Biologically Inspired Design

Ashok K. Goel · Daniel A. McAdams
Robert B. Stone
Editors

Biologically Inspired Design

Computational Methods and Tools

Springer

Editors
Ashok K. Goel
Georgia Institute of Technology
Atlanta, GA
USA

Daniel A. McAdams
Department of Mechanical Engineering
Texas A&M University
College Station, TX
USA

Robert B. Stone
School of Mechanical, Industrial and Manufacturing Engineering
Oregon State University
Corvallis, OR
USA

ISBN 978-1-4471-5247-7 ISBN 978-1-4471-5248-4 (eBook)
DOI 10.1007/978-1-4471-5248-4
Springer London Heidelberg New York Dordrecht

Library of Congress Control Number: 2013943573

Printed on acid-free paper

Springer is part of Springer Science+Business Media (www.springer.com)

In memory of my late parents, Urmil Goel and Satya Prakash Goel

Ashok K. Goel

I would like to thank the students I have had the opportunity and pleasure to teach. Their constant query of what, why, and how has kept me curious and energized to pursue my own line of what, why, and how

Daniel A. McAdams

Foreword: Curating Nature's Patent Database

The Moment of Truth

It is an odd hobby for a biologist, I admit, but I enjoy reading patents. Patents tell the meticulous story of how humans have solved the conundrums of their era, from catching mice to circling planets. This record of ingenuity is more than a legal necessity; it is an inspiration. Inventors read patents not just to avoid reinventing the wheel, but also to glimpse, from a mesa of inventive shoulders, the adjacent possible.

The volume you are about to read describes a Googlesque quest to develop another kind of patent database, one that describes nature's 3.8 billion years of adaptations. These adaptations are a record of life's long march to become well adapted to the particularities of this planet. While biologists ponder how adaptations help individual lilies, plankton, and pelicans survive, biomimics ask: "How might this adaptation, and the technology it inspires, help the human species fit in here over the long haul?"

In the last few decades, life's adaptations have inspired a series of game-changing technologies. A refrigeration-free vaccine inspired by the rugged Tardigrade, a coral-inspired way to sequester tons of carbon dioxide in concrete, and a material that captures fog as cleverly as a desert-dwelling beetle. Biomimics are working on ways to reduce pesticides in farmer's fields, ease traffic jams in cities, and prevent antibiotic resistance in our hospitals. Biomimetic products are doubling each year, and papers published in the field are doubling every 2–3 years, much faster than the 13-year doubling rate of other sciences (Lepora 2013). A 2010 economic study predicted that biomimicry could represent $1 trillion of global Gross Domestic Product by 2025,[1] and in 2012, biomimicry topped the

[1] Fermanian Business & Economic Institute, Point Loma Nazarene University. Global biomimicry efforts: an economic game changer, (2010).

Society of Manufacturing Engineers' annual list of "innovations that could change the way you manufacture."[2]

Though the prospects for bioinspired design have never been better, the discipline's moment of truth is here as well. A chasm still exists for many innovators, and unless we can cross it, biomimicry will remain the domain of the few innovators skilled and interested enough to decipher the primary biological literature. This leaves a knowledge divide for the millions of non-biologically trained engineers, architects, product designers, planners, chemists, material scientists, even policy makers for whom nature's strategies would be a revelation.

It is not that relevant biological information does not exist; we are, in fact, awash in it. If you can afford to access the full-text literature databases, and if you are fluent in the jargon, you have a chance of keeping up with the science. But as a designer, you are apt to have neither literature access nor a Rosetta Stone. For biomimicry to realize its potential, we need a Biological Information System (BIS) that is as ubiquitous and accessible as the Geographic Information System (GIS). That platform needs to deliver curated knowledge at the moment of creation, in a form tailored to fit the working styles of the people who invent our world. Like GIS, the success of BIS will depend on software tools that intelligently make sense of the raw data, augmented by apps that further extend its usefulness.

Transferable Ideas and Downloadable Beaks

Building a biological intelligence tool for inventors is the quintessential exercise in spanning disciplines. Biologists and inventors not only speak different languages—they ask different questions. Biologists might write a paper about the evolutionary significance of sharkskin that reduces biofouling, but often the "how" information—the dimensions and placement of the denticles—is buried in a paper devoted to the "why." Uncovering these gems of innovation from the continual blizzard of papers is a challenge, requiring enabling technologies like those collected in this volume: a way to describe biological phenomenon in machine readable language, an engineering-to-biology thesaurus, natural language query, near-clairvoyant search algorithms, and more.

Mining the literature, even clairvoyantly, is just the beginning. My biomimicry consulting colleagues at Biomimicry 3.8 have spent 15 years bringing biology to the design tables of companies such as Boeing, General Electric, General Mills, HOK, Nike, InterfaceFLOR, and Procter and Gamble. We have learned that most inventors are not interested in reading biological papers. They prefer that we synthesize and translate the papers into a taxonomy of promising mechanisms. Ultimately, they want a set of transferable ideas—*design principles* that will help them approach their challenge in a completely novel way. Our researchers can

[2] Society of Manufacturing Engineers http://www.sme.org/innovations12/#biomimicry.

easily read 10,000 papers to answer a question such as "How does nature contain liquids?" "How does nature manage vibration?" "How does nature store energy?" Building a taxonomy and extracting the design principles is a skill that takes years to master.

Once inventors are equipped with bioinspired design principles, there are still miles to go before these are translated into a product or process. This is where interactive tools could help, walking inventors through an iterative design process and giving them access to nature's ideas every step of the way. How, where, and when in the creative process this knowledge is delivered will mean the difference between inspiration and execution. Ideally, actionable plug-ins will be accessible right from the digital screens that designers, engineers, and architects use everyday, e.g., an AskNature button embedded in CAD/CAM or BIM tools.

While designing a roofing system, for instance, a building engineer would be able to visually browse reinforcement strategies in the natural world, and download actual truss designs based on this information. While laying out the HVAC system, he or she could run a branching algorithm to generate a fluid distribution system that keeps frictional losses to a minimum. Framing designs could be light weighted with the use of software that equalizes stress along surfaces, inspired by the growth of trees and bones. A genetic algorithm that uses natural selection protocols could optimize the entire design, all within the same program.

These digital modules are what our colleagues at Autodesk have described as the difference between "concept" and "content." Rather than read about a *concept*, inventors want to access biological information as *content* that they can use immediately. They would like to be able to download a biological library of forms—3D models of life's most streamlined, lightweight, or multifunctional designs. Imagine if Eiji Nakatsu, the JR West engineer who mimicked the kingfisher's beak to create Japan's Shinkansen train, had been able to download a 3D model of the beak before building a physical model. He could have attached the beak model to the train body, stretched and scaled it, even tested it *in silico* with computational fluid dynamic tools.

Building a biological library of forms would help biologists as well as inventors. With today's reality-capture software for cameras, it is possible to imagine "scan jams" where volunteers would digitize the artifacts of the world's natural history museums, freeing them from molding drawers so they can enliven the next generation of sustainable designs.

The internal blueprints of biomaterials will prove equally important, especially as we move to computer-controlled additive manufacturing (3D printing). Organisms add structure to common polymers to achieve extraordinary functionality, e.g., beetles layer chitin composites in a plywood hatch to achieve strength and toughness. A different structural design is used to create color, resilience, or water repellency, all from the same material. A biomimetic structure-function catalog could allow additive manufacturers to streamline their supply chain as nature does, using a small palette of easily recyclable polymers in unique architectures to achieve a wide range of functions.

Of course, each discipline will have different inventive needs. Chemists might prefer a "substitution engine" that allows them to replace an industrial chemical synthesis with a biochemical alternative, achieving similar effect without waste or toxic by-products. Organizational managers will want yet another slice of biological information, pertaining to topics like communication, cooperation, networks, or resilience. For each category of human endeavor, new user-centric applications will need to be created atop the BIS data.

Helping Innovators Meet Their Mentor

At the end of the day, even the cleverest information tools will not guarantee that a new invention, even one inspired by nature, will be sustainable in terms of energy and material use, toxicity, end-of-life fate, etc.

To help innovators create in ways that are deeply biomimetic, we find it useful to use systems-thinking tools such as Biomimicry 3.8's Life's Principles[3] in the scoping, creating, and evaluating phases of invention. These are meta principles common to most species on earth, and include reminders such as "build from the bottom up," "use a safe subset of the periodic table," "perform chemistry in aqueous solution," "embed feedback mechanisms to continually evolve your design," etc. Interactive software tools that screen for how well a design is meeting Life's Principles could help innovators solve problems without creating new ones.

If you look at all the ways that nature can influence decision-making, you realize that biomimicry is more than just a new way to innovate. It is a new way to think. University and professional training courses that prepare designers and engineers to ask "How would nature solve this?" are vital, as are techniques, described here, that help students make the all-important cognitive leap from design principle to application. The professors pioneering in this field are in a unique role; they have an opportunity to encourage the highest and best use of this new and powerful methodology, hopefully to solve the worthy conundrums of our era.

A prescient Steve Jobs said: "I think the biggest innovations of the twenty-first century will be at the intersection of biology and technology. A new era is beginning." If the age of biology is to keep its promise, the people who make our world will need to become biologically literate. But they will not want to become biologists themselves. Instead, they will want to know the key principles, the best practices, the operating codes of the natural world. They will want to understand ubiquitous patterns as well as the strange and wonderful curiosities in nature's patent database. Ultimately, they will want to understand how life has managed to enhance this planet, and how our innovations might do the same.

[3] http://biomimicry.net/about/biomimicry/lifes-principles/

A full-function tool to bring biological wisdom into human design is on its way, and the people in this volume will be instrumental in delivering it to us. They know that the key to wide-scale adoption of biomimicry is user-centric, curated knowledge, available at the moment of creation. Their efforts to help innovators learn from and emulate other species will one day be remembered as a pivotal leap in the evolution of our own.

Reference

Lepora, N. F., Ver schure, P. & Prescott, T. J. The state of the art in biomimetics. *Bioinspir. Biomim.* **8**, 013001 (2013)

Janine Benyus
Co-Founder and Institute Board President,
Biomimicry 3.8

Preface

Biologically inspired design is a promising paradigm for design innovation as well as sustainable design. The scientific challenge now is to transform it into a repeatable and scalable methodology. This requires addressing several big challenges, including the following four: the first and foremost of course is to use the paradigm to address increasing numbers of real problems that translate into real products in the market. A second challenge is to document the best practices of successful applications of the paradigm and develop a theory of biologically inspired design. A third challenge is to develop computational methods and tools that can make biologically inspired design repeatable and scalable. A fourth challenge is to educate new generations of would-be-designers in the paradigm of biologically inspired design. These four challenges are interconnected and build on one another: success at one likely will spur success at others.

This volume brings together a dozen chapters that together address all four of the above challenges at least to some degree, while emphasizing computational methods and tools for biologically inspired design. We are pleased to bring together these articles by some of the leading researchers and practitioners of biologically inspired design into a single volume.

Chapter 1 provides a brief review of two workshops sponsored by the United States National Science Foundation (NSF). These workshops served as the initial catalysis and formation of this book. Taken together, the two workshops brought together some 50 researchers in biologically inspired design, helped establish a stronger sense of research community, and led to the formulation of a research agenda outlined in the chapter.

Chapter 2 by Jon-Michael Deldin and Megan Schuknecht describe AskNature, Biomimicry 3.8 Institute's publicly available webportal that provides a functionally indexed database of biological design strategies and systems. Insofar as we know, this is the first scholarly article describing AskNature in detail, and thus adds an important piece to the growing literature on biologically inspired design.

In Chap. 3, Li Shu and Hyunmin Cheong describe a natural language approach to finding biological analogies and applying them to design problems. They review a decade long research program on developing the natural language approach to biologically inspired design, and also provide several examples of its application.

In Chap. 4, Jacquelyn Nagel presents an engineering-to-biology thesaurus, along with several examples of its use in addressing design problems. This kind of thesaurus can be a very useful tool for designers in finding biological analogies to their design problems.

Nagel, McAdams, and Stone in Chap. 5 describe the big picture of their several years of research on biologically inspired design. In particular, they present their information-processing theory of biologically inspired design, and illustrate it with several examples. They also describe a suite of tools that match several tasks in the process theory.

Goel, Swaroop Vattam, Bryan Wiltgen, and Michael Helms in Chap. 6 present their information-processing theory of biologically inspired design. They also compare their theory with similar theories such as Design Spiral and BioTRIZ, and examine what makes biologically inspired design different from other paradigms.

In Chap. 7, Jeannette Yen, Helms, Goel, Craig Tovey, and Marc Weissburg describe the evolution of a college-level interdisciplinary course on biologically inspired design. Their chapter reviews many lessons from teaching the course for several years. These lessons should be useful for potential teachers of similar courses.

In Chap. 8, Amaresh Chakrabarti focuses on analogical transfer from biology to engineering. He proposes guidelines for supporting this analogical transfer and describes an interactive tool that implements the guidelines. Comparative studies indicate that use of the tool increases the number of transferred designs.

Julie Linsey and Vimal Viswanathan in Chap. 9 study the cognitive challenges in biologically inspired design, focusing on design fixation. They describe several heuristics for addressing the challenges, including fixation. These heuristics should be useful for designing interactive tools for supporting biologically inspired design.

In Chap. 10, Wojciech Bejgerowski, John Gerdes, James Hopkins, Lengfeng Lee, Madusudanan Sathia Narayanan, Frank Mendel, Venkat Krovi, and Satyandra Gupta focus on bioinspired robotics. Building on several years of research, they present case studies ranging from bird-inspired robots to snake-inspired robots. They also describe a process by which biological features are selected and simplified for application using existing technologies for robot construction.

Julian Vincent in Chap. 11 specifies a need for identifying design principles that may help produce good technical designs without requiring biological expertise. He proposes four such principles derived from TRIZ: local quality, merging, dynamics, and prior cushioning. He calls for identification of design principles, especially those pertaining to information and material.

Frank Fish and John Beneski in Chap. 12 analyze the limits of design by evolution in biology and design by analogy in biologically inspired design. They argue that biological designs are not necessarily optimal with respect to any specific function. They advocate focusing on biological features that outperform currently available technologies for incorporation into technical designs.

We are grateful to Janine Benyus for writing the Foreword to this volume. Her work at Biomimicry 3.8 has long inspired the biomimicry movement, including our own efforts.

We are grateful also to the US NSF for sponsoring the workshops that led to this volume. In particular, we thank Christina Bloebaum for her support of this work as a Director of the Engineering Design and Innovation Program (now known as Engineering Systems Design) of the Civil, Mechanical, and Manufacturing Innovation Division of the Engineering Directorate at NSF.

We also thank the Springer publishing company, and especially Grace Quinn at Springer, for working with us on the preparation of this volume. We can only hope that scholars and practitioners of biologically inspired design, as well as design teachers and students more generally, will find this volume useful.

Ashok K. Goel
Daniel A. McAdams
Robert B. Stone

Contents

Chapter 1
Charting a Course for Computer-Aided Bio-Inspired Design

Robert B. Stone, Ashok K. Goel and Daniel A. McAdams

Abstract Bio-inspired design (BID) is an emerging research area in design, biology, computing, and engineering that seeks to systematically mine biological knowledge to solve design problems. To promote BID research, and especially research on computer-aided BID, the United States National Science Foundation (NSF) recently sponsored two workshops. These workshops served as the catalysis for this book. In this chapter, we review the discussions at the two workshops. We also sketch the outline of a research program on computer-aided BID that emerged from the workshops.

Keywords Biomimicry · Biologically inspired design · Computer-aided design · Engineering design · Design computing

1.1 Introduction

Bio-inspired design (BID or biomimicry or bionics) is an emerging research area in engineering design, computing, and biology that seeks to systematically mine biological knowledge to solve existing design problems. However, the community

R. B. Stone (✉)
School of Mechanical, Industrial, Oregon State University, 204 Rogers Hall, Corvallis 97331, USA
e-mail: rob.stone@oregonstate.edu

A. K. Goel
School of Interactive Computing, Georgia Institute of Technology, Atlanta 30332, USA
e-mail: goel@cc.gatech.edu

D. A. McAdams
Department of Mechanical Engineering, Texas A&M University, 411 Mechanical Engineering Office, 3123, College Station 77843-3123, USA
e-mail: dmcadams@tamu.edu

A. K. Goel et al. (eds.), *Biologically Inspired Design*,
DOI: 10.1007/978-1-4471-5248-4_1, © Springer-Verlag London 2014

of BID researchers at present is fragmented with no professional society, unifying funding source, or recurring conference.

As it stands, BID research is active across many disciplines and has had important and significant results. Nevertheless, BID remains largely a research activity contained in universities. BID is not yet an activity practiced by design engineers in the field. This "research-not-practice" status of BID exists because most BID is the result of researchers studying a biological entity or system such that their level of understanding allows an almost direct emulation but not necessarily new inspiration. This point solution status of BID sets a large, if a little ill-defined, scope for BID research: How do we transform BID from a point solution effort to fundamental theories and methods?

As a first draft of scoping BID research, some initial problems are apparent that served as motivation and agenda items for a workshop. How do we motivate and facilitate the interaction between engineering-oriented design researchers and science-oriented biology researchers—two fields with significantly different research cultures? How do we scope and scale biology in the context of abstracting engineering knowledge? How do we transmit biological knowledge to engineers and engineering knowledge to biologists without extensive discipline-specific training? How might computational theories of analogical reasoning inform the transfer of knowledge from biology to engineering and vice versa? How might artificial intelligence theories of knowledge representation and ontology support the construction of shared mental models among engineers and biologists? How might we teach biologically inspired design to engineers and biologists alike?

Of particular research interest to BID is how can we mine biology for solutions to problems for which we have no current solution. For example, some biological solutions exhibit superior sustainability to engineered solutions. Similarly, biological solutions often are complex both in their solution and the problem they solve. As the design problems solved become more complex, and the engineered solutions themselves also become more complex, biology may offer insights into how to solve the problem.

In this chapter, we provide an overview of two national science foundation (NSF)-sponsored workshops that started a discussion of these issues. The subsequent chapters in this volume discuss some of these issues in detail.

1.1.1 NSF Workshops

National science foundation funded two workshops to chart a course for BID. First, a one-day workshop in Palo Alto, California, was held on Sunday, March 20, 2011, in conjunction with the AAAI 2011 Spring Symposium on Artificial Intelligence and Sustainable Design.[1] A second, follow-up half-day workshop was held

[1] http://designengineeringlab.org/BID-workshop/Workshop_1.html

Table 1.1 Presenters and presentation titles at workshop 1

Speaker	Title
Janine Benyus Biomimicry guild/Biomimicry Institute	Biomimicry: sustainable design inspired by nature
Satyandra Gupta University of Maryland	Bio-inspired robotics
Daniel A. McAdams Texas A&M University	Bio-inspired design
Ashok Goel Georgia Institute of Technology	Computational methods and tools for biologically inspired design
Julian Vincent University of Bath	Methods of bridging biological information and engineering information
David Rosen Georgia Institute of Technology	SSS approach and creativity metrics for bio-inspired
Shapour Azarm University of Maryland	Bio-inspired design optimization
Thomasz Arciszewski George Mason University	Bio-inspiration
Jeannette Yen Georgia Institute of Technology	Center for biologically-inspired design
Amaresh Chakrabarti Indian Institute of Science	Biologically inspired design: an overview of research at ideas lab, IISc

in College Station, Texas, on June 5, 2012, in conjunction with the Fifth International Conference on Design Computing and Cognition.[2]

The first workshop was a good match to the AAAI 2011 Spring Symposium as the symposium had a topic on the related issue of biologically inspired and evolutionary models of sustainable design.[3] The objectives for the first workshop were to

1. Identify the research community.
2. Identify the topical coverage (overlap/gaps) of research.
3. Explore the gaps in current research efforts as it relates to a systematic application of biological information for engineering design.
4. Formulate major themes under the BID research banner.
5. Articulate steps toward a sustainable research community for BID.

Table 1.1 summarizes the workshop presenters and the title of their presentations. The workshop was broken into three sessions where leading researchers in the disciplines that intersect to form the BID research community presented the state of the art in bio-inspired research topics.

---

[2] http://designengineeringlab.org/BID-workshop/Workshop_2_%40_DCC.html

[3] http://www.vuse.vanderbilt.edu/~dfisher/AISD-Program.html

Table 1.2 Workshop 1 participants and affiliation

Name	Affiliation
Tom Arciszewski	George Mason University
Shapour Azarm	University of Maryland
Janine Benyus	The Biomimicry Group
Amaresh Chakrabarti	Indian Institute of Science (India)
Paul Egan	Carnegie Mellon
Douglas Fisher	Vanderbilt
Frank Fish	West Chester University
Robert J. Full	University of California, Berkeley
Erwin Gianchandani	Computing Research Association
Satyandra Gupta	University of Maryland
Norbert Hoeller	Sustainable Innovation Network
Chris Jenkins	Montana State University
Sangbae Kim	Massachusetts Institute of Technology
Venkat Krovi	University at Buffalo
Julie Linsey	Georgia Institute of Technology
Mary Lou Maher	University of Maryland
Julia O'Rourke	The University of Texas at Austin
David Rosen	Georgia Institute of Technology
Megan Schuknecht	AskNature.com
Justin Seipel	Purdue University
Jami Shah	Arizona State University
Jacquelyn Stroble	James Madison Univeristy
Srinivasan Arjun Tekalur	Michigan State University
Craig Tovey	Georgia Institute of Technology
Mohamed B. Trabia	University of Nevada, Las Vegas
Irem Tumer	Oregon State University
Swaroop Vattam	Georgia Institute of Technology
Julian Vincent	University of Bath (UK)
Anosh Wadia	Texas A&M
Jeannette Yen	Georgia Institute of Technology

Following the presentations, three breakout groups were formed to generate answers (and thus recommendations) for three general questions:

1. If BID is going to contribute to sustainability it needs to…?
2. If BID is going to contribute to complex system design it needs to…?
3. If BID is going to contribute to design education and pedagogy it needs to…?

The three breakout groups were led by Mary Lou Maher, Jami Shah, and Craig Tovey, respectively. The final session of the workshop featured debrief reports by each of the breakout groups and a general discussion session for all participants. The workshop included 30 researchers from the disciplines of engineering design, computer science, and biology in addition to the three organizers. The participants are listed in Table 1.2.

The second workshop, held in College Station, Texas, leveraged the computational design community at the 2012 Design Computing and Cognition

Table 1.3 Presenters and presentation titles at workshop 2

Speaker	Title
Rob Stone Oregon State University	Review of workshop 1 activities and findings
Ashok Goel Georgia Institute of Technology	Grounding bio-inspired design in cognition and computation
Julie Linsey Georgia Institute of Technology	Word tree express—a tool for design by analogy and bio-inspired design
Filipo Salustri Ryerson University	Analogy and systems are at the heart of BID
Marc Weissburg Georgia Institute of Technology	Pedagogical challenges to BID innovation
Rob Stone Oregon State University	Gathering designer feedback to generate requirements for intuitive biologically-inspired design tools

Conference to evaluate the first workshop's findings and generate a draft description of a program for research funding for BID. The objectives for the second workshop were to

1. Review the findings from the first workshop.
2. Introduce new ideas and research directions for BID through invited talks and a poster session with participants.
3. Draft a BID program description for a funding agency.

The outcome of the second workshop is captured in the later section entitled "Proposed NSF Program in BID." Table 1.3 summarizes the workshop presenters and the title of their presentations. The participants of the second workshop are listed in Table 1.4.

In addition to the individual presentations, the second workshop held breakout sessions with the goal of defining a potential NSF research program for funding BID research. The leaders of the breakout sessions included Alice Agogino, Tom Arciszewski, David Brown, Julie Linsey, and Marc Weissburg.

1.2 Current Research from the Disciplines

We begin with a brief summary of the state of the art from various subdisciplines of BID, with pointers to workshop presenters whose presentations elaborate each of those points.

Table 1.4 Workshop 2 participants and affiliation

Name	Affiliation
Alice Agogino	U. California at Berkeley
Kinda Al Sayed	University College London (UK)
Fernando Alvarez	Universidad de Bogotà (Columbia)
Tom Arciszewski	George Mason University
Ryan Arlitt	Oregon State University
David Brown	Worcester Polytechnic Institute
Christopher Earl	Open University (UK)
John Gero	George Mason University
Michael Helms	Georgia Institute of Technology
Julie Linsey	Georgia Institute of Technology
Mijeong Kim	Kyung He University (South Korea)
Pertti Saariluoma	University of Jyväskylä (Finland)
Filipo Salustri	Ryerson University (Canada)
Noe Vargas-Hernandez	U. Texas—El Paso
Pieter Vermaas	TU-Delft (Netherlands)
Marc Weissburg	Georgia Institute of Technology

1.2.1 Bio-Inspiration

One of the key aspects of BID is utilizing the similarities noted in nature for a particular problem to design and to bring inspiration to the designer. Inspiration through the forms of nature can come in three different types: visual, conceptual, and computational. Visual inspiration is widely used and understood. Pictures or other visuals of a biological system are used to create engineering systems that share the same visual appearance. Conceptual inspiration is the use of the knowledge found in biology to form design rules, heuristics, principles, or patterns. This type of inspiration requires an understanding in both nature and engineering. Algorithmic bio-inspiration is searching through nature to find algorithms like evolutionary computation and knowledge representations such as generative representations.

There are three different sources of bio-inspiration: evolution which is the gradual improvement in living systems in response to environmental stimuli, coevolution or coadaptation of a species in response to evolution of other species in the habitat, and morphogenesis—evolutionary development of an organism or its parts (Areiszewski).

Another search for bio-inspiration has been in the development of 3D manufacturing. While there are a large number of polymers used in engineering to complete various tasks, nature uses a small set. Taking inspiration from nature and using a smaller set more efficiently can be beneficial in manufacturing (Benyus).

Studies have shown that 70 % of engineering problems are solved by energy rather than by information. On the other hand, nature solves problems by using information rather that energy, making the system more efficient (Vincent).

Novelty is how unique the solutions are, while variety is how many of the solutions were new. A study containing a control group, a BID group, and an engineering group allowed the analysis of the groups' solutions using these metrics. The study showed that BID increases novelty but not necessarily variety. There was noticeable fixation in the engineering group that was not seen in the others (Rosen).

One can also use a creativity metric to analyze designs. By examining the novelty and the usefulness of a design, a creativity score can be determined. Biological systems may be a way to inspire more creative solutions when undergoing design problems (Chakrabarti).

1.2.2 Bridging Biology and Engineering

Historically, BID has been rather anecdotal rather than systematic; therefore, a way to make BID more systematic is needed to increase the benefits gained from BID. There are two main methods to enhancing BID, stimuli, and transfer guidelines. Stimuli is broken up into two different categories, structured and unstructured. Structured information allows for a simpler search and easier transfer of knowledge. However, the arrangement of the information into a structure can be difficult and time-consuming. Unstructured information requires no effort to arrange the information; however, the "search" carried out by the designer and the transferring of the idea tend to be more difficult and time-consuming. Transfer guidelines are broken into four general steps: formulate search objectives, search for biological analogs, analyze biological analogs, and transfer relevant knowledge to the target domain. Using the SAPPhIRE model, seven levels of abstraction can be obtained. These abstractions can be used to inspire ideas. The SAPPhIRE model excels at empirical findings and exploring the number of ideas which lead to a higher levels of SAPPhIRE and greater novelty. One can also use design creativity as a metric of the novelty and the usefulness of a design. Currently, not all levels of SAPPhIRE have been explored and studies are being done to examine these. Use of the SAPPhIRE model has shown that a systematic framework helps increase the overall number of ideas (Chakrabarti).

Bio-inspired design is responsible for many useful, innovative designs: The lotus leaf inspired self-cleaning water repellent surfaces, and the cocklebur inspired velcro. These are just a few examples of the power of BID. Engineering design is more problem driven, and the concepts are dominated by knowledge of similar systems. One can recast BID as a problem-driven effort by combining it with function-based design methods to create function-based BID. Function can be used as the analogical connection between what an engineering system needs to do and how the natural system completes that function. By using normal functional modeling techniques, naturel systems can be modeled. This modeling framework allows an analogous connection needed between engineering systems and natural systems. A practical challenge of this approach is that engineers and biologists use different terminologies when describing solutions to function; therefore, an engineering-to-biology thesaurus is needed. The current BID methodology uses the functional basis terms with biological function/flow correspondents. The main steps for this start with creating a functional model for the design problem using a terminology known as the functional basis. The next step is translating the functional model into the associated biological keywords. After this is done, one can search a specific biological knowledge repository or use Google, biology texts, or other publications (McAdams).

Goel has done work in developing information processing theories and computational methods for BID. This work starts with conducting empirical studies of BID, then developing general information processing accounts of BID, next constructing computational tools and techniques for aiding BID, and finally deploying and evaluating the tools in realistic settings. One of their findings is that BID often involves compound analogies, entailing intricate interaction between problem decomposition and analogical reasoning. A second finding is that BID engages not only in analogical transfer of functionally indexed mechanisms but also in transfer of problem decompositions. A third finding is that biological analogies are useful for several tasks of BID in addition to design concept generation, such as design analysis and explanation.

Other work bridges biological information with engineering information. Rosen has been researching and developing a strategy–state–structure ontology to create a formal language that represents designs. Strategy is function plus behaviors; therefore, this would include a taxonomy of functions and behaviors that could be used together to form strategies (Rosen).

At the University of Maryland, Gupta has been conducting studies with bio-inspired robots. Bio-inspired robots are robots with the main inspiration coming from a bird, animal, or fish. There are many applications of bio-inspired robots that include but are not limited to medical, reconnaissance, mine detection, entertainment, and space exploration. The traditional approach in creating these robots is taking inspiration from nature and adding modeling- and simulation-based optimization to create designs for these robots. Traditional manufacturing techniques are then examined to determine which designs can be executed. There are many successful robots larger than 100 mm; however, miniature robots (between 100 mm and 5 mm) usually come with limited capabilities. By finding new novel manufacturing concepts, highly capable yet miniature bio-inspired robots could be implemented. The first step is to approach design differently—to simplify by only retaining features of the biological creature that are useful and to identify high-value characteristics. Another need in the design phase is to amplify the useful biological characteristics to improve the performance of the robot. The second step is to approach the modeling and simulation of these robots differently using metamodel synthesis. The last is to approach assembly differently (Gupta).

1.2.3 Techniques and Tools for Bio-Inspired Conceptual Design

The development of design tools to help designers use BID effectively is crucial to being able to teach BID in a classroom and implement BID in the workplace. Asknature.org is an important tool for BID. Asknature.org takes organizes bio-literature by function and allows the user to search through. The Web site is public domain, part search engine and part social network, and helps people using BID connect with each other. As of now, asknature.org contains 1,300 strategies for design. These strategies contain links to biologist Web pages or Google Scholar articles to help the user be able to research the strategy. Support from Autodesk Inventor has been crucial as it implements BID into a commonly used design tool. A current redesign of AskNature is underway to follow the path of idea from nature–transferable idea–possible products–actual products–digital downloads–Autodesk Inventor. This would allow an easier transfer of flows from BID into products. Another possible tool would be implementing native ecosystem data in city planning. By setting new ecological performance standards, architects and city planners would use the ecosystem data to develop a city that is more sustainable and functions together (Benyus).

Another tool of importance is the engineering-to-biology thesaurus. This thesaurus helps engineers include bio-inspiration in the engineering design process by relating the engineering functions and flows in the functional basis to biological functions and flows. In this manner, the designer is allowed to use their functional model to relate key functionality to the way nature also solves it (McAdams). Another tool for BID is the creation of databases to organize information.

Goel described two knowledge bases: DANE and Biologue. Both DANE and Biologue use structure–behavior–function models of biological and engineering systems. DANE, which has already been released (http://dilab.cc.gatech.edu/dane/), is a functionally organized database of biological solutions. This is useful to designers because it allows them to look up the function they are trying to solve and relate that to biological solutions. The more recent Biologue system is a database that allows the indexing of biology-related documents using the structure–behavior–function model. Using this, a designer can compare the functions of what they are designing and also compare the structures and behaviors of their design to biological systems.

Of note, bio-inspired optimization techniques are also used in design. Genetic, ant, particle swarm, bee, and firefly algorithms all contain specific uses to help engineers optimize in design. By studying the ways grouped animals move and interact, techniques can be developed to mimic those interactions and create better optimization techniques and algorithms. These techniques can be used to help overcome challenges often found while trying to optimize designs. Such challenges include complexity, scalability, and convergence. There is a need to extend the applicability of genetic algorithms for design optimization with regard to uncertainty, system product design, and multi-objective genetic algorithms

(MOGA). Pros and cons exist for all of the BID optimization algorithms. The desirable traits are that the algorithm is population based, can optimize non-convex discontinuous functions, can handle discrete–continuous/combinatorial design problems, and obtain global optimums. However, sometimes these optimization techniques result in local optimums that can be undesirable. Nevertheless, there are complications with these algorithms that include complexity, scalability, and convergence (Azarm).

Vincent discussed another important BID tool, Bio-TRIZ. Bio-TRIZ is a reduced form of TRIZ that relates the TRIZ inventive principles to biological contradictions. Studying 5,000 examples, the conflict matrix was reduced from 39 conflict elements to 6 elements that appear in both biology and engineering, and a 6×6 contradiction matrix that contains all 40 of the inventive principles was created. These 6 conflict elements are substance, structure, time, space, energy/field, and information/regulation. Biological solutions were then studied to fill in inventive principles aspect of the matrix (Vincent).

1.2.4 Education in Bio-Inspired Design

The Biomimicry Institute (TBI) has been working on implementing BID into earlier education (K1-12) as well as into a professional masters program. An important part of featuring BID in a classroom setting is combining biologists, biomimetic scientists, engineers, and designers together. Academic settings tend to be very different from work settings. Therefore, working to close the difference between academic and work settings would lead to better transition from BID in the classroom to BID in the workplace (Benyus).

Since 2005, Georgia Institute of Technology has offered a BID interdisciplinary course for engineers, biologists, and other scientists. This course features ~30 engineering students and ~15 biology students that take part in semester-long, self-defined design projects that include the help of faculty mentors. This has been created to help promote BID practice and explore how BID aids in developing products. The main goal of this course is to encourage ways of thinking about and to explore nature that helps facilitate a designer's ability to implement biological strategies in engineered products and sustainable systems. The development results in a fifteen-week course that includes the principles of biologically inspired design, biomimetic materials, biologically inspired sensing and movement, system design and optimization, and green technology. These lectures contain both content and practice to help the students grasp the material. The final is a presentation by the students on a novel BID (Yen).

An important aspect of BID is implementing it in a classroom setting so students can not only learn and understand it but also implement it in industry. The University of Maryland has offered bio-inspired robotics undergraduate classes. These classes offer an excellent opportunity to teach students how to design a complex and modern mechatronics system. The content of the course includes the fundamentals of robotics, including kinematics, inverse kinematics, dynamics, trajectory generation and controls, bio-inspired robots where design concepts, sensors, and actuators are covered, and ending with robot building tutorials using mechanical design, servo motors, microcontrollers, and programming. The class is project based with bio-inspired robot outcome (Gupta).

1.3 Recommendations

Biologically inspired design has been practiced on an *ad hoc* basis throughout human history. The breakout sessions at the first workshop were designed to explore the research needed to move BID from *ad hoc* to *intentional* in its application. The breakout sessions at the second workshop were targeted toward defining an NSF program for funding BID research.

1.3.1 A Design Theory for Bio-Inspired Design

Bio-inspired design seeks to exploit biology for several different kinds of design such as sustainable design, creative design, and complex system design. Of course, these different kinds of design are mutually compatible and consistent: We can have complex systems that are sustainable, sustainable designs that are creative, and so on. Nevertheless, the three kinds of design have different emphases and foci.

The goal of bio-inspired sustainable design is to use biology as an inspiration for designing technological products that are ecologically sustainable. Although biological systems are not always optimal, the argument goes that they typically use only local and abundant resources and are often very efficient in terms of use of resources such as energy and water. Of course, this does not guarantee that bio-inspired products will be necessarily sustainable, but it promises that they may be more sustainable than equivalent products available in the market today. Consider the following specific cases:

The Biomimicry Institute's work on BID is driven by the growing need for sustainable design. An example of sustainable design at TBI is the novel design of a water bottle with "ribs" on its plastic surface mimicking the ribs on trees and providing strength to the bottle. This allows the bottle to use less plastic, which makes it lighter than similar water bottles (Benyus).

Recent work on BioTRIZ indicates that for many functions for which technological products typically use energy, equivalent biological systems use information instead. This suggests that we seek biological sources as inspiration to design a new generation of technological products that use information in place of energy to achieve as many functions as possible (Vincent).

Another line of research seeks to identify design patterns that biological systems use to achieve ecological sustainability. For example, the Namibian beetle uses an interleaved pattern of the biological effects of hydrophobia and hydrophylia for harvesting water from dry air. Bio-inspired designers have now used the same pattern for fog harvesting. This suggests building a classification of functionally indexed design patterns for sustainable design (Goel).

The goal of bio-inspired complex system design is to use the characteristically complex interactions found in nature as a guide to engineered systems that are complex and integrated among their constituent components. Although biologists welcome complexity, engineers typically are concerned about it and do what they can to avoid it. Approaching complex system design from a biologist's perspective that complexity could be a positive aspect and allow a mechanism for coping with product failures appears a promising avenue with the following observations:

On the engineering side of complex systems, one has relatively simple units leading to an emergent phenomenon that includes many interacting parts exhibiting nonlinear behavior, uncertainty, and multiple scales. The biology side of complex designs contains heterogeneity, multiple different parts that fit together and provide different outcomes dependent on initial conditions, and multi-functionality.

Biologists can help shift the paradigm of perceived complexity versus real complexity. This shift can help show engineers how to manage complexity in their systems which could lead to innovative products and elegant designs and help predict the performance of complex systems.

Nature has the ability to self-repair, adapt, add redundancies, accommodate failure, regenerate and reconfigure parts, as well as others. Engineering systems do not accept the notion of failure as a positive feature, but small, intentional failures to avoid a catastrophic failure can be a good thing.

1.3.2 A Funding Program for Bio-Inspired Design Research

As part of the second workshop, the participants reviewed the findings of the first workshop and produced variants of a proposed call for proposals. For purposes of context, it was assumed that the umbrella agency for such a program would be the United States NSF. The variant ideas were aggregated into the following brief description of an NSF program:

The NSF invites proposals for research in BID. Biologically inspired designs and processes have much potential to solve urgent and complex challenges faced by the United States and the planet such as those found in military, urban infrastructures, climate change, sustainability, and space exploration domains. Proposals must be from suitable multi-disciplinary teams (i.e., members might include biologists, computer scientists, engineers, or psychologists), addressing small- to medium-scale designs (such as household products or automotive systems), have demonstrated educational and computational potential, and be well evaluated. Suitable research areas include but are not limited to:

- BID system usability, interface design, visualization, and search;
- Identification of the role of biologically inspired tools and design methods for each stage of the design process, for example, problem framing, conceptual design, refinement, production (DfM for BIDs), marketing, reuse/recycle;
- Knowledge-base/database building and integration, ontologies (construction, use, and evaluation), and test beds (computational and physical);
- Evaluation of analogy utility (before, after, and during use) of BIDs, design methods, and the manufacturability of resulting designs;
- Roles of the scales (both spatial—from micro- to macro-, to systems of systems—and temporal—to promote desired emergent behavior over time) of biological knowledge for problem identification, design decomposition, generation, evaluation, and explanation;
- Impact of BID on communication in multiple disciplinary teams, for example, novice–expert studies, development of a community of practice and networks, sociology of disciplinary norms;
- Teaching approaches and curricular development.

The work must be extensible and must be shared in order to promote BID community building.

1.4 Summary

In summary, it is clear that recent research efforts across the disciplines of engineering design, computer science, and biology have attempted to address the various problems associated with not only developing biologically inspired designs, but also teaching students how to develop biologically inspired designs. However, it is also evident that there is a need for additional work on refining the proposed methods and tools as well as developing new methods to address current limitations. The recommendation of the workshop organizers is that a new cross-cutting NSF program in BID be established that seeks to fund transformative research as described in the program brief above. Such a program is expected to support high-risk/high-reward research that otherwise has no current home in the NSF.

Suggested Readings

Bio-Inspiration

Arciszewski T, Cornell J (2006) Bioinspiration: learning creative design principles. In intelligent computing in engineering and architecture. Lect Notes Comput Sci 4200:32–53

Bar-Cohen Y (2006) (ed) Biomimetics–using nature to inspire human innovation. Bioinspir Biomim 1(1):P1–P12. doi:10.1088/1748-3182/1/1/P01

Bar-Cohen Y (2011) (ed) Biomimetics: nature-based Innovation. CRC Press, Boca Raton

Benyus JM (1997) Biomimicry: innovation inspired by nature, 1st edn. Morrow, NewYork
Bonser R, Vincent J (2007) Technology trajectories, innovation, and the growth of biomimetics. In: Proceedings of the institution of mechanical engineers, part C: journal of mechanical engineering science, pp 1177–1180
Chakrabarti A, Shu L (2010) Biologically inspired design. AIEDAM 24:453–454
Helms M, Vattam S, Goel A (2009) Biologically inspired design: process and products. Des Stud 30(5):606–622
Lindemann U, Gramann J (2004) Engineering design using biological principles. Proc Int Des Conf Des 5:18–21 (Dubrovnik)
Mak T, Shu L (2008) Using descriptions of biological phenomena for idea generation. Res Eng Des 19:21–28
Shu, L, Ueda K, Chiu I, Cheong H (2011) Biologically inspired design. CIRP Ann Manuf Technol
Vincent JFV (2005) Deconstructing the design of a biological material. J Theor Biol 236:73–78
Vincent J, Mann D (2002) Systematic technology transfer from biology to engineering, philosophical transactions of the royal society. Physical Sci 360:159–173
Wilson J, Rosen D, Nelson B, Yen J (2010) The effects of biological examples in idea generation. Des Stud 31(2):169–186
Yen J, Weissburg M (2007) Perspectives on biologically inspired design: introduction to the collected contributions. J Bioinspiration Biomimetics 2

Bridging Biology and Engineering

Fish FE (1998) Imaginative solutions by marine organisms for drag reduction. In: Meng JCS (ed) Proceedings of the international symposium on seawater drag reduction. Newport, Rhode Island, pp 443–450
Fish FE (2006) Limits of nature and advances of technology in marine systems: What does biomimetics have to offer to aquatic robots? Appl Bionics Biomech 3:49–60
French M (1994) Invention and evolution: design in nature and engineering, 2nd edn. Cambridge University Press, Cambridge, 1988
Nagel JKS, Stone RB, McAdams DA (2010) An engineering-to-biology thesaurus for engineering design. In: ASME IDETC/CIE 2010 DTM-28233, Montreal, Quebec, Canada, 2010
Vattam S, Goel A (2011) Foraging for inspiration: understanding and supporting the information seeking practices of biologically inspired designers. In Proceedings of ASME DETC conference on design theory and methods, Washington DC, August 2011
Vogel S (1988) Life's devices. Princeton University Press, Princeton
Vogel S (1994) Life in moving fluids. Princeton University Press, Princeton

Techniques and Tools for Bio-Inspired Conceptual Design

Chakrabarti A, Sarkar P, Leelavathamma B, Nataraju B (2005) A functional representation for aiding biomimetic and artificial inspiration of new ideas. AIEDAM 19(2):113–132
Cheong H, Chiu I, Shu LH, Stone RB, McAdams DA (2011) Biologically meaningful keywords for terms of the functional basis. J Mech Des 132(2):02001-7–02001-11
Chiu I, Shu L (2007a) Biomimetic design through natural language analysis to facilitate cross-domain information retrieval. AIEDAM 21:45–59
Chiu I, Shu L (2007b) Using language as related stimuli for concept generation. AIEDAM 21:103–121
Goel A, Vattam S, Wiltgen B, Helms M (2012) Cognitive, collaborative, conceptual and creative—four characteristics of the next generation of knowledge-based CAD systems: a study in biologically inspired design. Comput Aided Des 44(10):879–900

Lenau T (2009) Biomimetics as a design methodology—possibilities and challenges, international conference on engineering design, ICED'09 24–27 August 2009. Stanford University, Stanford

Nagel JKS, Stone RB (2012) A computational approach to biologically-inspired design. AIEDAM 26(2)

Nagel RL, Midha PA, Tinsley A, McAdams DA, Stone RB, and Shu LH Exploring the use of functional models in biomimetic conceptual design. J Mech Des 130(12):121102-1–121102-13

Sarkar P, Chakrabarti A (2008) The effect of representation of triggers on design outcomes. Artif Intell Des Anal Manuf 22(02):101–116

Sartori J, Pal U, Chakrabarti A (2010) A methodology for supporting "transfer" in biomimetic design. AIEDAM 24:483–505

Shu L (2010) A natural-language approach to biomimetic design. AIEDAM 24:507–519

Srinivasan V, Chakrabarti A (2010) An integrated model of designing. ASME JCISE 10(3)

Vattam S, Helms M, Goel A (2008) Compound analogical design: interaction between problem decomposition and analogical transfer in biologically inspired design. In Proceedings of third international conference on design computing and cognition, Atlanta, June 2008. Springer, Berlin, pp 377–396

Education in Bio-Inspired Design

Bruck H, Gershon A, Golden I, Gupta S, Gyger L, Magrab E, Spranklin B (2007) Training mechanical engineering students to utilize biological inspiration during product development. Bioinspiration and Biomimetics 2:S198–S209

Glier MW, Tsenn J, Linsey JS, McAdams DA (2011a) Methods for supporting bioinspired design. In: Proceedings of the ASME 2011 international mechanical engineering congress and exposition, number IMECE2011-63247, Denver, Colorado

Glier MW, Tsenn J, Linsey JS, McAdams DA (2011b) Concepts in biomimetic design: methods and tools to incorporate into a biomimetic design course. In: Proceedings of the ASME 2011 international design engineering technical conferences and computers and information in engineering conference, number DETC2011-48571, Washington, DC

Helms M, Vattam S, Goel K (2010) The effect of functional modeling on understanding biological systems. In: Proceedings of ASME 2010 international design engineering technical conferences and computers and information in engineering conference (IDETC/CIE 2010). August 15–18, 2010, Montreal, Quebec, Canada

Nelson B, Wilson J, Yen J (2009) A study of biologically-inspired design as a context for enhancing student innovation. In: Proceedings of 39th ASEE/IEEE frontiers in education conference, 2009

Vakili V, Chiu I, Shu LH, McAdams DA, Stone RB (2007) Including functional models of biological phenomena as design stimuli. In: Proceedings of the 2007 ASME design engineering technical conferences and computers and information in engineering conference, number DETC2007-35776, Las Vegas, Nevada

Weissburg M, Tovey C, Yen J (2010) Enhancing innovation through biologically inspired design. Adv Nat Sci 3:145–167

Yen J, Weissburg M, Helms M, Goel A (2011) Biologically inspired design: a tool for interdisciplinary education. In: Bar-Cohen Y (ed) Biomimetics: nature-based innovation, Taylor & Francis

Yen J, Helms M, Goel A, Vattam S (2010) Evaluating biological systems for their potential in engineering design. Adv Nat Sci 3(2)

Design Theory for Bio-Inspired Design

Goel A, Bras B, Helms M, Rugaber S, Tovey C Vattam S, Weissburg M, Wiltgen B, Yen J (2011) Design patterns and cross-domain analogies in biologically inspired sustainable design. In: Proceedings of AAAI spring Symposium on AI and sustainable design, Stanford University, Palo Alto, March 2011, pp 45–51

Goel A, Vattam S, Wiltgen B, Helms M (2011b) An information-processing account of creative analogies in biologically inspired design. In: Proceedings of eighth ACM conference on creativity and cognition, Atlanta, Nov 2011, pp 71–80

Helms M, Goel A (2012) Analogical problem evolution in biologically inspired design. In: Proceedings of fifth international conference on design computing and cognition, College Station, Texas, July 2012, Springer

Nagel JKS, Nagel RL, Stone RB, McAdams DA (2010) Function-based biologically-inspired concept generation. Artif Intell Eng Des Anal Manuf 24(4):521–553

Vattam S, Helms M, Goel A (2010) A content account of creative analogies in biologically inspired design. AI Eng Des Anal Manuf Spec Issue Biol Inspired Des 24:467–481

Vincent J, Bogatyreva O, Bogatyrev N, Bowyer A, Pahl A (2006) Biomimetics: its practice and theory. J R Soc Interface 3:471–482

Chapter 2
The AskNature Database: Enabling Solutions in Biomimetic Design

Jon-Michael Deldin and Megan Schuknecht

Abstract Practitioners of biomimetic design express one consistent need—access to relevant biological information. This information is most useful when the transferable elements have been abstracted from the biological principle, are organized by design and engineering function, and are supported by contextual search functions. However, the fine details of how the information is organized and accessed are critical. In this chapter, we reflect upon AskNature.org, a biomimetic database created to address these issues, and its key elements, including a Biomimicry Taxonomy, biological strategy pages organized by function, search and tri-browse features, and bio-inspired product pages as examples of bio-inspired design successes.

2.1 Introduction

Biomimetic design professionals need access to relevant biological information expressed in common language (Bar-Cohen 2006). For example, both a biologist and a civil engineer can talk about "managing temperature" and have a basic understanding of what the other means, whether one of them is talking about an elephant seal managing temperature or the other about the need to regulate temperature in a building. However, if the biologist starts the conversation by talking about the intricate structure of a northern elephant seal's nasal turbinates, rather than their function, that information is not going to be of obvious interest or application to the civil engineer.

J.-M. Deldin · M. Schuknecht (✉)
Biomimicry 3.8 Institute, Missoula, MT, US
e-mail: megan.schuknecht@biomimicry.net

J.-M. Deldin
e-mail: jon-michael.deldin@biomimicry.net

A. K. Goel et al. (eds.), *Biologically Inspired Design*,
DOI: 10.1007/978-1-4471-5248-4_2, © Springer-Verlag London 2014

Table 2.1 AskNature.org website usage from November 12, 2008, through December 31, 2012

Years	Page views	Unique visitors
2008	125,568	22,386
2009	1,066,527	208,661
2010	1,195,928	236,117
2011	1,484,148	295,897
2012	1,790,709	478,486

Source Google analytics

We developed AskNature.org in an effort to translate biological information so it would be accessible to non-biologists, and to serve as a source of inspiration for biomimetic design. AskNature is a free, publicly available database of biological information and abstracts, most of which are drawn from peer-reviewed journals. The website catalogs biological information by function in "strategy" pages (described in Sect. 4.1) to bridge the gap between biology and fields like engineering, architecture, industrial design, chemistry, organizational development, and more. Since its launch at Greenbuild International Conference and Expo in 2008, the site has experienced rapid growth, recording almost 1.8 M page views in 2012 (see Table 2.1). Additionally, AskNature has been lauded by design professionals, winning an Earth Award in (2010) and becoming a finalist for an INDEX Award in 2011 (Biomimicry 3.8 Institute 2011).

Function is one way to interpret biological information in order to create a bridge of understanding across disciplines, and identifying function can be one key component to approaching biomimetic design (Helms et al. 2009, 2010; Stone and Wood 2000; Vattam and Goel 2011). The AskNature team developed the Biomimicry Taxonomy (taxonomy) as a means to organize biological data by function and present it to a design audience. The taxonomy represents the organizing schema for all of the biological strategy pages within the database. We will discuss the taxonomy in greater detail below.

2.2 Data

Trained biologists were responsible for gathering and generating the original almost 1,300 pages of biological data on AskNature. Researchers read scientific journals and books and perused scientific news, looking for leads on functional biology that might be of interest to innovators working to solve human challenges. Strategies were selected subjectively based on the researchers' assessment of whether the strategy held some potential for being useful within the field of bio-inspired design. AskNature's original data set represents a huge amount of human labor and was only possible due to generous funding from an independent investor.

Individuals continue to generate additional content for AskNature. Most of the biological data that has been added since 2008 has come from paid staff.

Qualified scientific curators are also able to add biological content. However, due to the rigors of maintaining the scientific integrity of the data and the need for strict adherence to the data's organizational structure, all curated content must be approved by AskNature's content editor.

2.3 The Biomimicry Taxonomy: Organizing Biology by Function

Once the strategy data were collected, the AskNature team began to look for patterns and to organize the data according to function. The result was a classification system we call the Biomimicry Taxonomy, shown in Fig. 2.1. The taxonomy categorizes strategies according to three levels: groups (highest level), subgroups, and function. Overall, the taxonomy includes 8 groups, 30 subgroups, and 162 functions. Individual strategies represent the next, most detailed level within the hierarchy.

As a specific example, an insect might face the challenge of protecting itself from other organisms that want to eat it. Its strategy to meet that challenge might appear like this within the taxonomy:

Group	Maintain physical integrity
Subgroup	Protect from biotic factors
Function	Protect from animals
Strategy	Nanoscale protrusions (AskNature 2008a).

2.3.1 *Compared to Other Taxonomies*

AskNature staff consulted with external design professionals when creating the taxonomy, particularly from chemical and materials science disciplines, but this schema is subjective and represents a small group of biologists attempting to organize a huge amount of biological data by function. Unlike other schemas that have been developed to assist bio-inspired design (Glier et al. 2011; Vattam et al. 2010; Vincent et al. 2006; Yen et al. 2011), the taxonomy has not been rigorously tested to assess its impact on its users.

2.4 A Tour of AskNature.org

AskNature consists of a number of components: biological strategy pages, biomimetic product case studies, a search engine, and a social network.

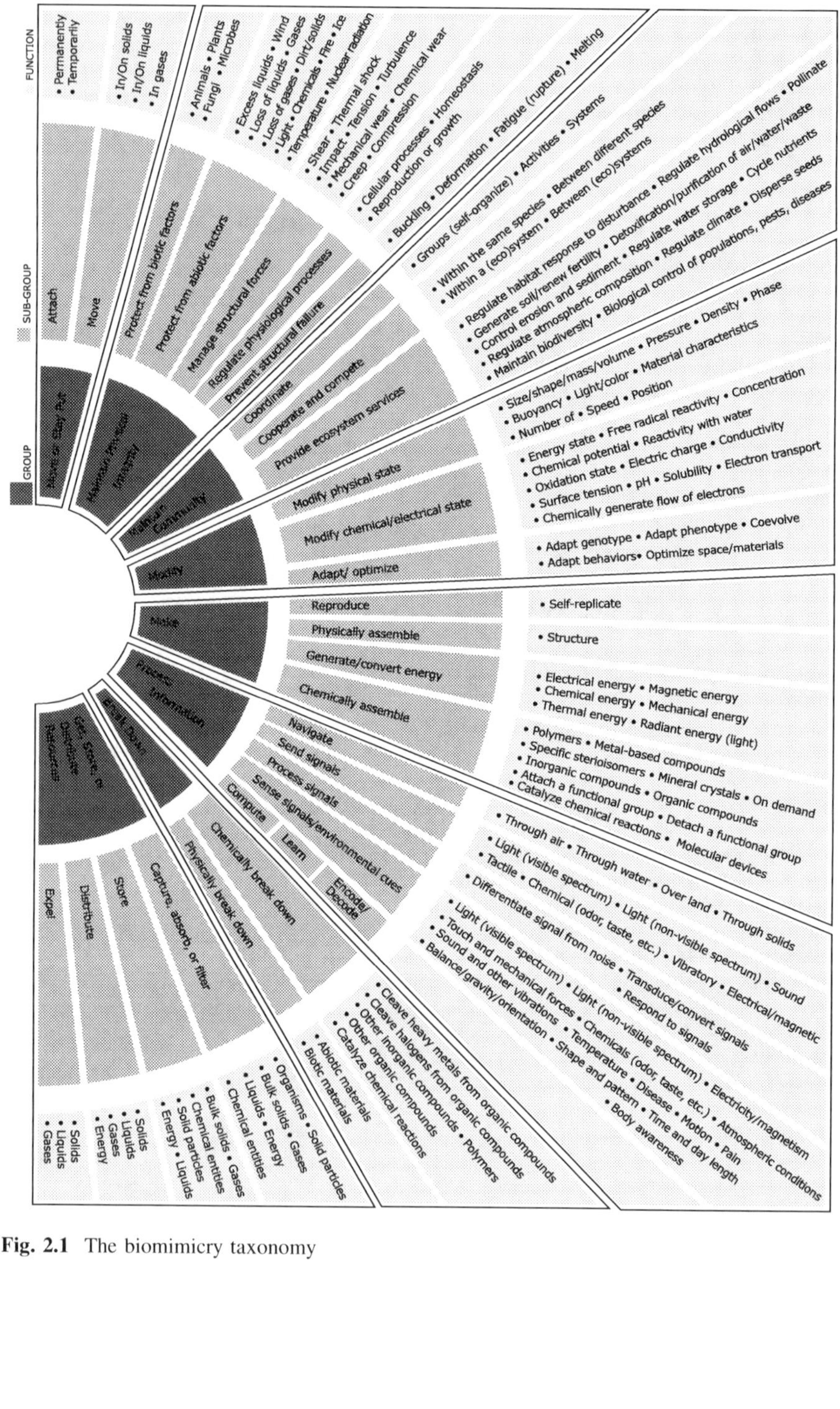

Fig. 2.1 The biomimicry taxonomy

2.4.1 Strategy Pages

All the biological content on AskNature is captured on strategy pages. As of January 2013, the AskNature database contains over 1,600 biological strategy pages. Strategies are solutions a given organism or ecosystem uses to meet functional challenges.

Each strategy page consists of a carefully crafted title, a short sentence explaining the essence of the biological strategy, the strategy's place within the Biomimicry Taxonomy, a scientific excerpt, biomimetic application ideas and/or links to biomimetic product pages, and links to scientific references via Google Scholar or Scirus. In addition, most of the strategy pages include photographs and/or illustrations of how the strategy works, basic natural history information (including IUCN Red List (International Union for Conservation of Nature and Natural Resources 2012) status number, if applicable), links to videos that provide further context on a given strategy, and links to scientific experts and/or laboratories.

For example, the title "Wing scales diffract and scatter light: Morpho butterflies," on the strategy page of the same name (AskNature 2008c), tells a user at a glance what function is being accomplished and by what organism (Fig. 2.2). The sentence below the illustration of the strategy provides more detail: "Wings of Morpho butterflies create color by diffracting and scattering light." On every strategy page, the structure of this sentence is the same: part of the organism, organism, what it does, and how it does it. This sentence represents one level of abstraction regarding the biology of the strategy; that is, it represents one interpretation of function at a given scale. Depending on the content, the scale discussed may vary.

The level of abstraction depends on a combination of the perspective of the original biological researcher, the interpretation by AskNature staff, the level of scientific detail known, and the framework of the database. In this particular case, much more detail is known about the architecture of the wings and how morpho butterflies refract and scatter light, that is, the mechanism used to accomplish the function, but that information and level of abstraction do not fit neatly into the summary sentence. A user must read further to find out, for example, that "nano-sized, transparent, chitin-and-air layered structures that, rather than statically absorb and reflect certain light wavelengths as pigments and dyes do, selectively cancel out certain colors through wavelength interference while reflecting others, depending on the exact structure and interspatial distance between diffracting layers" or to learn any exact dimensions used in this strategy.

While AskNature's strategies allow users to see a wide cross section of the ways organisms have met functional challenges, they generally does not provide a great level of detail about the mechanism of function. For example, what specifically about the nasal turbinates of northern elephant seals makes counter-current heat exchange so effective? However, the sheer number of strategy pages, the

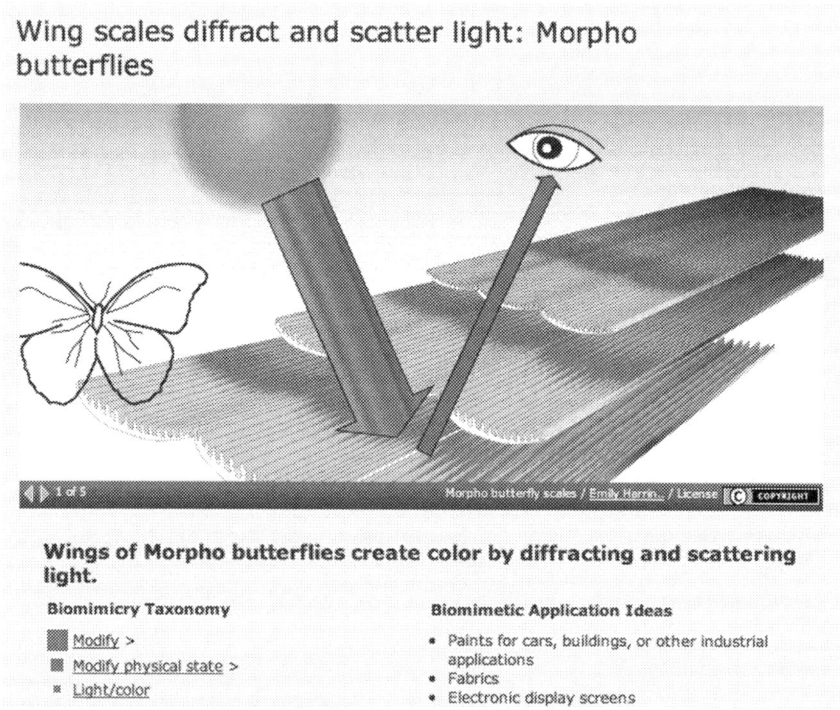

Fig. 2.2 Screenshot of the top of AskNature's morpho butterfly strategy page (AskNature 2008c)

organizational structure, and the fact that the database is free to the public have made AskNature a much more widely used tool than others that are accessible only to select users.

2.5 Biomimetic Products

AskNature features nearly 200 biomimetic product pages for users to see how others have applied a certain strategy. A product's page describes the history of the product, what challenges it solves, how it is different from existing products, and its bio-inspired principles. For example, Fig. 2.3 shows users how one design group was inspired by the whitebark pine's spiral fibers (AskNature 2008b) to develop a distinctive, 100 % recyclable polyethylene terephthalate (PET) bottle. As the site grows, we hope to add even more biomimetic products to our database.

Fig. 2.3 Screenshot of an AskNature product page (AskNature 2011) showing the product details, history, and more

2.6 Searching AskNature

Searching is one of the most vital features on AskNature. Without it, most users would never discover the wealth of biological strategies available on the site. We provide two mechanisms for finding content: a traditional search and a taxonomy browser.

As shown in Table 2.2, visitors use our search engine more than the browsing option. Every page on AskNature includes a full-text search box with the prompt "How would Nature…." The prompt is meant to encourage users to search by function; that is, by inserting a verb: How would nature cool? How would nature create color? How would nature distribute fluids? How would nature build community? The search function defaults to return strategy pages before any other type of page within the system, such as user profile pages.

Table 2.2 Search and tri-browse usage from November 12, 2008, through December 31, 2012 AskNature's search engine is used more often than the tri-browse feature

	Page views	
Years	Browse	Search
2008	5,806	10,332
2009	45,286	130,731
2010	44,241	173,183
2011	61,602	271,777
2012	62,106	303,861

Source Google analytics

There are numerous possible queries, but they must be plausible and recognizable by the database. For example, if a user types in "build an airplane," the search may not return any useful results. After all, Nature does not build airplanes, but it is a genius at flight. If the user refines his search to look for things such as "generate lift" or "reduce drag"—that is, functions he would like his design to accomplish—chances are he will find much more relevant results.

While finding strategies via search can provide immediate results, it may be useful for practitioners to view multiple strategies meeting similar functional challenges. To accommodate this, users can browse strategies and products by function in our "tri-browse" page shown in Fig. 2.4. The tri-browse page enables users to explore different strategies solving similar functional challenges, but users may find inspiration by browsing related subgroups as well.

2.6.1 Query Analysis

We have collected aggregate search queries since AskNature launched in 2008. Table 2.3 presents the top queries, and Table 2.4 presents a random selection. These tables reveal a number of observations. First, our suggested queries on each search result page attract a number of hits, indicating users are following them as a way of browsing. Second, users are searching without the Biomimicry Taxonomy and are instead using it as a generic search for organisms, environments, and other buildings. Third, literal questions are used as queries, which indicates inexperienced Web searchers (we ignore "how," "what," "the," and other common function words to mitigate this).

In summary, visitors are using our search engine like a traditional keyword Web search engine and not searching using the Biomimicry Taxonomy. Further research is needed to determine whether recasting a user's query in terms of the Biomimicry Taxonomy would increase user success rates.

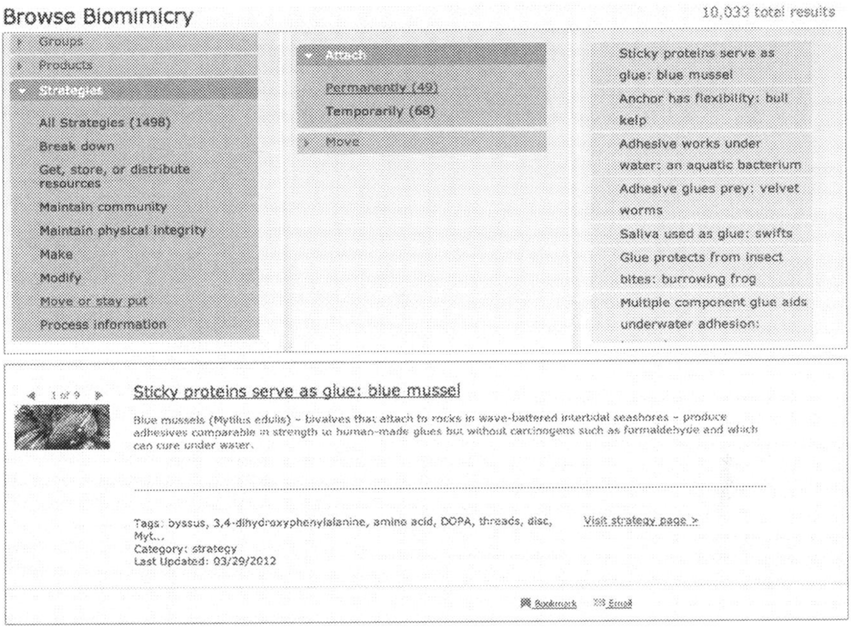

Fig. 2.4 Screenshot of the biomimicry taxonomy browser using group = modify, subgroup = attach, and function = permanently. Matching strategies appear in the rightmost column

Table 2.3 Top 15 search queries based on aggregate search logs from September 14, 2008, through February 20, 2012

Hits	Query
24964	Adhere to water
19359	How to purify water
18933	Water
18242	Structural color for painting
18201	Sticky berries adhere with strength and ease
18124	Capture water from fog in an arid environment
15046	Negative pressure used to suck moisture from soil
15035	Capture water from fogs
14096	Adhere to water
11896	Mutualism in nature
11719	Capture water from fog
11592	Hovering in mid-air
11233	Insects that capture water from fog
11177	How would nature purify water
10634	Structural color

The queries with a leading capital letter are suggested queries on each search result page

Table 2.4 Randomly selected queries from the same period as Table 2.3

Hits	Query
6172	A leaf capture water
5680	Green
4466	Plants that filter air
3267	How would nature resolve conflicts
2959	Solve green building challenges
2939	Green building
1698	Cool
1547	Improve air flow
1356	Organize an economy
1139	Build buildings
1207	Waterproof
1170	How to purify water, beavers
921	Solar
893	Glue
818	Desert

2.6.2 Social Network

AskNature is the execution of Benyus' vision (1997, p. 291) of a place where engineers and biologists can collaborate and share Nature's solutions to engineering problems. To support this, AskNature enables users to create profiles, comment on strategies and products, and discuss topics in forums. Users can connect with others in specific disciplines and countries also via the tri-browse page. We have not observed the level of collaboration desired, so we need to conduct a few user studies to determine how AskNature can be more conducive to collaboration.

2.7 Conclusion

In this paper, we have described AskNature.org and how we categorize its data according to the Biomimicry Taxonomy. We have learned four preliminary lessons from running a bio-inspiration database. First, if one is practicing biomimetic design, one needs tools to support it. AskNature is one such tool. Second, if a tool is going to support biomimetic design, it needs to provide a sizable corpus for inspiration. AskNature provides over 1,600 strategies. Third, it is important to provide a free and publicly available tool for anyone to use. AskNature's content is free and licensed under a Creative Commons Attribution-Noncommercial 3.0 License. Finally, AskNature is under development. It is an experiment, and we look forward to adapting AskNature to our user's needs.

Acknowledgements We thank Ashok Goel and our anonymous reviewers for their invaluable feedback.

References

AskNature (2008a) Eyes are anti-reflective: elephant hawk-moth. http://www.asknature.org/strategy/b389c40f5c8660fc082282bac3e00c3f

AskNature (2008b) Spiral fibers strengthen tree trunks: whitebark pine. http://www.asknature.org/strategy/e70b6b7753773a7177fe8358ec26ed2c

AskNature (2008c) Wing scales diffract and scatter light: Morpho butterflies. http://www.asknature.org/strategy/1d00d97a206855365c038d57832ebafa

AskNature (2011) Vitalis pet bottle. http://www.asknature.org/product/b5dc4030c48e41b03442d34aecdf46b4

Bar-Cohen Y (2006) Biomimetics–using nature to inspire human innovation. Bioinspir Biomim 1(1):P1–P12. doi:10.1088/1748-3182/1/1/P01

Benyus JM (1997) Biomimicry: innovation inspired by nature, 1st edn. Morrow, New York

Biomimicry 3.8 Institute (2011) Asknature selected as finalist for prestigious design competition. http://www.asknature.org/article/view/index2011

Glier MW, Tsenn J, Linsey JS, McAdams DA (2011) Methods for supporting bioinspired design. ASME Conf Proc 2011(54884):737–744. doi:10.1115/IMECE2011-63247

Helms M, Vattam S, Goel A (2009) Biologically inspired design: process and products. Des Stud 30(5):606–622

Helms M, Vattam S, Goel A (2010) The effect of functional modeling on understanding complex biological systems. ASME Conf Proc 2010(44137):107–115. doi:10.1115/DETC2010-28939

International Union for Conservation of Nature and Natural Resources (2012) The IUCN red list of threatened species. http://www.iucnredlist.org/

Stone RB, Wood KL (2000) Development of a functional basis for design. J Mech Des 122(4):359–370. doi:10.1115/1.1289637

The Earth Awards (2010) Systems winner. http://www.theearthawards.org/site/systems-win/

Vattam SS, Goel AK (2011) Foraging for inspiration: understanding and supporting the online information seeking practices of biologically inspired designers. ASME Conf Proc 2011(54860):177–186. doi:10.1115/DETC2011-48238

Vattam SS, Helms ME, Goel AK (2010) A content account of creative analogies in biologically inspired design. Artif Intell Eng Des Anal Manuf 24:467–481. doi:10.1017/S089006041000034X

Vincent JFV, Bogatyreva OA, Bogatyrev NR, Bowyer A, Pahl AK (2006) Biomimetics: its practice and theory. J R Soc Interface 3(9):471–482. doi:10.1098/rsif.2006.0127

Yen J, Weissburg M, Helms M, Goel A (2011) Biologically inspired design: a tool for interdisciplinary education. In: Bar-Cohen Y (ed) To appear in biomimetics: nature-based innovation. Newyork, Taylor & Francis

Chapter 3
A Natural Language Approach to Biomimetic Design

L. H. Shu and Hyunmin Cheong

Abstract Identifying relevant analogies from biology is a significant challenge for both designers who are interested in applying biological analogies for design and researchers who are developing general methods to support biomimetic design. This chapter discusses how a natural language approach facilitates the identification of biological analogies. We review methods developed to locate useful biological knowledge in natural language format, for example books and papers, as well as apply the identified knowledge to design solutions. We then discuss how these methods can bridge the gap between the increasing amounts of available information in biology and the goal of capturing computational design knowledge for biomimetic design. Finally, application examples of the natural language approach demonstrate that this approach can identify nonobvious analogies that are based on the transfer of abstract strategies, in addition to the direct mimicries that are used in most other biologically inspired design solutions.

Keywords Biomimetic design methods · Biomimetic design applications · Natural language processing · Computational linguistics · Information retrieval · Information extraction · Search keywords · Search categorization · Analogical transfer · Analogical mapping tools · Protocol analysis

---

L. H. Shu (✉) · H. Cheong
Department of Mechanical and Industrial Engineering, University of Toronto, Toronto, Canada
e-mail: shu@mie.utoronto.ca

H. Cheong
e-mail: cheong@mie.utoronto.ca

A. K. Goel et al. (eds.), *Biologically Inspired Design*,
DOI: 10.1007/978-1-4471-5248-4_3,

3.1 Introduction

Even as increasing examples of biomimetic or biologically inspired designs are developed and reported, we may have only scratched the surface in identifying and applying potential biological analogies. Designers who are interested in applying biological analogies are often limited by their own or others' bias, knowledge, or chance observation of relevant biological phenomena. However, much potentially useful biological information is already widely available but remains underutilized. Increasingly, books, papers, and other natural language resources that describe biological phenomena are becoming accessible in digital/online format. To enable a more objective and systematic approach to identifying possible analogies from the vast amount of such biological information, the lead author has chosen and pursued the natural language approach to biomimetic design since 2001. Objectives include developing strategies and techniques to first identify relevant biological analogies from natural language text, and second support designers in applying these analogies to develop design concepts. Shu (2010) and Shu et al. (2011) summarized the group's research work over the previous decade.

Other approaches in biomimetic design include modeling for, and compiling databases of, biological information. Such approaches may be particularly effective to aid in transferring knowledge from a chosen biological system to an engineering solution. However, a challenge that remains is identifying potentially useful biological knowledge to be modeled and entered into such databases.

This chapter first summarizes various approaches in biomimetic design to place our work in context. We then present the methods developed for the natural language approach, how computational techniques can help improve those methods, and the potential of the approach to support other biomimetic design research. Application examples are then presented, which demonstrate that the natural language approach is able to identify nonobvious analogies based on working strategy rather than superficial similarity.

3.2 Approaches to Support Biomimetic Design

Benyus (1997), a biological sciences writer who popularized the notion that humans emulate biological phenomena to design sustainable products and processes, founded the Biomimicry Institute, a clearinghouse for biomimicry researchers. A resource supported by the Biomimicry Institute is an online database of biological solutions, available at *AskNature.org*.

Vincent, a biologist working in engineering, is developing TRIZ (the Russian system for creative solution of problems) as the main tool of biomimetics. Vincent and Mann (2002) describe how TRIZ is adapted to support the transfer of knowledge from the biological to the engineering domain in a method they called 'BioTRIZ.'

In TRIZ, problems are first expressed in terms of contradictions or conflicts. Next, 'inventive principles' that typically solve those conflicts are identified and applied. TRIZ was developed by analyzing over two million Russian patent certificates. Vincent et al. (2006) created BioTRIZ using the same inductive approach as for TRIZ, but based on over 500 biological phenomena instead of patent knowledge. They found that the contradiction chart for BioTRIZ is only 12 % similar to the original TRIZ, suggesting that different principles are used to solve contradictions in biology. Notably, biological solutions rely less on energy, and therefore, the use of biologically inspired design is promising for the development of new and sustainable engineering solutions.

While Chakrabarti and Shu (2010) categorized biomimetic research as descriptive versus prescriptive, here we note that research on methods that support biomimetic design in general falls under the two high-level categories:

1. Methods to support search, retrieval, and representation of biological phenomena for design
2. Studies to better understand and therefore support the application of biological analogies to design

Clearly, the two categories are related and how the first category is implemented directly affects the second. Research under the above categories will be addressed in the order in which they are relevant during the biomimetic design process.

Helms et al. (2009) identify two directions in biomimetic design:

1. Solution driven, where an interesting biological phenomenon inspires the search for potential applications and
2. Problem driven, where a given problem motivates the search for biological analogies that could help solve the problem.

While a number of biologists and design researchers (Benyus 1997; Vincent and Mann 2002; Bar-Cohen 2006) have suggested potentially useful biological analogies for engineering applications, few methods have been developed to support systematic application of the solution-driven approach. The second direction, the problem-driven approach is more widely practiced, promoted, as well as studied. Since much of the general methods on biomimetic design support this approach, this section is devoted to the problem-driven approach.

One recurring message of this chapter is that there are several levels of biological organization, from the molecular, for example DNA, to the ecosystem/biosphere that can be exploited for biological analogies. Biological analogies are often recalled from the organ (e.g., lung, leaf) to organism (e.g., animal, plant) levels, evidenced by the many instances of existing biomimetic design based on organs and organisms. However, biomimetic design is more fully exploited by identifying analogies beyond the obvious ones that come to mind, which motivates more objective search approaches.

3.2.1 Problem Definition

The problem-driven approach requires that a problem be defined sufficiently to enable a meaningful search for biological analogies. The majority of work in biomimetic design suggests that functions are identified in this step. Functional modeling in engineering typically involves expressing the desired function to be fulfilled by a design solution. Complex high-level functions are decomposed into simpler functions, any one, or combination of which can be addressed using biological analogies. A primary goal in functional description is to avoid limitation to specific physical solutions for as long as possible.

Functions are typically represented by verbs. The use of nouns, except as objects of functional verbs, is usually discouraged as this may indicate that a physical solution is already in mind, and does not give other potential solutions fair consideration. However, the use of adjectives also has potential, as adjectives may describe desired qualities of the solution, for example flexible or transparent. Although to some extent, the use of adjectives also assumes certain physical entities, for example surfaces that are flexible or transparent, versus the need to allow for deflection or transmit light.

Other approaches may require redefining the original problem. In TRIZ and BioTRIZ, the problem is first reformulated as a conflict. Vincent et al. (2006) give an example where tire chains are needed on icy roads during winter to increase friction for safe driving. However, chains will damage ice-free road surfaces, and conventional alternatives such as changing between winter and summer tires are inconvenient. Therefore, a tire that can instantly adapt to road conditions is desired. The conflict becomes, 'How can friction between the tire and road increase, without increasing the weight (normal force) of the vehicle?'

Helms et al. (2009) and others suggest as a step between problem definition and solution search, to reframe or 'biologize' the problem, that is, by redefining problems in biological terms, often in the form of a question, for example, 'How do biological systems accomplish the desired function?' The example given re-frames or biologizes the function of 'stopping a bullet' into considering how biological entities 'prevent, withstand and heal damage.' However, this step must be undertaken with some caution, as it may predispose the designer to specific analogies or biological phenomena, and levels of biological organization.

Problem definition in biomimetic design is often iterative. Vattam et al. (2008) describe a complex interplay between problem decomposition and analogy retrieval and termed this process 'compound analogical design.' The authors note that multiple biological analogies could be used together to form a single design solution.

3.2.2 Search for Biological Analogies

There are a number of ways to search for biological phenomena that are relevant as possible analogies to a given problem, each with associated benefits and challenges. While the final goal is to identify an analogy that provides a working solution, the initial goal ought to be identifying a large variety of potential biological analogies in an objective manner.

Lindemann and Gramann (2004) propose a checklist of associations to translate between technical functions and biology terms. For example, the technical functions of 'change of the state of aggregation' and 'condense a gas' are associated with biological terms: nose passages, desert plants/animals, and leaves.

Helms et al. (2009) suggest several solution search heuristics including change constraints, identify champion adapters, examine variations within a solution family, and identify biological systems with multi-functionality.

With or without incorporating the above, below are approaches for finding biological analogies.

3.2.2.1 Ask Biologists Directly

The most obvious way of identifying potential biological analogies is to consult with biologists. AskNature (www.asknature.org) provides a social network of biologists specifically for this purpose. The advantage of simply asking biologists is that the engineer, generally untrained in biology, does not have to search for and interpret the relevance and potential application of information that may be unfamiliar. The disadvantage of this approach is that one must have access to biologists. In addition, biologists, as humans, may be biased toward their areas of expertise, rather than objectively recall and present a variety of phenomena as potential analogies.

3.2.2.2 Search Through Database

An obvious approach to address the limitations of asking biologists directly is to attempt to capture their knowledge in a database. AskNature (www.asknature.org) supports a database of potential biological analogies compiled by a network of biologists. One advantage of the database approach is that more focused search results are often possible. Frequently, the same keywords used to categorize biological phenomena and often, past engineering solutions developed based on them, are presented as search keywords. Therefore, the 'relevance' of information found is guaranteed since the same keywords used to categorize and enter information are used to then search the database.

Several engineering researchers propose modeling approaches to support the compilation of databases of biological knowledge for biomimetic design.

Chakrabarti et al. (2005) developed a model to represent causality of natural and artificial systems and used it to structure information in a database of systems from both natural and artificial domains. Bruck et al. (2007) created a design repository of bioinspired products and concepts to support bioinspired robotic projects for senior mechanical engineering students. Wilson et al. (2009) propose the use of reverse engineering and ontology to structure a database to support biomimetic design. Goel et al. (2011) constructed a functionally indexed knowledge base of structures and behaviors of biological and technological systems. Nagel et al. (2010) used functional modeling and a design repository to connect biological and engineering solutions.

The main disadvantage of searching through a database specifically developed to support biomimetic design is that the search results are limited to what was entered into the database. Depending on how the database is structured, bias may be imparted during the categorization of information as it is entered. A simple example is that while Velcro was developed from burrs, should the biological entity of a burr be categorized under the engineering function of fastening? If so, potential functions or strategies other than fastening that can be extracted from the burr may be lost.

3.2.2.3 Natural Language Based Search

In addition to building a database of biological phenomena categorized for engineering use, another approach takes advantage of the abundant biological information already available in natural language format (e.g., texts, papers) by searching them directly for relevant phenomena. This approach also avoids the enormous and likely subjective task of cataloging all biological phenomena for engineering. The natural language approach is supported by Vandevenne et al. (2011), who propose a scalable approach for the integration of large knowledge repositories in the biologically inspired design process. While this approach overcomes the many limitations of other approaches, there are also complex challenges, as detailed below.

3.3 Methods to Support the Natural Language Approach

This section summarizes methods developed to support the natural language approach. We also describe how tools from computational linguistics could be applied to these methods.

3.3.1 Source of Biological Information

The research group at the University of Toronto chose *Life, the Science of Biology* by Purves et al. (2001), a reference text for an introductory-level university biology course, as the initial source of natural language information in 2001. The text is written at a level that can be easily understood by engineers with little or no background in biology. In addition to ease of comprehension, introductory-level texts tend to be general and cover a wide range of organizational levels, from the molecular and cellular to the ecosystem, such that potential solutions are not limited to one or two familiar levels. The authors of the textbook kindly provided a digital searchable version to the University of Toronto to support this work. Although many other texts are available, this text was particularly well suited to this purpose.

Our group developed a search tool that identifies matches to keywords in the text. Other texts can be substituted or added as required for the initial or subsequent search. The challenging task is the initial identification of relevant biological phenomena. Once relevant phenomena are identified, further details can be found through more advanced texts, research papers, and traditional research methods. However, searching through more advanced sources initially will generate results that are in more technical language and thus more difficult to understand. Due to the designer frustration this may cause, such results are more likely to be overlooked even if they are relevant. Therefore, initial results from a more basic text may be more effective for introducing the subject and confirming relevance to the design problem. An understanding and confirmed relevance of the subject then motivates further research and effort to understand the possibly complex details needed to develop a solution.

3.3.2 Search Keywords

As the same keywords used to categorize phenomena are often presented as possible search keywords, searching databases specifically created to support biomimetic design tends to be more straightforward than searching natural language sources. Unlike with databases, finding relevant analogies in natural language text is subject to the word choices made by authors of the various natural language sources. That is, multiple terms can describe and thus be required to locate the same concept.

Vakili and Shu (2001) noted that synonyms are an obvious means to increase the number of matches for a given functional keyword. Chiu and Shu (2004) noted that troponyms, verbs that describe specific manners of another verb, for example, 'ambling' is a troponym of 'walking' as it is a particular manner of walking, also represent suitable alternative keywords.

Keywords used to search for biological analogies are verbs that describe the intended effect, or function, of desired solutions. Consistent with functional description, verbs are strongly preferred over nouns as keywords to initiate searches. Searching with nouns suggests preconceived analogies or solutions while searching for verbs that describe the desired action is more likely to objectively identify a wider range of biological forms that perform that action. For example, searching with the verb keyword, 'protect' will identify several biological phenomena that involve protection. However, searching using the noun keyword, 'shell' will only identify information related to shells, thereby limiting potential protection solutions to those that are based on shells.

WordNet (Miller 1995), an online lexical database, is used to identify synonyms or troponyms for originally formulated keywords. While WordNet uses semantic information to categorize its lexicon, VerbNet (Karin et al. 2002) uses both semantic and syntactic information to categorize verbs. Therefore, VerbNet can identify not only whether a particular verb has a similar meaning as another verb, but also whether both verbs can be used interchangeably in English without syntactic error. Therefore, VerbNet may be used to create more coherent groups of keywords.

3.3.2.1 Biologically Meaningful Keywords

Different lexicons, or vocabularies, used between engineering and biology to describe the same phenomena, motivated the identification of what Chiu and Shu (2005) refer to as biologically meaningful keywords. The motivating example involved the keyword 'clean,' which did not locate many useful matches for a problem involving cleaning. The advice of a domain expert was sought, who suggested 'defend' as an alternative keyword, since some organisms clean as a defensive mechanism. However, 'defend' is not intuitively related to 'clean' for most engineers and has no documented lexical relationship with 'clean,' for example, as a synonym. Therefore, Chiu and Shu (2005, 2007) developed a bridging method that identifies such highly relevant, but nonobvious, biologically meaningful keywords objectively and in a repeatable manner, without the need for domain experts.

The method uses word collocation and frequency analysis and is summarized in Fig. 3.1 for the 'clean/defend' example. First, hypernyms of 'clean' are found using WordNet. A hypernym is a more general form of a word, for example, 'remove' is a hypernym of 'clean.' WordNet is also used to identify related words, such as hyponyms/troponyms, of 'remove,' for example 'eliminate' and 'kill.' 'Eliminate' is a hyponym/troponym of 'remove,' as eliminating is a specific manner of removing. The original and additional keywords are then searched in the corpus, in this case, the Purves et al. (2001) text. Resulting text excerpts that contain these words are assessed for relevance using metrics that are described below.

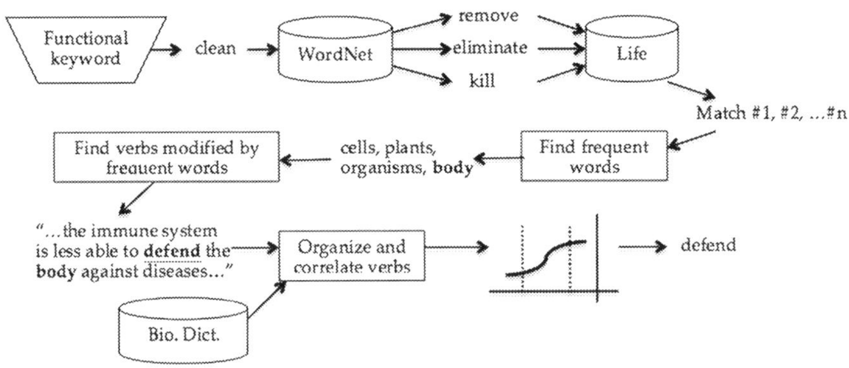

Fig. 3.1 Flow chart of the keyword bridging method, adapted from Chiu and Shu (2007), illustrates finding the biologically meaningful keyword ‘defend’ for the engineering keyword ‘clean’

Within the relevant matches, words, typically nouns, for example ‘cells,’ ‘plants,’ and ‘disease,’ which frequently occur, are then identified, along with the verbs that act on them. These verbs are then searched for in a biology dictionary. Verbs that are defined or part of a defined term are categorized as ‘biologically significant.’ For example, ‘abscise,’ meaning to naturally separate or fall off, such as by a dead leaf or ripe fruit, is part of the defined term ‘abscission,’ and therefore a biologically significant verb. Verbs that appear in the definition of terms, but are not defined themselves, are categorized as ‘biologically connotative.’ An example of a biologically connotative word is ‘defend,’ since it occurs in several definitions, but is not a defined term itself. Biologically significant and biologically connotative words combined are termed ‘biologically meaningful.’ Biologically meaningful keywords are then sorted by frequency of occurrence in a biology dictionary. Most useful bridge words are observed to occur between certain frequency cutoffs.

This method was able to identify ‘defend’ and other biologically meaningful keywords for ‘clean.’ Chiu and Shu (2007) used the method to also identify ‘survive’ as a biologically meaningful keyword for ‘encapsulate’ and ‘break’ or ‘breakdown’ as a biologically meaningful keyword for ‘release.’ A more obvious and straightforward approach to identify biologically meaningful keywords was to simply collect all verbs in the vicinity of the search verb. However, this method yielded far fewer useful, biologically meaningful words.

Chiu and Shu’s (2007) method involved a number of manual tasks, making it difficult to scale to larger amounts of text. Computational linguistic tools can be applied to automatize some of the processes. For example, when identifying frequent nouns, a part-of-speech tagger and stop list could be used to highlight more meaningful nouns. Also, finding the bridge verbs that modify frequent nouns can be automatically performed using the typed-dependency parser (de Marneffe et al. 2006), which can identify grammatical relationships between two words in a sentence, for example, between a verb and its object.

3.3.2.2 Biologically Meaningful Keywords for Verbs of the Functional Basis

Cheong et al. (2008, 2011) used the bridging method to identify biologically meaningful keywords for functional terms of the functional basis. The functional basis contains verb–object pairs intended to comprehensively represent the functionality of mechanical devices (Stone and Wood 2000). During the translation from functional to biologically meaningful keywords, Cheong et al. (2011) identified four categories where most biologically meaningful keywords appear. The four cases were used as criteria for identifying biologically meaningful keywords. Excerpts by Purves et al. (2001) illustrate these cases.

1. **Synonymous pair**: Words are used synonymously, often in the same sentence and adjacent to each other, for example, 'This information is received and *converted*, or *transduced*, by sensory cells into electric signals….'
2. **Implicitly synonymous pair**: Words are used synonymously, however, in separate clauses or sentences, for example, 'The xylem of tracheophytes *conducts* water from roots to aboveground plant parts. It contains conducting cells called tracheary elements, which undergo programmed cell death before they assume their function of *transporting* water and dissolved minerals.'
3. **Biologically specific form**: Biologically meaningful keywords describe a particular manner of the desired function, for example, '*Mutations* of one of the homeotic genes, bithorax, *transform* the third thoracic segment into a second copy of the second thoracic segment.'
4. **Causally related pair** (Causal relation): Biologically meaningful keyword is used to describe the enabling function for the desired function, for example, 'Humans *absorb* amino acids by *breaking down* proteins from food.'

A particularly interesting relation is the causally related pair, where the causal relation is a higher-order relation between the enabling and desired functions that may reveal a useful strategy for design-by-analogy. Cheong and Shu (2012) developed an algorithm, detailed in a later section, to automatically extract causally related functions from biological text. In addition, we believe that other semantic relations such as synonymous pairs or biologically specific forms could also be automatically extracted. For example, Hearst (1998) used lexico-syntactic patterns to automatically discover hyponym/hypernym relationships between nouns in a sentence. Perhaps, a similar method could be developed to identify synonyms/troponyms of verbs in a sentence, which could be applied to automatically identify synonymous pairs or biologically specific forms.

3.3.2.3 Adjectives as Keywords

While verb keywords can locate biological analogies that are functionally related to desired solutions, adjectives can be used to locate biological analogies that share similar qualities or working environments with the desired solution. For example, one can use the keyword ‘dry’ to find solutions that must work well in dry environments, or the keyword ‘wet’ to search for biological systems that enable moistness. Ke et al. (2010) used adjectives such as ‘wet,’ ‘humid,’ ‘dry,’ to search for biological analogies to inspire solutions for novel fuel cell bipolar plate concepts, where proper humidity is crucial to proper operation. Ke et al. (2010) explored searching for matches where the keyword is used as an adjective. Cheong and Shu (2012) implemented categorizing adjective search matches by the nouns modified by the adjectives.

3.3.3 Selection and Categorization of Search Matches

The biologically meaningful keywords described above often lead to a large number of matches. Not all matches are always useful and designers may find the large number of matches unmanageable. Therefore, techniques and strategies were devised to further identify and categorize relevant matches.

3.3.3.1 Abstract/Physical Phenomena

Hacco and Shu (2002) observed that keywords used in conjunction with non-physical phenomena contribute to many irrelevant matches. For example, in the following sentence, the verb ‘support’ acts on the abstract entity ‘theory’:

> Scientists have now found morphological evidence to **support** the theory. (Purves et al. 2001).

A designer wishing to find biological phenomena related to ‘supporting’ physical entities, may not find the above search result useful.

Ke et al. (2010) used the WordNet taxonomy to determine whether a particular noun is abstract or physical. At the highest level of the taxonomy, all nouns are classified as either an *abstract entity* or a *physical entity*. This information can be tagged on all nouns of the search corpus in advance to distinguish abstract objects. Creating this metadata is automatic and only needs to be performed once each time a new corpus is incorporated into the search tool.

3.3.3.2 Sense Disambiguation

Hacco and Shu (2002) noted that matches may contain search keywords used in different senses. For example, if the keyword ‘seal’ was used to look for biological analogies to sealing a joint, many matches that contain ‘seal’ in the aquatic mammal sense are clearly irrelevant.

Ke et al. (2010) explored using the word sense disambiguation algorithm developed by Pedersen and Kolhatkar (2009) to remove matches with keywords used in unintended senses. However, word sense disambiguation is still an ongoing challenge in computational linguistics, with the accuracy of many algorithms at only about 60 %.

3.3.3.3 Wikipedia Categorization

Ke et al. (2009) also used Wikipedia categories of biology-related entries to sort search results. This categorization can help designers skim through the search results based on the levels of biological organization or the types of organisms identified.

3.3.3.4 Subject/Object Categorization

Cheong and Shu (2012) used the typed-dependency parser (de Marneffe et al. 2006) to automatically identify the subjects and objects of verb keywords. Thus, search results for verbs can be sorted by both subjects and objects, which enable designers to more easily identify relevant search results for their design problems. For example, for the keyword 'prevent,' search results that contain 'water' as the object of 'prevent' may be especially useful to designers interested in waterproofing.

3.3.4 Evaluation and Application of Search Matches

Evaluating the relevance of potential analogies across domains is a more complex process than evaluating the relevance of matches to specific information sought. For example, in a traditional Web search for 'map of Toronto,' it is far more obvious which matches actually contain the map sought. However, with cross-domain analogies, the physical entities involved are likely to be different between analogy and problem. Thus, it is less straightforward to determine the relevance, as well as usefulness of a given match. Described below are challenges involved from the evaluation to the application of matches, and known ways of addressing these challenges.

3.3.4.1 Types of Similarity

Mak and Shu (2004) observed how students applied descriptions of biological phenomena functionally related to a given problem to solve that problem. Specifically, students were asked to develop concepts that result in 'clean clothes,' by

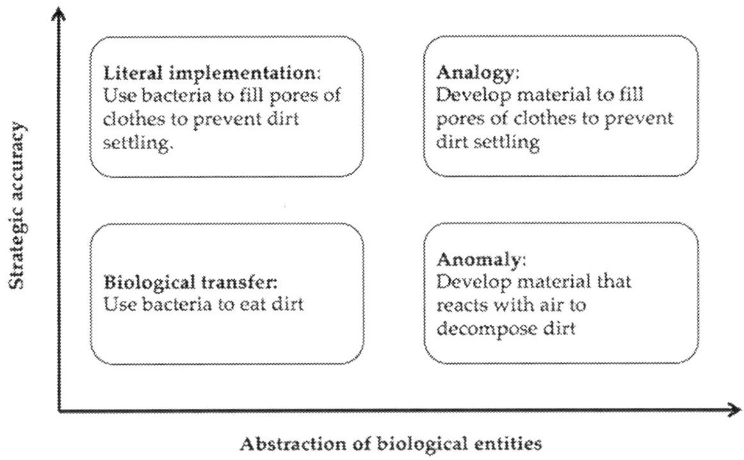

Fig. 3.2 Types of similarity comparisons between biological phenomena and developed concepts, adapted from Mak and Shu (2004)

using the following description of a biological phenomenon by Purves et al. (2001):

> Barriers and local agents defend the body—skin is a primary innate defense against invasion. The bacteria and fungi that normally live and reproduce in great numbers on our body surfaces without causing disease are referred to as normal flora. These natural occupants of our bodies compete with pathogens for space and nutrients, so normal flora are a form of innate defense.

Four similarity types were observed between the description of the biological phenomenon and the concepts developed using them as stimuli. These similarity types are shown with examples in Fig. 3.2, whose axes are strategic accuracy and abstraction of biological entities. Details on the four similarity types are as follows:

1. *Literal implementation*: A literal implementation involves using biological entities, for example bacteria, directly to solve the engineering problem, for example, by filling clothing pores to prevent dirt from settling. Here, the biological entities are not abstracted, but rather, used directly, with the same strategy between source and problem domains.
2. *Biological transfer*: A biological transfer involves transferring the biological entities, for example bacteria, into the solution domain, but without applying the strategy presented in the biological domain. For example, bacteria are used to provide the solution of clean clothes by eating dirt.
3. *Anomaly*: An anomalous solution involves neither the entities nor the strategy from the biological phenomenon. Some anomalous concepts are due to lack of understanding. Other anomalous concepts are likely due to fixation on a few words, for example 'motor' in motor proteins, in the text description while disregarding the overall strategy or principle presented.

4. *Analogy*: The intended analogous solution accurately applies the strategy from the biological phenomenon to the concept without transferring the biological entities, for example bacteria, into the solution.

3.3.4.2 Analogical Mapping Tools

Although identifying the undesired categories of similarity reduced the number of misapplied biomimetic design, two persisting difficulties observed by Mak and Shu (2008) in students using descriptions of biological phenomena were as follows: inability to transfer information from biology to engineering and fixation on specific phrases of the description of biological phenomena. This motivated further studies where students were provided outlines of strategies to be applied to both biological and engineering domains to facilitate analogical mapping. Compared to results without the mapping tool, the quality of generated concepts improved. However, although participants who used the mapping tool extracted strategies consistent with the biological phenomena presented, they continued to apply strategies to specific attributes of the given problem. Because generated concepts varied with selected attributes, a possible method to reduce fixation and increase the variety of solutions developed is to instruct the designer to list the problem attributes, for example 'clothing,' 'dirt,' 'detergent,' and then develop ways in which the biological strategies can be applied to each item in the list of attributes.

Further work by Cheong and Shu (2009) and Cheong et al. (2010) support that a template helps designers to recognize causal relationships in descriptions of biological phenomena and transfer this relationship across domains. Figure 3.3 shows an example of the template, which can be used to transfer functions of the relevant strategy, while abstracting and mapping the associated entities from the biological domain to the solution domain.

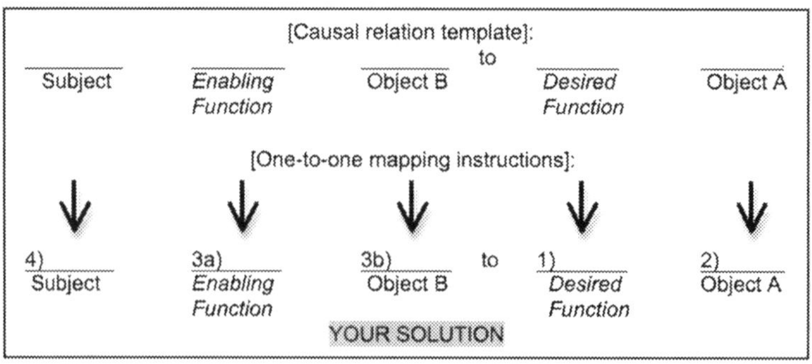

Fig. 3.3 Template for abstracting and mapping causally related functions from descriptions of biological phenomena, adapted from Cheong et al. (2010)

3.3.4.3 Understanding Analogical Reasoning in Biomimetic Design

The challenges observed in novice designers while identifying and applying biological analogies motivated us to study analogical reasoning from a different approach. In particular, we were interested in understanding why designers tend to fixate on certain features of biological phenomena and fail to identify the relevant analogy, as observed by Mak and Shu (2008), Cheong and Shu (2009), and Helms et al. (2009).

Therefore, Cheong et al. (2012) studied verbal protocols of fourth-year mechanical design students working in groups to solve one of three biomimetic design problems. Each problem had a corresponding description of a biological phenomenon to be used as a source of inspiration. The discourse of all group members was recorded. Both qualitative and quantitative analyses were performed on the verbal protocol.

We noted that participants made three types of similarity comparisons, based on Gentner's (1983) work on analogical reasoning:

- **Entity**: A comparison to superficial characteristics of entities of the biological phenomenon.
- **Function**: A comparison to functions of the biological phenomenon.
- **Strategy**: A comparison involving a higher-order relation (strategy) from the biological phenomenon.

Qualitatively, we noted that:

- Similarity between analogous elements at lower levels of comparison, for example superficial (entity) and functional, prevented novice designers from detecting the higher, overall strategy of the analogy.
- Domain knowledge can provide readily available associations at low levels of comparison and induce fixation.
- Novice designers focus on mapping multiple features of the source analog, instead of projecting multiple inferences (solutions) from the identified analogy, perhaps due to lack of confidence in the analogy.

Quantitative analysis was performed to evaluate the types of similarity comparisons made over time. Figure 3.4 shows the trend of similarity comparisons over time for one of the groups. This representation enables an objective assessment of the analogical reasoning process occurring during biomimetic design sessions. Such visualization of design protocols would be useful not only for researchers studying the protocols, but also for designers to better understand their own performance.

Applying computational linguistic techniques such as word frequency/co-occurrence count and lexical chain analysis may enable more fine-grained and automatic analysis of the analogical reasoning process. The application of natural language processing techniques to analyze verbal protocols has been widely used in other design research. Similar studies of verbal protocols in the context of

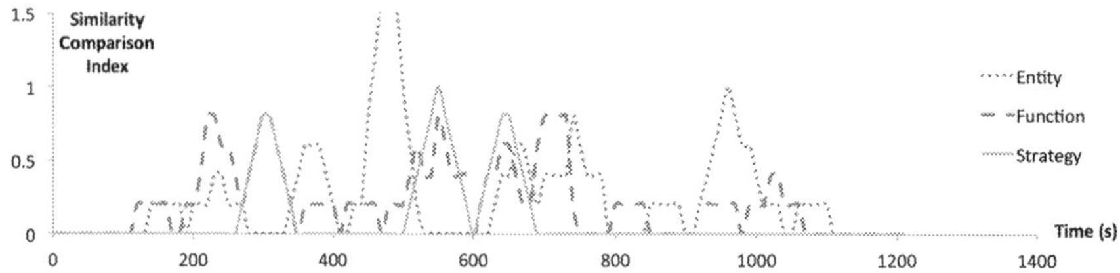

Fig. 3.4 Graph depicting the types of similarity comparison made over time, adapted from Cheong et al. (2012). The similarity comparison index on the y-axis is the rolling average of instances of similarity comparisons over five-second time segments

biomimetic design, which requires complex analogical reasoning, may warrant more focus in the future.

3.3.5 Automatic Extraction of Causally Related Functions

This section summarizes an algorithm developed and applied by Cheong and Shu (2012) to automatically extract causally related functions from text. As motivation, Cheong and Shu (2009) and Cheong et al. (2010) had noted that causally related functions in descriptions of biological phenomena may help designers identify and apply relevant analogies. Gentner (1983) puts forth that higher-order relations, such as causal relations, provide a framework for correct analogical transfer. Other researchers in biomimetic design, including Goel et al. (2009) and Chakrabarti et al. (2005), have modeled biological systems using a causality framework.

Computational linguistics may be applied to extract causal information from text. Some of the most promising algorithms (Khoo et al. 2000; Girju 2003) use a set of explicit causative words, for example 'causes' or 'because,' and lexico-syntactic patterns to identify cause and effect from text. While these algorithms are able to extract explicit causal relations between concepts, our work focused on extracting causally related functions that are implicit and may be more relevant for biomimetic design.

3.3.5.1 Extraction Method

A set of linguistic patterns that represent causally related functions was first manually identified from seven chapters, covering a variety of topics, from our search corpus, *Life* (Purves et al. 2001). Table 3.1 shows the linguistic patterns identified.

For each pattern, a particular grammatical relation is held between two causally related verbs that can be automatically detected from text parsed with Stanford typed-dependency parser v1.6.7 (de Marneffe et al. 2006). For example, in

'Lysozymes destroy bacteria to protect animals,' the parser is able to identify the grammatical relation *open clausal complement* between 'destroy' and 'protect,' that is, *xcomp*(destroy-2, protect-5). The numbers within the parentheses indicate the position of the tagged word in the sentence parsed. The definitions of the identified grammatical relations are shown in Table 3.1.

The extraction algorithm compares the part-of-speech tags and dependency relations of the corpus to rules defined based on the linguistic patterns of Table 3.1. The algorithm mainly looks for types of dependency relations, for example *xcomp*, *prepc_by*, or *purpcl*, and uses the part-of-speech tags to test additional rules, such as the exception listed under Pattern #1 in Table 3.1.

The extraction algorithm was implemented in our search tool, which takes a verb keyword as input and automatically retrieves enabling functions for the keyword verb. In other words, the search tool returns causal relations that are formed between the search keywords (desired functions) and the identified enabling functions. Figure 3.5 shows the results of an example search.

Some matches that contained nonmeaningful verbs were removed. These verbs include light verbs, simple causative verbs, and other frequently appearing verbs in the corpus. A light verb has little semantic meaning on its own but becomes more meaningful when combined with an object, for example 'use' versus 'use heat.' Simple causative verbs, including 'cause,' 'lead to,' 'allow,' are used to identify explicit causality, but on their own, do not tend to provide functional information. Manning and Schutze (1999) note that the most frequently appearing words in a corpus are likely to be semantically weak, that is, not meaningful. Cheong and Shu

Table 3.1 Typed-dependency relations and corresponding syntactic rules used to extract causally related functions, adapted from Cheong and Shu (2012). 'DR' stands for *dependency relation*

1. Lysozymes *destroy* bacteria to *protect* animals

   The verb 'protect' is an open clausal complement (DR: *xcomp*) to the verb 'destroy,' that is, 'protect' does not have its own subject, but has the same subject as 'destroy'

   Exception: When the first verb is intransitive, that is, does not have an object, the verbs are usually not causally related. For example, 'I like to swim' does not express any causality although 'swim' is defined as an open clausal complement to 'like'

2. Bacteria are *destroyed* to *protect* animals

   Similar to Pattern #1, the verb 'protect' is an open clausal complement (DR: *xcomp*) to the verb 'destroy.' In this case, however, the main verb 'destroy' is in the passive voice and the exception rule for Pattern #1 is ignored

3. Lysozymes *destroy* bacteria, *protecting* animals

   The verb 'protect' is an open clausal complement (DR: *xcomp*) to the verb 'destroy'

4. By *destroying* invading bacteria, lysozyme *protects* animals

   The gerund 'destroying' is a prepositional clausal modifier of the verb 'protect,' linked with the preposition 'by' (DR: *prepc_by*)

5. To *protect* animals, lysozymes *destroy* bacteria

   The verb 'protect' is part of a purpose clause modifier 'To protect animals,' which specifies the purpose of the following clause 'lysozymes destroy bacteria' (DR: *purpcl*)

6. *Destroying* bacteria *protects* animals

   The gerund 'destroying' acts as a clausal subject for the verb 'protect' (DR: *csubj*)

4 match(es) in which the enabling function of move is beating :
Section 31_12: Small ribbon worms **move** by **beating** their cilia .
Section 31_13: Flatworms (phylum Platyhelminthes) have no body cavity , lack organs for oxygen transport , have only one entrance to the gut , and **move** by **beating** their cilia .
Section 31_7: They **move** by **beating** these cilia rather than by muscular contractions .
Section 42_2: The tiny gametes of males , called sperm , are mobile and **move** by **beating** their flagella .

2 match(es) in which the enabling function of move is extending :
Section 43_3: Once they bulge into the blastocoel , they **move** by **extending** long processes called filopodia along an extracellular matrix of proteins that is laid down by the ectodermal cells lining the blastocoel .
Section 47_1: Amoebas **move** by **extending** lobe-shaped projections called pseudopods and then seemingly squeezing themselves into those pseudopods .

Fig. 3.5 Enabling functions for the keyword 'move' extracted and categorized in search tool using Purves et al. (2001), adapted from Cheong and Shu (2012)

(2012) confirmed that the most frequent verbs in the corpus included many light and simple causative verbs and therefore created a procedure to remove matches with such semantically weak verbs.

3.3.5.2 Algorithm Accuracy

Development testing was performed to examine the accuracy of the extraction algorithm and identify sources of errors to be removed. The performance of the algorithm was compared against the manual coding of causal relations by a single coder. The test was performed on three randomly selected chapters that were not used for algorithm development. Precision, recall, and F-measure, which are defined below, were used to assess the accuracy.

- Precision = (# of correctly retrieved causal relations)/(# of causal relations retrieved)
- Recall = (# of correctly retrieved causal relations)/(# of causal relations in text)
- F-measure = 2 * (Precision) * (Recall)/[(Precision) + (Recall)].

The extraction algorithm scored precisions of 0.800–0.933, recalls of 0.762–0.778, and F-measures of 0.780–0.848. For all three chapters, high precision and moderate recall were obtained. The narrow range of performance measures suggests the consistency of the extraction algorithm over different topics of text.

The majority, that is, 69 %, of the sources of errors involved parsing errors. These include the parser incorrectly identifying the part of speech of a relevant verb, not being able to find dependency relations between causally related verbs, and incorrectly identifying the dependency relation between causally related verbs. The most frequent nonparsing error was when the algorithm was unable to identify a complex causal relation.

3.3.5.3 Challenges

Most of the processing tasks required for the causal relation extraction are automated. Also, the extraction rules for the algorithm use domain-independent linguistic patterns. Therefore, the process is scalable to extracting information from a large amount of text. On the other hand, the algorithm does not completely capture causally related functions in biological text, as described below.

Causally related functions from multiple sentences. Currently, the parser only identifies grammatical relations between words within a single sentence, which limits the extraction algorithm to find causally related functions from a single sentence. Identifying causally related functions across multiple sentences would require anaphora resolution. Anaphora resolution refers to determining the antecedent or referent of an anaphor, such as the pronouns 'them' or 'it.' Hobbs (1978) first tackled the problem by analyzing a parse tree, while Soon et al. (2001) recently used a set of syntactic/semantic conditionals to evaluate the candidate referent of anaphors. However, the performance of these algorithms still requires improvement for practical applications.

Causally related functions involving light verbs. The current algorithm removes matches with instances of light verbs. A technique called *light verb construction* reduces a light verb and its following object into a 'heavy' verb that has more semantic context on its own, for example, by reducing 'to take a walk' into 'to walk.' The simplest technique identifies whether the complement noun itself can be used in a verb form. Stevenson et al. (2004), however, observed various limitations in light verb construction. For instance, original nuances may be lost, for example, the meaning of 'to use heat' differs from the meaning of 'to heat.'

3.4 Examples Applying the Natural Language Approach

The natural language approach enables designers to retrieve biological analogies that are not limited to those compiled in databases. Our research group applied the natural language approach to a number of case studies, including design for remanufacture (Shu and Flowers 1999; Hacco and Shu 2002), authorized disassembly (Saitou et al. 2007), microassembly (Shu et al. 2003, 2006), sensing (Lenau et al. 2008), redesign of fuel cells (Currie et al. 2009; Ke et al. 2009, 2010), protection during hobbies (Cheong et al. 2008), and protection from lunar regolith (Davidson et al. 2009). Below, three of these case studies, incorporating insights as they were gained over the course of a decade, are presented.

3.4.1 Design for Remanufacture

Snap fits are often used as a fastening method due to their ease of assembly. However, snap fits frequently break and are difficult to repair during remanufacture. For example, a toner cartridge with failed snap fits would likely be discarded rather than repaired or reused. Hacco and Shu (2002), therefore, searched for biological analogies that could inspire solutions to redesign snap fits for easier remanufacture.

3.4.1.1 Keywords and Biological Phenomena Identified

Initial keywords 'remanufacture' and 'refurbish' did not retrieve any matches in the biological text. The authors therefore used direct and related synonyms of the initial keywords for the search, which included keywords such as 'renew,' 'restore,' 'repair,' 'correct.'

The keywords 'repair' and 'correct' located useful biological phenomena. The first phenomenon is based on the strategy used by plants to replace damaged parts:

> The defense systems of plants and animals differ. Animals generally **repair** tissues that have been infected. Plants, on the other hand, do not make repairs. Instead, they seal off and sacrifice the damaged tissue so that the rest of the plant does not become infected. This approach works because most plants, unlike most animals, can replace damaged parts by growing new stems, leaves, and roots (Purves et al. 2001).

Relevant to the design problem, plants do not expend resources to repair damaged parts but instead grow new parts to replace damaged ones. This strategy can be implemented as a product component designed such that the failed feature can be easily removed and replaced.

Another useful phenomenon located was based on fainting:

> Blood must be returned from the veins to the heart so that circulation can continue. If the veins are above the level of the heart, gravity helps blood flow, but below the level of the heart, blood must be moved against the pull of gravity. If too much blood remains in the veins, then too little blood returns to the heart, and thus too little blood is pumped to the brain; a person may faint as a result. Fainting is self-**correct**ing: A fainting person falls, thereby moving out of the position in which gravity caused blood to accumulate in the lower body (Purves et al. 2001).

The biological phenomenon of fainting suggests that a form of defensive failure can be an effective strategy in preventing more serious failure.

3.4.1.2 Solution Developed

Incorporating the preemptive failure strategy to the redesign of snap fits, Hacco and Shu (2002) specified predetermined breakpoints, shown in Fig. 3.6a, that may cause earlier failure, but the part containing the snap fit feature can be more easily reused.

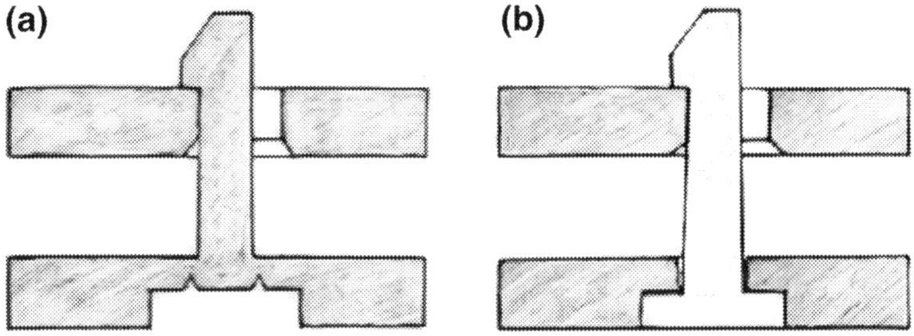

Fig. 3.6 Snap fit redesigned to facilitate easy refurbishment, adapted from Hacco and Shu (2002). **a** Re-designed snap fit. **b** Replacement snap fit installed

Incorporating the strategy to sacrifice and replace parts, a possible planned replacement part is shown in Fig. 3.6b that can be used once the sacrificial feature in Fig. 3.6a fails.

3.4.2 Microassembly

Size effects complicate the handling and assembly of micromechanical parts. Specifically, surface-related forces, for example, electrostatic, van der Waals and surface tension forces, dominate gravitational forces. As a result, sticking between the micropart and the gripping device during release hinders the automation of microassembly operations. Shu et al. (2006) developed biologically inspired solutions to overcome sticking in microassembly.

3.4.2.1 Keywords and Biological Phenomena Identified

Keywords used include ‘remove’ and ‘release.’ In addition, ‘defend’ was selected as a keyword, because biological systems often ‘remove’ parts as a defensive mechanism.

The keyword ‘defend’ led to a match with DNA transcription as the basis for how proteins are selectively synthesized to defend against infections, which led to a concept where features with different geometries would be used to maximize or minimize surface contact with the microobject as needed. However, this complex interaction mapped into a relatively complex solution compared to one that was developed based on the abscission principle, as described below.

Multiple keywords, including ‘defend,’ ‘remove,’ and ‘release’ led to the biological phenomenon of abscission. Abscission is the process by which leaves, petals, and fruits separate from a plant. A hormone called auxin is strategically released in plants to direct growth. When auxin is no longer produced, other hormones are released. The combined effect of these hormones breaks down parts of plants, for example, stalks of leaves damaged through infection or are no longer needed, as in the winter season, such that they become completely detached from the plant. The base of some leaves contains the abscission zone, which is a special layer of cells. Without auxin, these cells swell and form a cork-like material, which cuts off nutrients to the leaf, forms a seal between the leaf and the plant, and protects the plant once the leaf separates.

3.4.2.2 Solution Developed

The abscission principle was applied abstractly to overcome difficulties associated with 'sticking' as follows. The microobject is released together with a part of the tool designated as sacrificial, which can be of significant mass to take advantage of gravity. The object can thus be easily released, and the sacrificial part of the tool can then either remain with the microobject or be subsequently removed.

For the specific application of inserting a 0.6 mm metallic microscrew into a plastic counterpart, the abscission zone is physically implemented as a polypropylene rod of 4 mm diameter that is easily gripped and positioned by a small industrial robot with six degrees of freedom and a specified repeatability of $\pm$0.02 mm (see Fig. 3.7). The tip of the polypropylene rod is locally melted by heating and then pressed over the head of the screw. Upon solidification of the polypropylene, a solid bond between the rod and the screw is formed, and a robot can then manipulate the screw into the plastic counterpart. Once the screw is tightened into its final position, the resulting increased torque will break the bond between screw and rod.

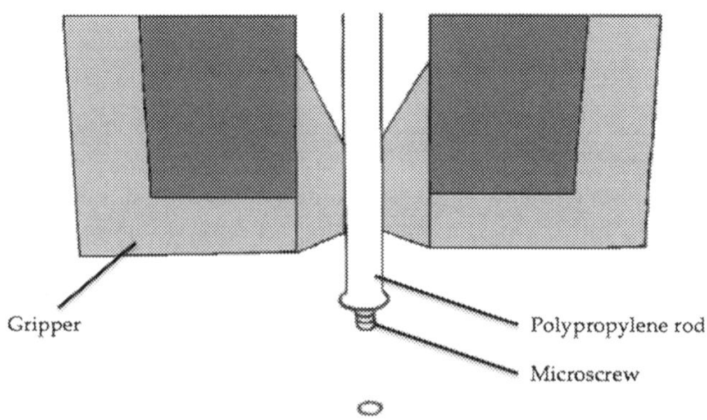

Fig. 3.7 Implementation of the abscission strategy for micropart release, adapted from Shu et al. (2006)

3.4.3 Protection from Lunar Regolith

The Light Detection and Ranging (LIDAR) device is an optical instrument that can detect particle concentrations kilometers above the instrument itself. A LIDAR consists of a high-powered laser that points upward and an optical receiver. A cover, when closed, protects both the laser beam canister and the receiver lens. Improved protection is sought for the lens both during operation (when the cover is open) and while idle. Lunar regolith, or lunar dust, causes significant problems in lunar exploration because the dust is pervasive, abrasive, and has been known to damage and cause premature failure of lunar exploration equipment.

Two components of the lunar environment limit the choice of materials. First, the thin lunar atmosphere approximates a hard vacuum and eliminates the specification of materials such as polymers, which will outgas in a vacuum and result in severe physical degradation. Second, in the most common lunar exploration areas, temperatures range from 120 °C during the day to −150 °C at night. Solutions therefore must accommodate large and rapid temperature swings that occur during the change from day to night and vice versa, further limiting the selection and arrangement of materials.

Both mechanical and electrostatic aspects contribute to lunar regolith adhering strongly to all surfaces. Without an atmosphere, no wind rounds regolith particles. The sharp, jagged edges of regolith cause the mechanical aspect of adhesion, as well as abrasion of mechanical seals. Furthermore, the small size of most regolith particles (below 70 μm) leads to infiltration of almost all mechanical systems. Positive charging of particles by solar wind during the lunar day and negative charging by plasma electron currents at night enable particles to cling to ungrounded surfaces. Finer regolith grains will levitate under this electrostatic charging and thus, dust is present at the instrument level, even without mechanical disturbance of regolith.

3.4.3.1 Keywords and Biological Phenomena Identified

'Protect' and its corresponding biologically meaningful keywords 'cover,' 'surround,' 'inhibit,' 'destroy,' 'change shape,' 'bind,' 'repel,' etc., are used as the search keywords.

Several biological phenomena were found to be relevant for potential design solutions. The keyword 'protect' identified bivalve shells that deflect sand away from mating surfaces and have a ligament joint that is more resistant to particle fouling than a rotational hinge. The keyword 'repel' identified phenomena such as waxy coating on hair that repels water and the use of like charges to repel. The keyword 'destroy,' although seemingly antonymous to 'protect,' identified phenomena describing proteins destroying invading bacteria to protect animals.

3.4.3.2 Solutions Developed

Strategies from the first two biological phenomena, bivalve shells and repelling charges, were implemented in the chosen two-piece cover system concept, shown in Fig. 3.8 that mimics the curvature and ligament joint of the bivalve shell. The curvature of the lids reduces the amount of lunar dust falling onto the inside lens area during opening, and the ligament hinge avoids the relative motion found in rotating joints, increasing resistance to invasive dust particles. When the cover is open during operation of the LIDAR, high-voltage DC electromagnetic fields repel dust particles away from the lens surface.

3.4.4 Conclusions from Examples

Below are retrospective observations from first-hand experience in the selection of biological analogies, extraction of biological strategies and their application in engineered systems described in our examples.

3.4.4.1 Identify Other Preventive Solutions in Biology

Many of the biological analogies discussed in our examples are based on preventive strategies in biology. For the remanufacture and LIDAR protection example, this is not surprising because of the nature of the problems and the keywords used for the searches. On the other hand, it is interesting that a preventive strategy, that is, abscission in plants, was also identified and finally chosen to address the microassembly release problem.

In biology, the consequence of failure or damage may be dire. Therefore, strategies such as abscission and fainting are used before more critical damage to

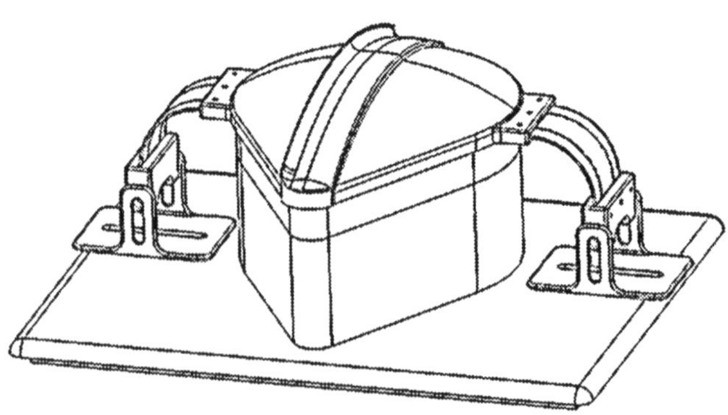

Fig. 3.8 CAD model of the bivalve-inspired lid assembly, adapted from Davidson et al. (2009)

the organism occurs. There is likely an abundance of preventive strategies in biology that are highly relevant to engineering, particularly the development of more sustainable products and processes.

3.4.4.2 Implement Strategies Passively to Enable/Replace Active Solutions

Although not detailed in this chapter, earlier reports of both the remanufacture and microassembly examples include biological analogies that were not as obviously or successfully implemented in the engineering solution. For example, identified using the keyword ‘repair’ for the remanufacture problem are DNA repair mechanisms. Specifically, excision repair targets damaged sections of a DNA molecule, including that which occurs during the life of the cell. Chemically damaged abnormal bases are excised and replaced with functional bases. Clearly, this phenomenon is highly analogous to the repair of damaged parts during remanufacture. The text on excision repair by Purves et al. (2001) follows:

> For example, in excision repair, certain enzymes ‘inspect’ the cell’s DNA. When they find mispaired bases, chemically modified bases, or points at which one strand has more bases than the other (with the result that one or more bases of one strand form an unpaired loop), these enzymes cut the defective strand. Another enzyme cuts away the bases adjacent to and including the offending base, and DNA polymerase and DNA ligase synthesize and seal up a new (usually correct) piece to replace the excised one.

While the above description confirms relevance, it is difficult to apply the active strategies given. For instance, who or what can be used in place of enzymes in an engineering solution to ‘inspect, find, cut, synthesize, seal up, and replace’ the defective parts, activities needed in excision repair? Relegating these duties to a human repair technician confers no improvement over the current situation in remanufacture. Further details on excision repair from a more advanced source mention conformation changes that result from the interaction of enzymes and contribute to the above activities in the repair sequence. The conformation change strategy was then matched to the engineering strategy where failure induces a conformation change that helps release the part for replacement, for example, self-removal during failure, somewhat implemented already in the chosen solution described.

Another biological analogy identified for the microassembly example also involves complex molecular interactions. This phenomenon was developed into a potentially feasible physical concept, but was far more complex to implement than the polypropylene rods used to represent the abscission zone in the chosen solution described. However, as engineering operates at increasingly smaller scales, it is possible that these molecular strategies can be better implemented directly.

In the lunar regolith example, the charge-based solution, although also ‘active’ in that energy is required, can be implemented without the need for a physical actor, for example, an astronaut, a robot, or a wiper to clear the lens of regolith while the LIDAR is being used, all of which would be more complex as well as potentially unsafe.

3.4.4.3 Consider Analogies by Environment in Addition to Function

In addition to searching for and selecting biological analogies based on functional similarity, as suggested by many biomimetic design researchers, the lunar regolith example highlights the role of environmental similarity. Specifically, while the clam was identified more by superficial than functional characteristics, it turned out to provide a wealth of transferable strategies. The clam's physical characteristics that are suited to its particulate-laden environment, for example, shell shape that sheds sand and ligament hinge less susceptible to particle fouling, are also well suited to protection from lunar dust. Conversely, strategies from a more obvious eye protection analogy, though more functionally similar, turned out to be much less directly transferable to the lunar environment. For example, tears/fluids are unsuitable in the lunar environment, and wiping (by the eyelid) is unsuitable for abrasive particles. Therefore, we now support the use of adjective keywords and sort matches by the objects the adjectives describe, in our search tool.

Helms et al. (2009) indirectly refer to the identification of biological analogies by environment as well as function though champion adapters—organisms that survive in the most extreme of the environment of interest.

3.4.4.4 Explore Transferring Strategic in Addition to Superficial Similarity from Biological Analogies

While many of our example solutions, for example, for remanufacture and microassembly, apply strategies from biological phenomena that share little superficial similarity, many biomimetic design examples mimic biological phenomena that share both functional and superficial similarity with the design solution. For example, many legged robots are based on, and appear superficially similar to, insects; Velcro shares both functional and physical similarity to the burrs on which it is based; even lotus-effect-based solutions share physical and functional similarities at the micro- and possibly nano-level with the lotus surface.

Several design researchers (Bonnardel 2000; Hon and Zeiner 2004; Tseng et al. 2008) report that cross-domain analogies, which often involve functional but not superficial similarities, can evoke creative design solutions. Cross-domain analogies are frequently more challenging to apply than within-domain analogies because designers must identify similarities at the functional level, while there may be little similarity at the superficial level. However, because it is easier for humans to recall and observe information based on superficial similarity, designers left to identify biological analogies based on their own or others' observation or recollection of knowledge may be more likely to develop solutions that directly mimic biological systems/phenomena based on superficial similarities.

Our natural language search approach has focused on using keywords, for example verbs and adjectives, to more objectively identify potential analogies based on functional and environmental similarity. Our examples have demonstrated the ability of this approach to identify analogies that seem unrelated, but are in fact functionally or environmentally relevant to the design problems. Our approach is consistent with artificial intelligence systems that make comparisons between designs at functional, behavioral, or causal levels to support analogical design (Goel 1997).

3.4.5 Conclusions on the Natural Language Approach

Our work on the natural language approach to support biomimetic design has progressed to incorporate more computational linguistic tools and to automatize the processes involved. Our latest work on the automatic extraction of causally related functions in particular creates the potential to extract useful information from a large amount of text sources. At the same time, it is essential to question whether the information extraction techniques are capturing the appropriate type of information, and the completeness of the captured information.

In addition, difficulties experienced by designers when applying analogies found in natural language sources must be considered. Natural language text, or language in general, is a flexible yet rich medium to represent and communicate knowledge. How designers interpret text descriptions of biological knowledge and form appropriate analogies to design solutions should be studied in more detail. Interestingly, computational linguistic tools can also assist in such cognitive studies.

Described below are other possible benefits of the natural language approach, including its potential to bridge the gap between the increasing amount of biological information available and capturing meaningful information to support biomimetic design.

3.4.5.1 Translate Natural Language Text to Other Formats that Support Biomimetic Design

An extensive amount of information in biology is becoming available in digital/ online format. For example, Rebholz-Schuhmann et al. (2005) observed that the number of articles available from Medline (Medical Literature Analysis and Retrieval System Online) is exponentially increasing.

Although indexing biological information for the purpose of biomimetic design is extremely useful, we cannot expect either biologists or designers to compile all the available information (Bar-Cohen 2006). AskNature (www.asknature.org), a popular database of biological information for biologically inspired design, has been able to index about 1,400 strategies over the past five years, with most strategies compiled before 2009. Vandevenne et al. (2011) noted that manually compiling information would not efficiently integrate the vast majority of biological knowledge into design.

Importantly, most of the growing digital information is in the form of natural language text. The field of natural language processing has existed since the 1950s (Turing 1950). Therefore, plenty of tools exist that can be used to capture meaningful information from the biological domain. This chapter referenced a number of these tools that support the natural language approach to biomimetic design.

One promising direction is to automatically translate natural language text into a form that more directly supports other approaches to biomimetic design. Many modeling frameworks developed for biomimetic design, for example, SBF models by Goel et al. (2009) and SAPPhIRE by Chakrabarti et al. (2005), may benefit from techniques developed to automatically translate text information to the specific representation format used in modeling. Databases to support biomimetic design become increasingly useful according to the amount of knowledge they contain.

Limitations of this approach exist because English, for one, is a highly ambiguous language. Many areas of natural language processing, such as word sense disambiguation, anaphora resolution, and speech recognition, are actively being researched. The accuracy of the information extraction and translation efforts may be limited by the performance of the state-of-the-art natural language processing techniques. On the other hand, increased application of natural language processing in research domains other than computational linguistics will likely support the advancement of existing techniques.

3.4.5.2 Determine Linguistic Relations Relevant to Biomimetic Design

An important step in the translation of natural language text to other useful formats is identifying which syntactic or semantic information must be captured. For example, the causal relation extraction algorithm used syntactic relations, such as *open clausal complement*, *prepositional clause modifier*, etc., shown in Table 3.1, to identify causally related verbs. Most computational models in biomimetic design are based on formal frameworks to index and reason with biological information. Chakrabarti et al. (2006) defined different relations of verbs, nouns, and adjectives to express each SAPPhIRE construct. Goel et al. (2009) applied well-defined syntax and semantics to specify SBF models. Nagel et al. (2010) used functional basis lexicons to form verb–object pairs to represent biological systems. Because these approaches have already defined specific frameworks to index relevant biological information, researchers could then determine whether linguistic relations relevant to the corresponding frameworks can be identified in natural language text.

Since patterns of linguistic relations are present in all natural language text, once techniques and strategies are developed to identify specific sets of useful linguistic relations, they could be easily shared or extended to other information extraction tasks. In a sense, the linguistic relations found in natural language text could be considered an abstract framework that can be developed into specific frameworks suitable for specific approaches of modeling biological information.

One challenge in any computing task is ensuring completeness. For instance, the syntactic relations listed in Table 3.1 may not capture all causally related verbs in a corpus. In fact, another syntactic relation that could be used to identify causally related verbs was located while testing the relations. Therefore, the framework required for information extraction may be learned while examining the source of the information sought.

While many researchers in biomimetic design are working toward developing frameworks for indexing/reasoning, fewer are working on locating appropriate information. The examination of linguistic relations and corresponding computational linguistic techniques may thus advance biomimetic design.

3.4.5.3 Identify Patterns in Biological Phenomena

One benefit that comes with the ability to process a large amount of information is the chance to identify patterns in information. TRIZ was developed based on manually observing patterns of innovation in over 40,000 patents and has been used as an effective problem-solving tool. Such a procedure is obviously difficult to replicate, given the significant undertaking required. BioTRIZ developed by Vincent et al. (2006) was based on about 500 biological phenomena. Son et al. (2012) have recently investigated the patterns of biological transformations in 113 biological systems/organisms.

With highly automated information extraction techniques, such as the causal relation extraction algorithm by Cheong and Shu (2012), patterns of strategies used by biological systems could be identified. These patterns may reveal ubiquitous solutions employed in nature, which could be mapped to solve a variety of problems in engineering. Hoeller et al. (2007) propose that capturing recurring solutions in biology would inspire effective designs of sustainable products or services. Yen and Weissburg (2007) observe that the generality and robustness of a particular biological strategy may be determined by whether that strategy is implemented across many organisms. In addition to advancing biomimetic design, significant findings on patterns of biological strategies may also advance our understanding of biology.

3.5 Summary

This chapter summarized previous work on the natural language approach to biomimetic design and discussed some possible directions for future work. Three key points are as follows:

1. The natural language approach can enable designers to identify nonobvious, previously unused analogies as demonstrated by our group's application examples.
2. Various computational linguistic tools can be used to develop and automate the natural language approach.
3. An automated natural language approach could be applied to translate the enormous amount of biological information in natural language format to knowledge that more directly informs design, thereby supporting other work in biomimetic design.

References

Bar-Cohen Y (2006) Biomimetics—Using nature to inspire human innovation. Bioinspiration Biomimetics 1(1):1–12

Benyus JM (1997) Biomimicry: Innovation inspired by nature. William Morrow & Co, New York

Bonnardel N (2000) Towards understanding and supporting creativity in design: Analogies in a constrained cognitive environment. Knowl-Based Syst 13(7–8):505–513

Bruck H, Gershon A, Golden I, Gupta S, Gyger L, Magrab E (2007) Training mechanical engineering students to utilize biological inspiration during product development. Bioinspiration Biomimetics 2(4):S198–S209

Chakrabarti A, Shu LH (2010) Biologically inspired design. Spec Issue Biologically Inspired Des AIEDAM 24(4):453–454

Chakrabarti A, Sarkar P, Leelavathamma B, Nataraju BS (2005) A functional representation for aiding biomimetic and artificial inspiration of new ideas. Artif Intell Eng Des Anal Manuf 19(2):113–132

Cheong H, Shu LH (2009) Effective analogical transfer using biological descriptions retrieved with functional and biologically meaningful keywords. Proceedings ASME International design engineering technical conference, San Diego, CA, 30–2 Aug–Sept 2009, DETC2009-86680

Cheong H, Shu LH (2012) Automatic extraction of causally related functions from natural-language text for biomimetic design. Proc. ASME Int. Design engineering technical conference, Chicago, IL, 12–15 Aug 2012, DETC2012-70732

Cheong H, Shu LH, Stone RB, McAdams DA (2008) Translating terms of the functional basis into biologically meaningful keywords. Proceedings ASME international design engineering technical conference, NYC, NY, 3–6 Aug 2008, DETC2008-49363

Cheong H, Chiu I, Shu LH (2010) Extraction and transfer of biological analogies for creative concept generation. Proceedings ASME international design engineering technical conference, Montreal, Canada, 15–18 Aug 2010, DETC2010-29006

Cheong H, Chiu I, Shu LH, Stone RB, McAdams DA (2011) Biologically meaningful keywords for functional terms of the functional basis. J Mech Des 133:021007

Cheong H, Hallihan G, Shu LH (2012) Understanding analogical reasoning in biomimetic design: an inductive approach. Design computing and cognition DCC'12, College Station, TX, 7–9 June 2012

Chiu I, Shu LH (2004) Natural language analysis for biomimetic design. Proceedings of ASME design engineering technical conference, Salt Lake City, UT, 28–2 Sept–Oct 2004, DETC2004-57250

Chiu I, Shu LH (2005) Bridging cross-domain terminology for biomimetic design. Proceedings ASME international design engineering technical conference, Long Beach, CA, 24–28 Sept 2005, DETC2005-84908

Chiu I, Shu LH (2007) Biomimetic design through natural language analysis to facilitate cross-domain information retrieval. Artif Intell Eng Des Anal Manuf 21(1):45–59

Currie J, Fung K, Mazza AG, Wallace JS, Shu LH (2009) A comparison of biomimetic design and TRIZ applied to the design of a proton exchange membrane fuel cell. 6th International conference on innovation and practices in engineering design and engineering education, Hamilton, Ontario, 27–29 July 2009

Davidson M, Bligh D, Maloney N, McKnight C, Young W, Shu LH, Potvin M-J, Warkentin A (2009) Biomimetic design of a multi-layered dust protection system for optical instruments operating in the lunar environment. 6th International Conference on Innovation and practices in engineering design and engineering education, Hamilton, Ontario, 27–29 July 2009

de Marneffe M-C, MacCartney B, Manning CD (2006) Generating typed dependency parses from phrase structure parses. Proceedings of the international conference on language resources and evaluation, pp 449–454

Gentner D (1983) Structure-mapping: a theoretical framework for analogy. Cogn Sci 7(2):155–170

Girju R (2003) Automatic detection of causal relations for question answering. Proceedings of the ACL 2003 workshop on multilingual summarization and question answering, vol 12. pp 76–83

Goel AK (1997) Design, analogy, and creativity. IEEE Expert 12(3):62–70

Goel AK, Rugaber S, Vattam S (2009) Structure, behavior, and function of complex systems: The structure, behavior, and function modeling language. Artif Intell Eng Des Anal Manuf 23(1):23–35

Goel AK, Vattam S, Wiltgen B, Helms M (2011) Cognitive, collaborative, conceptual and creative—Four characteristics of the next generation of knowledge-based CAD systems: a study in biologically inspired design. Comput Aided Des 44(10):879–900

Hacco E, Shu LH (2002) Biomimetic concept generation applied to design for remanufacture. Proceedings ASME international design engineering technical Conference, Montreal, Canada, 29–2 Sept–Oct, DETC2002-34177

Hearst M (1998) Automated discovery of WordNet relations. In: Fellbaum C (ed) WordNet: An electronic lexical database. MIT Press, Cambridge

Helms M, Vattam SS, Goel AK (2009) Biologically inspired design: process and products. Des Stud 30(5):606–622

Hobbs JR (1978) Resolving pronoun references. Lingua 44(4):311–338

Hoeller N, Salustri F, DeLuca D, Pedersen Zari M, Love M, McKeag T, Stephens E, Reap J, Sopchak L (2007) Patterns from Nature. Proc. of the 2007 Society for Experimental Mechanics Annual Conference and Exposition on Experimental and Applied Mechanics, Springfield, MA, 4–6 June 2007

Hon KKB, Zeiner J (2004) Knowledge brokering for assisting the generation of automotive product design. CIRP Ann 53(1):159–162

Karin K, Martha P, Owen R (2002) Extending PropBank with VerbNet semantic predicates. Workshop on applied interlinguas, held in conjunction with AMTA-2002, Tiburon, CA, 8-12 Oct 2002

Ke J, Wallace JS, Shu LH (2009) Supporting biomimetic design through categorization of natural-language keyword-search results. Proceedings ASME international design engineering technical conference, San Diego, CA, 30–2 Aug–Sept, DETC2009-86681

Ke J, Chiu I, Wallace JS, Shu LH (2010) Supporting biomimetic design by embedding metadata in natural-language corpora. Proceedings ASME international design engineering technical conference, Montreal, Canada, 15–18 Aug 2010, DETC2010-29057

Khoo C, Chan S, Niu Y (2000) Extracting causal knowledge from a medical database using graphical patterns. Proc. of the 38th Annual Meeting of the Association for Computational Lingustics

Lenau T, Cheong H, Shu L (2008) Sensing in nature: using biomimetics for design of sensors. Sensor Rev 28(4):311–316
Lindemann U, Gramann J (2004) Engineering design using biological principles. Proceedings 8th international design conference design 2004, pp 355–360
Mak TW, Shu LH (2004) Abstraction of biological analogies for design. CIRP Ann 53(1):117–120
Mak TW, Shu LH (2008) Using descriptions of biological phenomena for idea generation. Res Eng Design 19(1):21–28
Manning C, Schutze H (1999) Foundations of Statistical Natural Language Processing. MIT Press, Cambridge
Miller GA (1995) WordNet: a lexical database for English. Commun ACM 35(11):39–41
Nagel JKS, Nagel RL, Stone RB, McAdams DA (2010) Function-based, biologically inspired concept generation. Special Issue Biologically Inspired Des AIEDAM 24(4):521–535
Pedersen T, Kolhatkar V (2009) WordNet::SenseRelate::AllWords—A broad coverage word sense tagger that maximizes semantic relatedness. Proceedings of the North American chapter of the association for computational linguistics—Human language technologies conference, Boulder, CO, 1-3 June 2009, pp 17–20
Purves WK, Sadava D, Orians GH, Heller HC (2001) Life, the science of biology, 6/e. Sinauer Associates, Sunderland
Rebholz-Schuhmann D, Kirsch H, Couto F (2005) Facts from text—Is text mining ready to deliver? PLoS Biol 3(2):e65. doi:10.1371/journal.pbio.0030065
Saitou K, Shalaby M, Shu LH (2007) Bioanalogous mechanical joints for authorized disassembly. CIRP Ann 56(1):33–36
Shu LH (2010) A natural-language approach to biomimetic design. Special Issue Biologically Inspired Design J AIEDAM 24(4):483–505
Shu LH, Flowers WC (1999) Application of a design-for-remanufacture framework to the selection of product life-cycle fastening and joining methods. Int J Robot Comput Integr Manuf (Special Issue Remanufacturing) 15(3):179–190
Shu LH, Lenau TA, Hansen HN, Alting L (2003) Biomimetics applied to centering in microassembly. CIRP Ann 52(1):101–104
Shu L, Hansen H, Gegeckaite A, Moon J, Chan C (2006) Case study in biomimetic design: handling and assembly of microparts. ASME international design engineering technical conferences & computers and information in engineering conference, Philadelphia, PA, 10–13 September 2006, DETC2006-99398
Shu LH, Ueda K, Chiu I, Cheong H (2011) Biologically inspired design. CIRP Ann 60(2):673–693
Son JJ, Raulf C, Cheong H, Shu LH (2012) Applying patterns of transformation in biology to design. Proceedings ASME international design engineering technical conference, Chicago, IL, 12–15 Aug 2012, DETC2012-71296
Soon WM, Ng HT, Lim DCY (2001) A machine learning approach to conference resolution of noun phrases. Comput Linguist 27(4):521–544
Stevenson S, Fazly A, North R (2004) Statistical measures of the semi-productivity of light verb constructions. Proceedings of the ACL 2004 workshop on multiword expressions: Integrating processing, pp 1–8
Stone RB, Wood KL (2000) Development of a functional basis for design. J Mech Des 122(4):359–369
Tseng I, Moss J, Cagan J, Kotovsky K (2008) The role of timing and analogical similarity in the stimulation of idea generation in design. Des Stud 29(3):203–221
Turing AM (1950) Computing machinery and intelligence. Mind 59:433–460
Vakili V, Shu LH (2001) Towards biomimetic concept generation. Proceedings ASME design technical conferences, Pittsburgh, PA, 9–12 Sept 2001, DETC2001-21715 (DTM)
Vandevenne D, Verhaegen P-A, Dewulf S, Duflou JR (2011) A scalable approach for the integration of large knowledge repositories in the biologically-inspired design process.

International conference on engineering design, Technical University of Denmark, 15–18 Aug 2011

Vattam SS, Helms ME, Goel AK (2008) Compound analogical design: Interaction between problem decomposition and analogical transfer in biologically inspired design. Design computing and cognition DCC'08, Atlanta, GA, pp 377–396

Vincent JFV, Mann DL (2002) Systematic technology transfer from biology to engineering. Philos Trans R Soc Phys Sci 360(1791):159–173

Vincent JFV, Bogatyreva OA, Bogatyreva NR, Bowyer A, Pahl A-K (2006) Biomimetics: its practice and theory. J R Soc Interface 3(9):471–482

Wilson J, Chang P, Yim S, Rosen DW (2009) Developing a bio-inspired design repository using ontologies. Proceedings ASME international design engineering technical conference, San Diego, CA, DETC2009-87272

Yen J, Weissburg M (2007) Perspectives on biologically inspired design: introduction to the collected contributions. Bioinspiration Biomimetics doi: 10.1088/1748-3190/2/4/E01

Chapter 4
A Thesaurus for Bioinspired Engineering Design

Jacquelyn K. S. Nagel

Abstract Biological systems provide insight into sustainable and adaptable design, which often leads to designs that are more elegant, efficient, and sustainable. There are, however, significant hurdles to performing bioinspired design. This chapter presents a design tool, the engineering-to-biology thesaurus, that addresses several challenges engineers may encounter when performing bioinspired design, allowing engineers without advanced biological knowledge to leverage nature's ingenuity during engineering design. Along with the thesaurus tables, detailed information on the thesaurus model, structure, population, term placement, term placement review, and limitations is provided. Applications of the design tool are discussed. Examples are provided to demonstrate the goals and applications of the design tool followed by a review of integration with computational design tools.

Keywords Thesaurus · Function-based design · Design tools · Functional modeling · Functional basis · Translation · Analogies · Concept generation · Identification · Inspiration search · Brainstorming · Dialogue facilitation

4.1 Introduction

Engineering design is considered a creative endeavor involving many activities, which encourages the use of engineering principles, imagination, prior knowledge, stored knowledge, and a designer's intuition to create engineering solutions. The resulting solution may or may not be innovative, novel, or what some would call creative; however, the design should fulfill a purpose or answer a need (Hyman 1998;

J. K. S. Nagel (✉)
Department of Engineering, James Madison University Harrisonburg,
Harrisonburg, VA, USA
e-mail: nageljk@jmu.edu

A. K. Goel et al. (eds.), *Biologically Inspired Design*,
DOI: 10.1007/978-1-4471-5248-4_4, © Springer-Verlag London 2014

Otto and Wood 2001; Dym and Little 2004; Ulrich and Eppinger 2004; Voland 2004; Ullman 2009; Dieter and Schmidt 2009; Cross 2008). To arrive at a solution, it is not uncommon for engineers to use design methods and tools, make analogies among different engineering disciplines (i.e., an electrical resistor and mechanical damper are mathematically analogous) during ideation to find solutions, or use metaphors to frame or assist in defining the design problem (Hey et al. 2008). The use of design tools and methods is recognized as standard practice in both industry and academia. These tools and methods play a pivotal role in many of the design activities from concept generation to detailed design. The leap made between engineering disciplines using analogies is to be expected as one gains more experience. Making a leap between domains, however, is less likely to occur without an impetus. Take for instance Velcro©, if it were not for the curiosity of George de Mestral that caused him to investigate how the tiny burrs he and his dog accumulated from walking through wooded areas, modern day hook and loop may never have been invented or it may not be as effective. George de Mestral's chance observation of a biological phenomenon resulted in a very simple, reusable material (de Mestral 1955) that has been used for securing everyday items such as shoes to mission-critical items needed for exploring space.

The natural world provides numerous cases for inspiration in engineering design. Though biological organisms, phenomena, and strategies, herein referred to as biological systems, provide a wealth of elegant and ingenious approaches to problem solving, there are challenges that prevent designers from leveraging the full insight into the biological domain. The leap from engineering to biological science and back has posed a challenge. Engineers often struggle with how to best use the vast amount of biological information available from the natural world around them. Often, it is because there is a knowledge gap or terminology is difficult or terminology takes different meanings. Moreover, the time required to learn and become fluent in biology poses too large a hurdle. This reveals a fundamental problem of working across the engineering and biology domains. The effort and time required to become a competent engineering designer creates significant obstacles to becoming sufficiently knowledgeable about biological systems (the converse can also be said).

This chapter presents a design tool, the engineering-to-biology thesaurus, that addresses the main barriers to bioinspired design and aims to lower the hurdle, allowing engineers without advanced biological knowledge to leverage nature's ingenuity during engineering design. The three key goals of this thesaurus are to (1) lessen the burden when working with knowledge from the biological domain by providing a link between engineering and biological terminologies; (2) assist designers by establishing connections between the two domains; and (3) to facilitate bioinspired design.

From the perspective of engineering design, biological information that can interface with existing design methods will further lessen the burden on the designer and increase the probability of looking to nature for inspiration. One approach to interfacing biological information with engineering design is through terminology or, more specifically, through grouping of synonyms and related

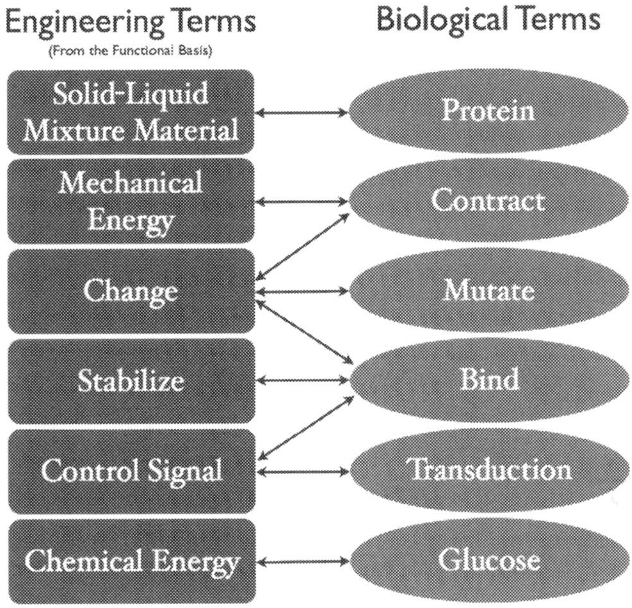

Fig. 4.1 Graphical representation of the engineering-to-biology thesaurus

concepts in a classified form as a thesaurus. Linking biology terms to engineering modeling terms (used for functional representation and abstraction) assists in reframing the biological terminology in an engineering context. Thus, the biological information is accessible to engineering designers with varying biological knowledge, but a common understanding of engineering design methodologies. The engineering-to-biology thesaurus design tool, described in this chapter and graphically depicted in Fig. 4.1, addresses several challenges engineers may encounter when performing bioinspired design, which include:

- Terminology differences.
- Understanding how a biological system works or functions.
- Discovering relevant biological systems.
- Abstracting biological principles/solutions for inspiration.

Acknowledging the difficulties of utilizing biological systems in engineering design, the engineering-to-biology thesaurus serves as a versatile design tool that affords design engineers, with a limited biological background, a means for developing connections between nature and engineering. The primary applications include assisting engineers with translation of biological information, communication with biologists, and concept generation with biological information. Building upon multiple design language research efforts, the engineering-to-biology thesaurus design tool has the potential to play a key role in several bioinspired design activities.

4.2 Background and Related Work

It is evident that nature can inspire innovative engineering designs (Toko 2000; Bar-Cohen 2006a, b; Benyus 1997; Stroble et al. 2009a; Vincent et al. 2006; Brebbia 2006, 2008; Brebbia and Carpi 2010; Brebbia and Collins 2004; Brebbia et al. 2002; Forbes 2006; The Biomimicry Institute 2009). Biological systems provide insight into sustainable and adaptable design, which often leads to designs that are more elegant, efficient, and sustainable. For engineering designers, however, to adopt such a practice, design tools, techniques, and methods are needed. Utilizing biological information during the engineering design process has taken many forms. Inspiration for solving or finding direct solutions to engineering problems has been obtained through chance observances (Hill 1995; Nachtigall 1989, 2000, 2002), functional keyword searches (Chiu and Shu 2007a, b; Vakili and Shu 2001; Bruck et al. 2007), systematic reverse engineering (Wilson and Rosen 2007; Lindemann and Gramann 2004), use of function–structure–behavior terms to search a database (Sarkar et al. 2008; Wilson et al. 2009; Design and Intelligence Laboratory 2010), TRIZ (Vincent et al. 2006), analogical reasoning (Tsujimoto et al. 2008; Mak and Shu 2004, 2008; Helms et al. 2009; Vattam et al. 2010a), and functional representation through functional models (Nagel et al. 2008, 2010a, b; Vakili and Shu 2007; Shu et al. 2007; Stroble et al. 2009b; Vattam et al. 2010b). Although each method has a different procedure, they all share one thing in common; the promising biological system or phenomena must be abstracted to capture the functional principle.

The research presented in this chapter explores the structure and purpose of an engineering design thesaurus and how it enhances an existing design lexicon for bioinspired design. Researchers at many universities are working on the knowledge transfer problem between the engineering and biological domains by further developing function- or function–behavior–structure-based design languages. Research into the engineering-to-biology thesaurus builds upon multiple design language research efforts.

The formal idea of a function-based design language, a standard set of engineering function and flow terms for systematically creating function structures, was originally proposed by Pahl and Beitz (2007). A function represents an operation performed on a flow of material, signal, or energy. Numerous researchers further evolved this set of generally valid functions and flows. Hundal proposed a further refined set of function and flow classes in (Hundal 1990); however, flows were excluded. Little et al. (1997) developed a set of function and flow terms, which classified both functions and flows at class and basic levels. Szykman et al. (1999) created a standardized taxonomy of function and flow terms, separated into classes down to the fourth level, for the purpose of computer-based design. Separately, but at the same time, Stone and Wood developed a well-defined standardized modeling lexicon comprised of defined function and flow sets with definitions and examples, entitled the Functional Basis (Stone and Wood 2000). Hirtz et al. (2002) later reconciled the Functional Basis into its most current set of terms, with research efforts from the National Institute of Standards and Technology (NIST), two universities, and their industrial partners. Within the Functional Basis, there exist eight classes of functions and three classes of flows, both having an increase in specification at the secondary and tertiary levels. There are 21 secondary and 24 tertiary functions, accompanied by correspondent terms to aid the designer in choosing the correct function. Similarly, there are 20 secondary and 22 tertiary flows accompanied by correspondent terms.

In a similar vein, function–behavior–structure languages were being explored and developed to provide designers a way to uniformly represent system behavior. Umeda et al. (Umeda et al. 1990) postulated that structure correlates with a system's state, or with the physical description of an entity in a design, while behavior was defined as the change in the state, and function was defined as the realization of the behavior through the use of the design. The work by Goel et al. uses structure to represent a physical description for components, function is the pre- and post-conditions for the behavior of the system, and behavior is the transition between states (Goel et al. 2009; Goel and Chandrasekaran 1992). To capture environmental interactions with a system, Gero developed the situated function–behavior–structure framework (Gero and Kannengiesser 2002). Building on this prior work, Chakrabarti et al. (2005) developed an approach to describe natural and artificial systems and their functionality. The representation is implemented in a software package entitled Idea-Inspire that allows one to search a database comprised of natural and artificial complex mechanical systems with a function–behavior–structure, or verb–noun–adjective, set. Each entry's motion or process is described functionally by behavioral language in the form of a function–behavior–structure model, which the user chooses from a predefined list of terms. The Idea-Inspire software yields seven behavioral constructs following the SAPPhIRE model—state change, action, parts, phenomenon, input, organ, and effect—for each search result that adequately fits the chosen function–behavior–structure set (Srinivasan and Chakrabarti 2009).

Researchers at the University of Toronto worked to provide designers with biologically meaningful words that correspond to engineering functions. To identify biologically meaningful words, the strategy developed by Chiu and Shu for searching biological literature using functional keywords for design inspiration (Chiu and Shu 2007a, b) was employed. Cheong et al. (2008) used the search strategy in conjunction with the terms of the Functional Basis to create the listing of biologically meaningful words. Only the Functional Basis functions in the secondary, tertiary, and correspondent levels were analyzed. Based on semantic relationships, the engineering function terms of the Functional Basis were used to generate a list of biologically significant and connotative functional keywords. In a similar vein, Stroble et al. (2009c) worked to provide designers with biological terms that correspond to engineering flows of the Functional Basis. Flow-type biological correspondent terms were collected utilizing an organized verb–noun search that extracts collocated words from a biological text. The Functional Basis flows in the class, secondary, and tertiary levels were analyzed. The macrorelevant biological flow terms identified were mapped to engineering flow terms through dictionary cross-referencing.

The work by Cheong et al. concentrated on finding specific connections between a subset of engineering functions and biological words, resulting in only a partial thesaurus. Stroble et al. concentrated on finding specific connections between engineering flows and biological terms, which also resulted in a partial thesaurus that was later expanded to include biological function terms by Nagel et al. (2010c). Both works share a common starting point, but diverged in approaches to filling out a thesaurus that maps engineering terms to biology terms as well as to what depth the thesaurus was populated. Additionally, the work of Nagel et al. (2010c) combined multiple design language efforts by (Chakrabarti et al. 2005; Srinivasan and Chakrabarti 2009; Cheong et al. 2008; Stroble et al. 2009c; Nagel et al. 2010c) into a single thesaurus. The result of that work is presented in this chapter.

4.3 Engineering-to-Biology Thesaurus

The engineering-to-biology thesaurus design tool (Nagel et al. 2010c) aims to facilitate collaboration between biologists and engineers and the discovery and creation of bioinspired engineering solutions. Applications of the engineering-to-biology thesaurus include, but are not limited to, (1) translation of biological information to increase comprehension or develop connections, (2) concept generation with biological inspiration through function-based approaches, and (3) dialogue facilitation between the engineering and biology communities. To achieve the stated applications, the structure of the thesaurus was molded to fit the knowledge and purpose of an engineering design perspective; biological synonyms and related concepts to the engineering terms of a well-established modeling lexicon are grouped together. The engineering-to-biology thesaurus, provided in Tables 4.1, 4.2, 4.3 and 4.4, does more than arranging terminology of one domain side by side with terminology of another; it serves as the intermediary between the biology and engineering domains. Furthermore, this design tool increases the interaction between the users and the knowledge resource (Lopez-Huertas 1997) and aims to increase a designer's efficiency when working across the engineering and biology domains. Efficient information retrieval through the engineering-to-biology thesaurus allows an engineering designer to cross into the biology domain and gain functional knowledge without becoming overwhelmed by unfamiliar biological systems and terminology. This is, and is envisioned to be, a work in progress that slowly and steadily bridges the terminology gap between the two domains based on contributions of a community of researchers and practitioners.

In this section, the thesaurus model, population methods of the biological flows and functions, term placement, term placement review, and limitations are explained.

Table 4.1 Engineering-to-biology thesaurus function terms

Functional basis terms			Biological function correspondent terms
Class	Secondary	Tertiary	
Branch	Separate		Bleaching, meiosis, abscission, mitosis, segment, *electrophoresis*, dialysis, denature, free, detach, release
		Divide	Divide, *prophase*, *metaphase*, *anaphase*, cleave, cytokinesis
		Remove	Deoxygenate, filtrate, liberate, expulsion, evacuate
	Distribute		*Circulate*, *diffusion*, *exchange*, disperse, scatter, spread, spray
Channel	Import		*Absorb*, attract, consume, inhale, intake
	Export		*Bind*, block, *breakdown*, excrete, inactivate, repel
	Transfer		Migrate, transfer
		Transport	*Circulate*, conduct, diffuse, pump, shift, displace, fly, swim, jump, bounce
		Transmit	*Communicate*, *transduce*
	Guide		Orient, position, slide, tunnel
		Translate	*Synthesize*, transcribe
		Rotate	Oscillate, spin, turn, swivel, roll
		Allow DOF	Articulate
Connect	Couple		Recombination, mate, build, *phosphorylate*, *bond*, *synthesis*, latch, lock, extend, link, overlap
		Join	*Bind*, adhere, bond, fuse
		Link	Clamp, *activate*, *bind*, project
	Mix		Blend, *contract*, *exchange*, fragment
Convert	Convert		Polymerize, *synthesize*, burn, gluconeogenesis, metabolize, grow, *transduction*, *transduce*, fermentation, glycolysis, hydrolyze, hydrolysis, respiration, ionize, decompose, *degrade*, *develop*, *mutate*, *photosynthesize*
Control	Actuate		*Activate*, induce, trigger
Magnitude	Regulate		*Electrophoresis*, gate, organogenesis, respire, sustain, preserve, remain, stabilize, maintain, regulate
		Increase	Hyperpolarize, *pinocytosis*, *grow*, expand, multiply, replicate
		Decrease	Compress, coil, *divide*, fold, shorten, *wrap*
	Change		*Pinocytosis*, *degrade*, alter, *bind*, catalyze, *contract*, hydrolysis, twist, *mutate*, radiate, charged, slip, acclimatize, alternate, fluctuate
		Decrement	Decarboxylation, *constrict*

(continued)

Table 4.1 (continued)

Functional basis terms			Biological function correspondent terms
Class	Secondary	Tertiary	
		Shape	Elongate, *stretch*, attach, *spread*
		Condition	Osmosis, *constrict*
	Stop		Clog, extinguish, halt, interphase, seal, suspend
		Prevent	Constrain, obstruct
		Inhibit	Cover, destroy, *inhibit*, repress, *surround*
Provision	Store		Conserve, *hold*, *convert*, deposit, *photosynthesize*
		Contain	*Absorb*
		Collect	*Absorb*, catch, *breakdown*, concentrate, digest, reduce
	Supply		Feed, lactate
Signal	Sense	Detect	Detect, locate, see, smell, *transduce*
		Measure	Observe, monitor, gauge, watch
	Indicate		Fluoresce, *communicate*, *react*, mark
	Process		Learn
Support			*Develop*, *wrap*
	Stabilize		Homeostasis, cling, *hold*, *bind*, *connect*
	Secure		*Surround*, envelope

Table 4.2 Engineering-to-biology thesaurus material flow terms

Functional basis terms			Biological function correspondent terms
Class	Secondary	Tertiary	
Material	Human		Being, *body*
	Gas		Oxygen, nitrogen, chlorine
	Liquid		Acid, chemical, water, *blood*, *solution*, base, buffer, fluid, plasma
	Solid	Object	Fiber, *body*, substrate, microfilament, microtubules, structure, chain, *organ*, nucleus, *tissue*, muscle, cilia, flagella, tube, vein, heart, plant, somite, apoplast, stem, kidney, egg, ovary, leaf, embryo, bacteria, chloroplast, carbon, adipose, angiosperm, meristems, mineral, stoma, shoot, seed, capillary, receptors, hair, bone, tendon, neuron, sporangium, photoreceptor, mechanoreceptor, chromosome, petiole, lysosome, archaea, cone, strand, centriole, spore, zygote, sulfur, lipoprotein, nephron, hyphae, plasmodesma, conifer, plasmid, plastid, xylem, pigment, sperm, hippocampus, phloem
		Particulate	Cytokinin, pyruvate, nicotine, opium, glycerol, carotenoid, GTP, ATP, urea, *RNA*, tRNA, mRNA, *DNA*, *glucagon*, parathormone, cryptochromes, *ligand*, promoter, gene, exon, intron, molecule, enzyme, *lipid*, *hormone*
		Composite	Enzyme, virus, *ribosome*, prokaryote, macromolecule, polymerase, nucleotide, polypeptide, organelle, symplast, mesophyll, brood, codon, messenger, *DNA*, *RNA*, cytoplasm, *organ*, *tissue*
	Mixture	Gas–gas	Air
		Liquid–liquid	*Hormone*, melatonin, thyroxine, calcitonin, thyrotropin, estrogen, somatostatin, cortisol, *glucagon*, adrenocorticotropin, testosterone, auxin, insulin, intracellular fluid, extracellular fluid, spinal fluid, poison, urine, peptide, *solution*, steroid
		Solid–solid	Adenosine, glomerulus, blastula, monosaccharide, membrane, phosphate, *ribosome*, centrosomes
		Solid–Liquid	Algae, synapse, cell, glia, phytochrome, retina, protein, receptor site, repressor, hemoglobin, *blood*, membrane, bacterium

4.3.1 Thesaurus Model

The purpose of a thesaurus is to represent information in a classified form to group synonyms and related concepts. A thesaurus of the English language has classes and categories with an index of terms directing the user to the correct instance (i.e., noun, verb, adjective) of the term under examination. The engineering-to-biology thesaurus design tool has a unique structure and classification; it is merged with the reconciled Functional Basis as a set of biological correspondent terms. In the Functional Basis lexicon, a function represents an action or transformation (verb) being carried out, and a flow represents the type (noun), material, signal, or energy,

Table 4.3 Engineering-to-biology thesaurus signal flow terms

Functional basis terms			Biological flow correspondent terms
Class	Secondary	Tertiary	
Signal	Status		Change, variation, lateral, swelling, catalyzed, translation, exposed, active, separated, cycle, formation, reaction, redox, deficient, saturated, diffusion, broken, hybridization, orientation, resting, cue, magnetic, volume, under, organized, fruiting, fatty, *anaphase*, *metaphase*, *prophase*, conjugation, osmolarity, senescence, signal
		Auditory	Sound
		Olfactory	Smell
		Tactile	Pain
		Taste	Gustation
		Visual	Length, shortened, long, dark, full, double
	Control		Place, *inhibit*, release, excrete, *develop*, match, induce, *digest*, integrate, translation, *transduction*, equilibrium, grow, splice, capture, *distribute*, *phosphorylation*
		Analog	*Binding*, center, synthesis, photosynthesis
		Discrete	Flower, translocation

Table 4.4 Engineering-to-biology thesaurus energy flow terms

Functional basis terms			Biological flow correspondent terms
Class	Secondary	Tertiary	
Energy	Acoustic		Echolocation, sound wave
	Chemical		Calorie, metabolism, glucose, glycogen, *ligand*, nutrient, starch, fuel, sugar, mitochondria, *lipid*, gibberellin
	Electrical		Electron, potential, feedback, charge, *field*
	Electromagnetic	Optical	*Light*, *infrared radiation*
		Solar	*Light*, sunlight, ultraviolet light
	Hydraulic		*Pressure*, osmosis, osmoregulation
	Magnetic		Gravity, *field*, wave
	Mechanical		*Muscle contraction*, pressure, tension, stretch, depress
	Pneumatic		*Pressure*
	Thermal		Temperature, heat, *infrared radiation*, cold

passing through the functions of the system. There exist eight classes of functions and three classes of flows, both having an increase in specification at the secondary and tertiary levels. Both functions and flows have a set of engineering correspondent terms that aid the designer in choosing Functional Basis terms during model creation. The complete function and flow lexicon can be found in (Hirtz et al. 2002).

The engineering-to-biology thesaurus provides biological correspondent terms for engineering functions and flows at the class, secondary, and tertiary levels, which follows the structure and classification of the Functional Basis. This grouping highlights the synonyms and related concepts across the two domains.

The thesaurus does not include an exhaustive set of engineering and biology terms; rather, it contains a representative set of engineering and biology terms. Biological correspondent terms to the Functional Basis functions and flows are provided in place of the original engineering correspondent terms, as shown in Tables 4.1, 4.2, 4.3 and 4.4. Only biological verbs and nouns that are synonymous to terms of the Functional Basis are considered. The thesaurus does not include adjectives nor does it include an index. The Functional Basis class-level terms, however, do emulate the classes of a traditional thesaurus. Furthermore, the secondary- and tertiary-level Functional Basis terms emulate the categories of a traditional thesaurus. Thus, the classification is predetermined according to the Functional Basis model; however, it remains the intermediary between the biology and engineering domains.

A tool such as the engineering-to-biology thesaurus increases the interaction between the users and the knowledge resource (Lopez-Huertas 1997) by presenting the information as a lookup table. This simple format fosters one to make associations between the engineering and biological lexicons, thus strengthening the designer's ability to utilize biological information.

4.3.2 Biological Functions

The majority of biological information is written in such a way that correlating biological verbs with Functional Basis functions is relatively straightforward. However, there are always exceptions. Well-known functional terms that appear in a biological corpus may not have the meaning an engineer would typically know. For instance, the term bleaching outside of the biological domain means to clean, sterilize, or whiten, as most know. The biological meaning, however, refers to the process of separation between the retina and opsin in vertebrate eyes and causes the retinal molecule to lose its photosensitivity (Campbell and Reece 2003). It is these types of exceptions that researchers were cognizant of when compiling the set of biological correspondent function terms for the engineering-to-biology thesaurus. Keyword searches of a biology textbook using the automated information retrieval tool (Stroble et al. 2009d) and Functional Basis functions were performed to gather a list of collocated verbs that occur within the same sentence as the search word. To signify which function terms are utilized in both domains, the Functional Basis term is repeated in the biological correspondent list. It should be noted that some of the biological correspondent function terms are nouns that name a process corresponding to a Functional Basis function. Identified biological functions were cross-referenced in the Oxford American dictionary (McKean 2005), Henderson's dictionary of biological terms (Lawrence and Holmes 1989), and the Oxford Dictionary of Biology (Matrin and Hine 2000) before placement in the thesaurus, which was at the discretion of the author. All other function terms were obtained from research performed at the Indian Institute of Science and University of Toronto, which are made explicit in (Nagel et al. 2010c).

Functional terms from the Indian Institute of Science were collected from the Idea-Inspire software. Every natural system entered into the software's database was indexed using the predetermined list of verbs, nouns, and adjectives. Analyzing the list of verbs by cluster (Srinivasan and Chakrabarti 2009) revealed scientific terms applicable to biological systems grouped with engineering terms exactly matching those of the Functional Basis. Utilizing multiple dictionaries as in the prior analysis, the verbs of Idea-Inspire were mapped to the Functional Basis functions as biological correspondent function terms.

Functional terms from the University of Toronto were collected from the work of Cheong et al. (2008). As background work was already performed on the semantic relationships of the biologically meaningful words to Functional Basis functions, further investigation was not performed. Rather, the terms were directly added to the thesaurus as biological correspondent function terms following the classification in Cheong et al. (2008).

4.3.3 Biological Flows

In the authors' experience, understanding biological terms that were considered flows (material, signal, and energy) when utilizing biological systems for idea generation or design inspiration posed the most difficulty. Determining whether a biological material is liquid, solid, or a mixture by its name typically requires domain knowledge that most engineers do not have, which can cause biological information to be perplexing. Similarly, needing a reference to look up biological terms each time, a potential biological system was found made the design process tedious and disrupted thought patterns leading to decreased efficiency. Thus, the inclusion of biological correspondent flow terms strengthens the ability of the design tool to address the challenges engineers may encounter when performing bioinspired design.

Identification of engineering-to-biology thesaurus biological correspondent flow terms was achieved through keyword searches of a biology textbook using the automated information retrieval tool (Stroble et al. 2009c, d). Functional Basis functions were used for the keyword searches to extract biological nouns that an engineering designer interested in function-based design might encounter. The nouns that were collocated within the sentence to the search word were counted and sorted by frequency of appearance. All nouns that appeared more than two times were considered macrorelevent. Each macrorelevent term was determined if it was of signal, material, or energy type through cross-referencing in the new Oxford American dictionary (McKean 2005) and Henderson's dictionary of biological terms (Lawrence and Holmes 1989) before being placed, which was at the discretion of the author.

4.3.4 Term Placement Review

Mapping engineering terms to the biological domain, which the author is not an expert in, requires a review of the relationships by a biologist. A biologist in two instances reviewed the thesaurus: (1) when the biological correspondent flow listing was generated and (2) when the biological correspondent function listing was generated. The terms of the thesaurus in Tables 4.1, 4.2, 4.3 and 4.4 represent the complete set of terms that are the result of the biologist's reviews. Review of the placement, type, and structure of the thesaurus terms was initially performed by a biology student at Missouri University of Science & Technology after flows were placed (Linsey 2008a, b). A large group of tertiary terms under discrete control signal were moved to the secondary level as they could be either discrete or analog control signals. Other misplaced biological correspondent terms were moved between the tertiary-level material classifications. After the first-term placement review, 32 (10.9 %) of the flow terms were moved to a more appropriate mapping, 20 (6.8 %) were removed completely, and 11 new terms were added, thus changing the total biological correspondent flow term count to 285 from 294.

A second review of the placement, type, and structure of the thesaurus terms was performed by a professor of zoology at Oregon State University after functions were placed (Brownell 2010a, b). The professor of zoology reviewed both the function and flow biological correspondent terms and offered his insight. To better map the terminology to the engineering domain, biological terms of a similar type or related concept that were scattered were moved to the same classification and the multiple meanings of terms are emphasized through repetition of the terms across classifications. For example, all the terms representing molecules across multiple material tertiary terms were collected and placed into the particulates' classification. A similar change was made for terms representing hormones, which were placed in the composites' classification. The term organ is found under multiple classifications as it can be thought of as an object or a composite of tissues. Other changes include removing terms due to ambiguity and changing term tense. After the second-term placement review, 34 (11.9 %) of the flow terms were moved, 24 (8.4 %) were corrected to the proper tense, and 22 (7.7 %) were removed. Also, 7 new terms were added and 5 existing terms were repeated, thus changing the total biological correspondent flow term count from 285 to 275. For the biological correspondent function terms, the changes consisted of moving 2 (0.01 %) of the terms, correcting 1 (0.005 %) to the proper tense, and removing 5 (2.4 %) of the terms. Also, 1 new term was added and 3 existing terms were repeated, thus changing the total biological correspondent function term count to 206 from 207.

4.3.5 Thesaurus Details

Key challenges to the approach for populating the thesaurus were (1) the time required searching each term to generate a listing of collocated terms and (2) understanding the definitions provided by the three dictionaries used in the analysis. To determine the material, energy, or signal type of the flow term in question, generally multiple biological dictionary entries were referenced. Considering biological processes that perform a specific function within the system revealed many macrorelevant terms that would have been overlooked if only verbs were analyzed.

The Functional Basis offers a definition and example for each class, secondary, and tertiary terms. Definitions of the correspondent terms are, however, not provided. Rather, the correspondent terms are synonyms to the Functional Basis terms. This is also true for the biological correspondent terms. Biological terms that correspond to multiple functions or flows are repeated and are italicized to designate the special case of those terms. This treatment is similar to the repeated words of the engineering correspondent terms.

4.3.6 Limitations

As with all engineering design tools, limitations exist. For the engineering-to-biology thesaurus, the limitations are the terms available and the focus on mapping biological terms to the functions and flows of an existing engineering modeling language. Setting the boundaries on the small but representative set of engineering terms of the Functional Basis is a major limitation. This directly affects the design tool and limits the biological terms that can be correlated with the engineering domain. For example, the biological function of *protect* loosely aligns with the Functional Basis tertiary term *prevent*, which is under the secondary-level term *stop*. This relationship, however, is not intuitive because a shell of an animal protects the animal from harm or death, but it does not prevent or stop the attack from occurring. Therefore, there are relevant biological function terms that are often performed by biological systems that are currently not included. While the thesaurus does not include a comprehensive list of all biology and engineering terms, it does, however, contain a representative set that can guide the designer to make informed judgments on terms that are not included. Depending on the level of confidence by the designer, this may or may not be achievable. Both the engineering and biology term sets could be expanded to include other relevant terms, as they are made available.

Translating biological information to the engineering domain is possible, as shown in Sect. 4.4.2; however, it is not as intuitive to an engineer as translating information from one engineering subdomain to another (i.e., fluid resistance to electrical resistance). Designers must make creative leaps through reliance on prior knowledge, a knowledge base, a design tool, or communication with a biologist to perform translation across domains. The current state of the art, which includes the engineering-to-biology thesaurus, allows engineers to translate a function in nature to an equivalent, or analogous, engineering function. What the current state of the art does not do is automatically identify what the biological system performing the function is or translate the fundamental mechanism of the biological system to the engineering domain. Therefore, the engineering-to-biology thesaurus requires that the user has a good understanding of engineering function, which will assist in making the connections between the domains to correlate biological systems with engineered systems.

An additional limitation of the design tool is that it is a manual design tool. A designer must manually look up terms in the thesaurus tables. To assist the manual process, an alternate form of the thesaurus, organized and alphabetized by biological term, was created to better assist the manual translation from biology to engineering. It can be found at www.designengineeringlab.org. Although two versions of the thesaurus have been compiled, a computational version that would take input terms from either domain and automatically look up the corresponding term or, if the term does not exist, suggest a relevant corresponding term would be beneficial. Development of a biology-to-engineering translator for automatic machine translation of biological information into an engineering context is underway.

4.4 Design Tool Applications

Mapping terms between the biology and engineering domains not only reduces the terminology barrier and addresses other challenges engineers face when working across the two domains, but also supports several engineering design applications. Applications of the engineering-to-biology thesaurus include, but are not limited to, (1) translation of biological information to increase comprehension or develop connections, (2) concept generation with biological inspiration through function-based approaches, and (3) dialogue facilitation between the engineering and biology communities. The concept generation application activities supported by the thesaurus are the identification of relevant biological terms to use during brainstorming or searching for biological inspiration, functional modeling of biological systems, and identification of analogies between the domains. Figure 4.2 depicts how the applications of the engineering-to-biology thesaurus design tool fit within a typical function-based engineering design process. Communication and translation applications can occur at many points during the engineering design process, affording versatility of the design tool. Additional versatility is provided during the early phases of design through the multiple concept generation application activities. Integration and prescriptive use of the engineering-to-biology thesaurus design tool into an engineering design process to facilitate bioinspired design are described in Chap. 5.

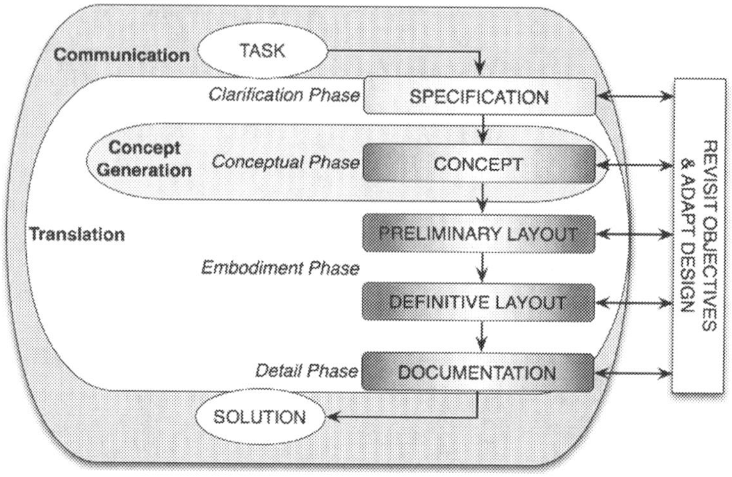

Fig. 4.2 Thesaurus applications mapped to a typical function-based design process (adapted from Pahl et al. 2007)

All the design activities lead to, or can be incorporated into, concept development, which is an overarching application of this design tool. Additionally, the majority of these applications capture the biological system through an abstraction, which is very valuable in solving design problems and can assist designers with learning design principles from nature. The following subsections describe the inherent and plausible applications of the design tool. Application examples are also provided along with references to in-depth examples.

4.4.1 Communication

The design tool presented in this chapter can aid the communication and collaboration between engineers and biologists. There are distinct relationships that the two domains share. For example, function in engineering is analogous to physiology in biology, and structure in engineering is analogous to morphology in biology. Similar relationships exist among the subject-domain-specific terms of biology and engineering. Many are captured in the thesaurus tables. Recognizing those relationships can help an engineer or biologist understand the other's viewpoint, which, subsequently, can aid communication across the domains. Specifically, engineers get glimpse of how biological terms have multiple meanings and levels of interpretation. Being aware of the terminology differences and levels of abstraction can ease communication with biologists.

The engineering-to-biology thesaurus can assist engineers with asking better questions of biologists. For example, asking "how does a biological system transmit information?" is very open-ended and may not lead to a useful answer. A better question to ask would be "what role does transduction play in communicating information?" By replacing transmit with transduction and rewording, the question to be more specific is likely to reduce the time clarifying what is meant by certain terms or phrases, and more time is spent discussing topics of interest, resulting in an intellectually stimulating conversation versus a frustrating discussion of the differences between the domains. Using the terms of the thesaurus can similarly aid a biologist to reciprocate and give an answer in a context that an engineer is more likely to understand.

4.4.2 Translation

Lopez-Huertas wrote that a thesaurus "… is thought of as a way of easing communication between texts and users in order to increase the interaction in information retrieval, and thus facilitate information transfer" (Lopez-Huertas 1997). The engineering-to-biology thesaurus can aid engineering designers with the translation process that changes biological information into engineering "speak" to facilitate information transfer. Translation aims to effectively change the context of the information from biology to engineering. The fundamental activity of translation is to increase comprehension of biological information. Empirically, translation has led to a greater understanding of the inspiring biological system and impacts the time required to develop a bioinspired concept. Consequently, increased comprehension positively affects other engineering-to-biology thesaurus applications as well as the bioinspired design process. Other reasons for translation include exploration of analogies, assistance with concept generation, and communication across the domains.

Translation is achieved by manually substituting biological terms that appear in the thesaurus with their corresponding Functional Basis terms. Substitution of terms can be approached incrementally or all at once. Essentially, this will rewrite the biological information in engineering "speak" and increase the likelihood of a designer making connections between the two sets of information. Describing a biological system in engineering terms of the Functional Basis is advantageous. Not only does it increase the likelihood of a designer understanding the biological system, but it also lends itself to formulating connections between the biological and engineering domains, easy comparison to other abstractions, and easy integration with function-based design methods. Efficient information retrieval through the engineering-to-biology thesaurus allows an engineering designer to cross into the biological domain and gain functional knowledge without becoming overwhelmed by unfamiliar biological systems and terminology.

The following example serves as a qualitative measure of the engineering-to-biology thesaurus to show that the design tool can assist in translating biological information into engineering "speak" without requiring the designer to learn deep biological knowledge. Consider the text excerpt describing insect olfaction through antennae taken from the section Chemoreception in The Encyclopedia of Insects (Mitchell 2003). The contextual difference between the original and translated forms aims to, "eas[e] communication between texts and users in order to increase the interaction in information retrieval, and thus facilitate information transfer" (Nagel et al. 2010c). The original text from Mitchell is as follows.

> In insects, odor molecules first contact the cuticular surface, and because it is waxy, they easily dissolve. From here they move in two dimensions, and some find their way into the opening of a pore canal. .... Eventually, however, before it arrives at the receptor surface of a dendrite, the hydrophobic odor molecule will encounter water. .... The other type binds less specifically [to] a variety of nonpheromone molecules (e.g., food odors) and are called general odor binding proteins (GOBP). The odorant binding proteins (OBP) act as shuttles and carry odor molecules through the aqueous medium to the surface of the dendrite. In the membrane of the sensory cell are receptors for various odors, depending on the specificity of the cell.

Mappings used to translate the text on insect chemoreception are with function- and flow-type terms. Many biological terms in the text excerpt were found in the thesaurus tables and manually swapped with the corresponding engineering terms. Examples are molecule, protein, receptor, dendrite, shuttles, and cell. While the biological process of olfaction through antennae described might be clear, the types of materials and signals involved in the process are less clear. Translation of the flow-type terms assists in determining the types of materials and signals involved. Biological terms to translate that were not found within the thesaurus, such as dendrite, were addressed through translation of the definition of the biological term. By manually identifying unclear biological terms and substituting Functional Basis terms, the translated text presents the information in a more generalized context. The translated insect olfaction text is as follows:

> In insects, odor particulates first contact the solid material surface, and because it is waxy, they easily dissolve. From here the odor particulate moves in two dimensions, and some find their way into the opening of a pore canal. ... Eventually, however, before it arrives at the solid object surface of a an object that transfers electrical energy, the hydrophobic odor particulate will encounter a liquid material. ... The other type joins less specifically [to] a variety of nonpheromone particulates (e.g., food odors) and are called general odor joining solid-liquid mixtures. The odorant joining solid-liquid mixtures act as guides and transfer odor particulates through the liquid material to the surface of the object that transfers electrical energy. In the solid-solid material of the sensory solid-liquid material are solid objects for various odors, depending on the specificity of the solid-liquid material.

The biological information is now presented in a more generalized, engineering context, which can be used to facilitate a range of bioinspired design activities. One could further translate the text if the biological system is still unclear. Each engineering designer will have different background knowledge of biology; therefore, iterations are recommended until the designer comprehends the biological information.

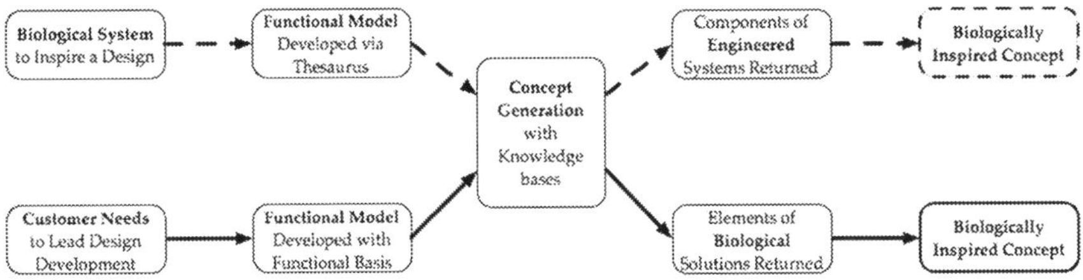

Fig. 4.3 Typical concept generation processes utilizing the thesaurus

4.4.3 Concept Generation

Concept generation—manual or computational—aims to generate several conceptual design variants. During this process, engineers engage in many activities and draw on their prior knowledge or search knowledge bases such as specialized repositories, design catalogs, patents, and technical literature (Voland 2004; Otto and Wood 2001; Ulrich and Eppinger 2004; Cross 2008). Nature is another resource available to engineers during concept generation, and the subject of biology serves as the body of knowledge engineers can interface with to gain inspiration. By mapping biological terms to the Functional Basis modeling language terms, the engineering-to-biology thesaurus integrates with existing function-based concept generation activities to facilitate bioinspired design. The concept generation application activities supported by the thesaurus are identification of relevant biological terms to use during brainstorming or searching for biological inspiration, functional modeling of biological systems, and identification of analogies between the domains. Figure 4.3 depicts the process for the concept generation application activities that fit within the conceptual phase of the design process given in Fig. 4.2.

The process flow in Fig. 4.3 with a dashed line relies on the use of biological functional models to derive bioinspired conceptual designs. In this concept generation approach, the biological system is known or chosen prior to concept generation. The biological functional model is used to query a knowledge base indexed by engineering function, such as the Design Repository[1] discussed in Sect. 4.5.1, to retrieve engineered components and subsystems that perform the same functions as the biological system. The MEMIC tool described in Sect. 4.5.3 supports an automated morphological matrix approach that interfaces with the Design Repository. A manual morphological matrix approach could also be taken by manually searching patents, design catalogs, and literature for solutions to the biological functions. Once engineered solutions are determined, the designer must draw analogies or connections to make the leap to a bioinspired concept.

[1] Design Repository www.designengineeringlab.org.

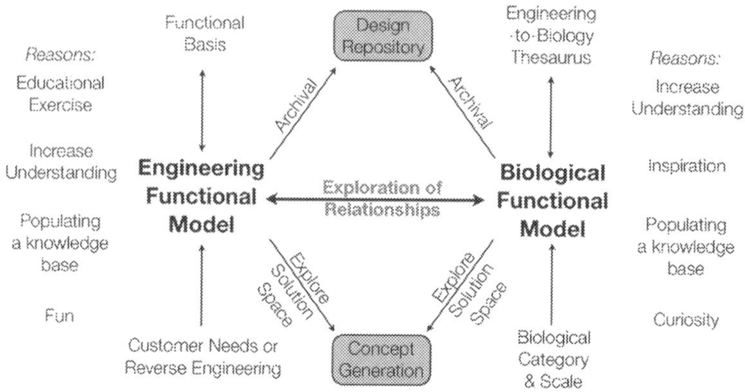

Fig. 4.4 Relationship of biological functional models to engineering functional models

The process flow in Fig. 4.3 with a solid line relies on the use of engineering functional models to derive bioinspired conceptual designs. In this concept generation approach, the inspiring biological systems are not known prior to concept generation and are identified through searching a biological knowledge base. Biological knowledge bases that integrate with the thesaurus include the Design Repository and the biological corpora that are made available to the organized search tool discussed in Sect. 4.5.2. Once biological solutions are identified, the designer must learn more about the inspiring systems to draw analogies or connections to make the leap to a bioinspired concept. Translation of biological information or communication with biologists may be necessary.

Considering biological systems through generalized engineering terms allows connections to be made between the domains, which facilitates knowledge transfer, and allows the biological information to be used during the engineering design process. Prior knowledge of a broad range of engineered systems and processes is not required for concept generation of bioinspired designs; however, that knowledge provides the impetus for readily recognizing the connections between systems of two dissimilar domains. Concepts are formulated directly and indirectly from biological inspiration. Using the thesaurus could result in conceptual designs that partially (i.e., one or two components) or completely (i.e., entire design) mimic a biological system. Although the engineering-to-biology thesaurus assists in making the leap from biology to engineering, to arrive at the final concept, the designer is still required to make the leap within the engineering domain.

The following subsections describe the concept generation application activities in detail.

4.4.3.1 Functional Modeling of Biological Systems

The engineering-to-biology thesaurus provides direction when choosing the best-suited function or flow term to objectively model a biological system. A wide

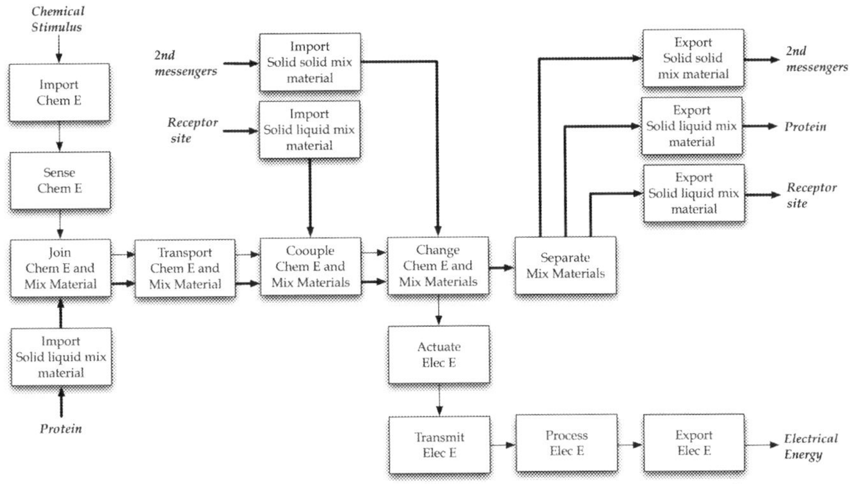

Fig. 4.5 Biological functional model of chemoreception

range of biological terms have been collected and placed into the thesaurus, which can accommodate a designer when developing functional models of well-known to lesser-known biological systems. Functional modeling of biological systems allows representation of solutions to specific engineering functions and direct knowledge discovery of the similarities and differences between biological and engineered systems, as viewed from a functional perspective (Nagel et al. 2010d). Figure 4.4 summarizes the relationships between biological and engineering functional models. Detailed instructions and examples of creating biological functional models can be found in (Nagel et al. 2010b, d, 2011).

A biological functional model could be thought of as an additional form of translation, which allows a designer to qualitatively model a system to comprehend and compare functionality. Representing biological, functionally using the lexicon of the Functional Basis, also allows biological solutions to be stored in an engineering design knowledge base for future reuse, such as for concept generation or for educational purposes. The archived biological solutions can then be recalled and adapted to engineered systems. The creation of engineered systems that implement strategies or principles of their biological counterparts without reproducing physical biological entities is an additional benefit to biological functional models.

Continuing the example from Sect. 4.4.2, a biological functional model of insect olfaction through antennae is developed and provided in Fig. 4.5. The information contained within the prior text excerpt is further expanded below to assist in understanding details of insect olfaction or chemoreception.

Antennae are made of a chitin–protein complex referred to as cuticle, which are porous, and covered in a waxy layer to prevent desiccation (Mitchell 2003). Multiple parts of the insect body, particularly the antennae, are covered in cuticular protrusions in the form of sensilla (e.g., hairs, pegs) that house the chemically sensitive cells for olfaction (Mitchell 2003; Møller 2003; Eguchi and Tominaga 1999; Klowden 2008). In order to detect the chemical stimulus, the odor molecules must make contact with the waxy layer of a sensillum and travel through the porous cuticle. Once inside, odor molecules encounter an aqueous medium containing odor-binding proteins and receptor sites on the dendrite surface (Mitchell 2003; Møller 2003; Eguchi and Tominaga 1999). As the name implies, the odor-binding proteins bind to the odor molecules and essentially shuttle one odor molecule at a time to a receptor site. The dendrite is connected to a sensory cell that, in most cases, is activated by specific odor types (e.g., food, pheromones) through the receptor sites at the dendrite surface. Regardless, once an odor molecule comes into contact with a receptor site, a signal is generated, the signal is amplified, and the odor-binding protein then causes hydrolysis to separate the odor molecule from the receptor site and the protein itself (Mitchell 2003; Møller 2003; Eguchi and Tominaga 1999). The odor-binding protein is responsible and required for receptor site activation and deactivation (Klowden 2008). Binding to a receptor site causes activation and conformational change and leads to the generation of an action potential (electrical signal), which is summarized as signal transduction. This is achieved through second messengers, typically cyclic adenosine monophosphate, which increases the sensory cell's permeability to sodium ions and alters the electrical potential of the cell membrane (Mitchell 2003; Møller 2003; Eguchi and Tominaga 1999; Klowden 2008). After the signal has been generated and separation by hydrolysis is complete, esterase enzymes breakdown the odor molecule and the odor-binding protein is recycled.

Using the terms within Tables 4.1, 4.2, 4.3 and 4.4, the biological information is mapped to engineering terms, similar to the translation process described in Sect. 4.4.2. Where the translation example of Sect. 4.4.2 had an emphasis on biological flows, the biological functional model has an emphasis on biological functions. Combined, the two create a comprehensive translation of biological information into an engineering context. The following paragraph demonstrates the distilled biological information with the substituted engineering terms in italics used to create the biological functional model in Fig. 4.5.

When the *chemical energy*, odorant, enters the insect cuticle, the odor-binding proteins immediately sense their presence and begin the detection process. The function of *join* represents the protein binding to the chemical stimulus, which is then carried to the receptor site noted by the function of *transport*. The *couple* function denotes binding of the odor molecule and odorant-binding protein to the receptor site. *Change* represents the activation, conformational change in the receptor site, and generation of an action potential and is why the flows of chemical energy and mixture materials are all present for that function. Signifying the receptor site deactivation in parallel with the electrical signal that is sent to the nervous system to be identified is the function of *separate* and *actuate*, respectively. The final portion of the chemoreception process is transmission of the electrical signal to the brain to produce a response.

Following the process outlined in Fig. 4.3, the biological functional model of chemoreception could be used to inspire an innovative chemical sensor. Additionally, the model could be entered into a specialized knowledge base for future bioinspired design reuse. An example of modeling the symbiotic biological system lichen with the help of the thesaurus can be found in (Nagel et al. 2010b, 2011). The lichen biological functional model resulted in a concept for an innovative solar energy system that is adaptable to different climates to improve efficiency of electricity generation.

4.4.3.2 Inspiration Brainstorming and Search

Identification or discovery of relevant biological systems to take inspiration from when addressing an engineering problem can be a challenge. The biological terms of the engineering-to-biology thesaurus can assist in discovering biological systems to use for design inspiration by offering more functions for brainstorming and relevant keywords to use when searching for inspiration. It is often advantageous to reframe the design problem to relate it to biology (Helms et al. 2009). Searching a biology corpus, such as a textbook, for biological inspiration based on engineering terms typically produces results that are mixed. Results containing the search word often use the search word out of context, not at all or in a different sense than the designer intended. By utilizing the biological correspondent terms of the thesaurus when searching for specific functions or flows that solve the engineering problem, search results improve (Nagel and Stone 2012) and become more focused on the desired biological systems, consequently, aiding concept generation.

Inspiration brainstorming and searching integrates with the concept generation processes depicted in Fig. 4.3 and can be applied with manual or computational function-based methods. Two applications include identification of biological systems that address customer needs and identification of a biological system for functional modeling.

Applying the thesaurus during brainstorming biological solutions or searching for biological systems that offer solutions to engineering problems can assist in achieving relevant results. For example, consider the design of a braking system for a pedaled vehicle. A simple way to relate the problem to biology is to consider "How do biological systems brake?" This approach, however, will not yield results that relate to the given design problem. To increase relevant results, the designer can develop a black box or detailed functional model to determine engineering functions that closely relate to the braking system design. One result is the Functional Basis engineering term *stop*. Looking for the engineering term *stop* in the engineering-to-biology thesaurus (Table 4.1) leads to the biological correspondent terms of *extinguish*, *halt*, *clog*, *seal*, and *suspend*. Using the identified set of biological correspondent function terms provides a wide range of terms to assist in brainstorming or searching to identify biological systems that address the design problem.

Identified biological systems include spider webs, fainting goats, giraffe throat valves, blood clots, scabs, clam shells, and puffer fish.

With several biological systems identified, a designer, depending on the level of understanding, can begin to develop bioinspired conceptual designs. If comprehension is low, the designer could look up information on each system to perform translation or functional modeling or analogy discovery to become inspired. Considering fainting goats and puffer fish leads to two interesting braking system concept variants. Learning that fainting goats' muscles seize when they become over-excited leads to a concept of suspending bike movement to prevent input from the rider and slow down. Essentially, the entire bike acts as the braking system. Learning that puffer fish expand to halt predators leads to a concept that has tires that act as brakes by expanding to create more friction and stop the bike from moving.

4.4.3.3 Discovery of Analogies

The established terminology relationships of the engineering-to-biology thesaurus provide a foundation for discovering analogies between the engineering and biology domains. Abstractions of biological functions, processes, materials, signals, and energies, as presented in the engineering-to-biology thesaurus, allow a designer to make connections with engineering abstractions, principles, components, and systems. Connections can be formulated directly and indirectly from biological inspiration. Thus, connections and analogies do not need to be one to one, as in the case of directly copying a biological system. Discovering analogies between biology and engineering can happen at any time during exploration of a biological system for inspiration; however, after the translation of biological information into an engineering context, the connection may become clearer.

Considering biological systems in a generalized engineering context allows connections to be made between the domains, which facilitates analogy discovery. For example, consider the southern three-banded armadillo, which has three bands along its back that spread and allow it to roll up into a ball and cover itself from predators in the event of an attack. The armor surrounds the animal's tail, head, feet, and back. Similar to the development of biological functional models, the biological information is manually mapped to engineering terms using the thesaurus tables to assist the discovery of analogies. Looking for the biological terms of *surround*, *cover*, and *spread* in the thesaurus tables leads to the corresponding engineering terms of *inhibit* and *shape*. Inhibit is a tertiary term under stop, and shape is a tertiary term under change, with both falling under the class of control magnitude. Using the identified engineering terminology, the designer can identify analogous engineered systems to the armadillo that involve in changing shape to inhibit an external input. Engineered systems that perform the functions corresponding to *surround*, *cover*, and *spread* are stadiums with retractable roofs and convertible automobiles. Both engineered systems change shape and inhibit the "enemy" of bad weather.

Querying the Design Repository facilitates a computational approach to discovering analogies. With both biological and engineered systems in the knowledge base, one could retrieve biological and engineered components when searching for solutions, thus allowing for comparison and analogy discovery through function. Both the manual and computational approaches to analogy discovery can lead to an "Aha" moment that assists in concept generation.

4.5 Integration with Computational Design Tools

Computational design tools promise engineers a faster realization of potential design solutions based on previously known products and implementations, or databases of design information. The engineering-to-biology thesaurus design tool, which is a stand-alone tool, has been integrated with established function-based computational design tools to support bioinspired engineering design. The following subsections describe the integration efforts.

4.5.1 Design Repository

The Design Repository is an engineering design knowledge base that contains descriptive product information such as functionality, component physical parameters, manufacturing processes, failure, and component compatibility of over 130 consumer products. Each consumer product was decomposed and functionally modeled using the Functional Basis. Each repository entry is designated as an artifact or assembly of artifacts, whether it performs a supporting function (secondary to the product's operation) and the class of the artifact when entered into the repository database. Additionally, several artifact attributes are captured and stored in a relational database.

The integration of the thesaurus with the Design Repository has been with the population of a biomimetic design repository, which enables the storage of biological knowledge indexed by engineering function. Storing the biological information based on the function the biological system solves allows quick access to principle solutions. There are a total of 30 biological entries in the Design Repository that comprise the biomimetic design repository. The Design Repository facilitates computational concept generation and comparison of biological and engineered components. The designer chooses from resulting computational concept generator suggestions, engineered and biological, to develop a complete conceptual design.

4.5.2 Organized Search Tool

The organized search tool (Stroble et al. 2009d) was developed for retrieving relevant biological systems that perform functions of interest. Specifically, the organized search tool is designed to work with non-engineering subject-domain-specific information, such as biology. The majority of biological information is written in natural language format, which prompted the investigation of using both a Functional Basis function and flow term when searching for solutions. Realizing how the topic of the text is treated increases the extensibility of the organized verb–noun search algorithm.

The verb–noun combination search strategy incorporates the terms of the engineering-to-biology thesaurus into the search algorithm and provides two levels of results: (1) associated with verb only, of which the user can choose to utilize or ignore, and (2) the narrowed results associated with the verb–noun pair. This search strategy requires the designer to first form an abstraction (e.g., functional model) of the unsolved problem using the Functional Basis lexicon. The verbs (functions) of the abstraction are input as keywords in the organized search tool to generate a list of biological matches. The search algorithm swaps the engineering function term for the corresponding biological function terms in the engineering-to-biology thesaurus. The biological corpora are then searched for the biological function, and all sentences containing the function are extracted for further processing. Each match is stored for display to the user. When multiple biological correspondent function terms are present, the search is executed recursively until all corresponding biological functions have been searched. The noun listing is then used in combination with the search verb results for a second, more detailed search to locate specific text excerpts from the biological corpora that describe how the biological systems perform the functionality with the desired flows.

This search strategy is embodied in an automated retrieval tool that allows an engineering designer to selectively choose which corpora to search and to upload additional searchable information as it is made available. The user interface initially presents the designer with a function (verb) entry field and search options. Search options prompt the designer to choose from exact word, derivatives of the word, and partial word. Once the biological corpora are searched for the function term, the designer is presented with a listing of flows (nouns) that occur in proximity to the searched verb for each corpus searched followed by a group of sentences that include the function and listed flows. The resultant biological information is more relevant and focused due to the integration of the thesaurus in the search algorithm.

4.5.3 MEMIC

Computational concept generation is an efficient way to generate several conceptual design variants. Also, it adds the benefit of providing lists of engineering components that may be used to solve a particular function. The morphological evaluation machine and interactive conceptualizer (MEMIC) was created for use during the early stages of design to produce design solutions for an engineering design from a functional model using knowledge of existing engineered products (Bryant Arnold et al. 2008; Bryant et al. 2007). The concept generator software MEMIC accepts an input functional model and uses functionality and compatibility information stored in the Design Repository to generate, filter, and rank full concept variants. The software returns a listing of engineering component solutions for each function–flow pair of the input functional model, which allows a designer to easily choose between multiple solutions for a given function and interactively build a complete conceptual design.

To extend the MEMIC computational concept generation approach to support bioinspired design, the organized search tool algorithm, which utilizes the engineering-to-biology thesaurus, is merged (Nagel and Stone 2012). Integrating biological information with an established, computational method for concept generation enables designers to consider taking inspiration from biology without having to expend extra effort to learn a new method. Computational concept generation of bioinspired designs through this approach requires the designer to input desired functionality, in the form of a functional model. Functionality is a useful metric for defining a conceptual idea, as functional representation has been shown to reduce fixation of how a product or device would look and operate (Otto and Wood 2001; Pahl et al. 2007). Each function/flow pair of the model is then searched in the engineering and biological knowledge bases, the Design Repository and the biological corpora made available to the organized search tool, respectively, to identify solutions to the function/flow pairs.

This computational approach expands the biological knowledge base beyond that of the biomimetic design repository to provide a greater range of biological solutions for engineering inspiration. Multiple solutions from both domains, to each function/flow pair, are returned and presented to the designer. The biological solutions are not indented for physical use, but are intended for spurring creative ideas or connections to the engineering domain that could be implemented in an engineered system that partially (i.e., one or two components) to completely (i.e., entire design) mimic a biological system. This computational approach assists in identifying biological solutions to engineering functions, analogies, and engineering components that map to biological system attributes. To arrive at the final concept, however, the designer is required to identify principles, components, materials, and/or systems within the engineering domain that support what the biological solution suggests. Therefore, this approach lends itself more toward innovative design problems where novel solutions tend to dominate.

4.6 Conclusion

The engineering-to-biology thesaurus presented in this chapter affords engineers, with limited biological background, a tool for leveraging nature's ingenuity during many steps of the design process. An engineering-to-biology thesaurus (1) lessens the burden when working with knowledge from the biological domain by providing a link between engineering and biological terminologies; (2) assists designers in establishing connections between the two domains; and (3) facilitates bioinspired design through many applications. The thesaurus is presented as a versatile design tool that can be used during many points in the bioinspired design process and is customizable to the designer based on needs and prior biological knowledge.

Biological terms in the thesaurus are correlated with the engineering domain through pairing with a synonymous function or flow term of the Functional Basis lexicon. The engineering-to-biology thesaurus is a work in progress and is not a comprehensive list of all biological and engineering terms. Rather than encompassing all engineering terminology, the confined set of generalized engineering terms of the Functional Basis was chosen as an initial starting point. Through this research, biological correspondent function and flow (material, signal, and energy) terms were mapped to the engineering domain. The mapping groups related terms and concepts in a classification structure that supports bioinspired design activities in the engineering domain. Additionally, the terminology mappings of the engineering-to-biology thesaurus facilitate recognition of related terms and concepts between engineering and biology by providing a structured framework. The engineering-to-biology thesaurus increases the interaction between the users and the knowledge resource and fosters one to make associations between the engineering and biological lexicons, thus strengthening the designer's ability to utilize biological information.

Listing biological correspondent terms that an engineering designer interested in function-based design might encounter is among the first step to bridging the terminology gap between the biology and engineering domains. Furthermore, the engineering-to-biology thesaurus is a subject-domain-oriented, intermediary structure, which can be updated as needs are identified.

The work to develop and formulate the engineering-to-biology design tool makes a fundamental contribution to the fields of engineering design and biomimicry. The thesaurus is envisioned to enable the engineering and biology communities to better collaborate, create, and discover. Furthermore, it facilitates many bioinspired design activities as a stand-alone tool, but also supports and enables computational bioinspired design activities.

Future work for improving the engineering-to-biology thesaurus includes examining potential terms through clustering and analyzing terms contained within the glossary of a collegiate entry-level biological textbook. While collocated terms provide an indication for macrorelevant terms, clustering analysis could be utilized to find less obvious, but equally important, biological terms for thesaurus population. Additionally, biological texts that focus on a topic of interest (i.e., insects, fungi) should be analyzed for relevant biological terms that an introductory text does not include. Further work also includes adopting a hierarchy for the mapped biological terms. Many of the repeated biological flows currently in the thesaurus indicate that the biological flow has different scales. For example, an organ is a composite of tissues, and tissue is a composite of cells. A hierarchy could help make the terminology mappings between the domains more clear.

Beyond the terms within the thesaurus, there are future work opportunities for the implementation of the tool as well. The terms have successfully been integrated with other computational design tools; however, a stand-alone computational version of the thesaurus would still be useful. Development has begun on biology-to-engineering translator for automatic machine translation of biological information into an engineering context. A long-term goal is implementation of the thesaurus as a dynamic Web entity that would allow collaboration where those with different expertise can contribute to the terms as well as support dissemination of the research in real time.

References

Bar-Cohen Y (2006a) Biomimetics biologically inspired technologies. CRC/Taylor & Francis, Boca Raton

Bar-Cohen Y (2006b) Biomimetics—using nature to inspire human innovation. J Bioinspiration Biomimetics 1:P1–P12

Benyus JM (1997) Biomimicry innovation inspired by nature. Morrow, New York

Brebbia CA (2006) Design and nature III: comparing design in nature with science and engineering, vol 3. WIT, Southampton

Brebbia CA (2008) Design & nature IV: comparing design in nature with science and engineering. WIT, Southampton

Brebbia CA, Carpi A (2010) Design & nature V: comparing design in nature with science and engineering. WIT, Southampton

Brebbia CA, Collins MW (2004) Design and nature II: comparing design in nature with science and engineering, vol 3. WIT, Southampton

Brebbia CA, Sucharov LJ, Pascolo P (2002) Design and nature: comparing design in nature with science and engineering, vol 1. WIT, Southampton

Brownell P (2010a) Engineering-to-biology thesaurus term check meeting. Nagel JKS. 8 Apr 2010, Corvallis

Brownell P (2010b) Engineering-to-biology thesaurus term check and validation meeting. Nagel JKS. 4 May 2010, Corvallis

Bruck HA, Gershon AL, Golden I, Gupta SK, Gyger LS Jr, Magrab EB, Spranklin BW (2007) Training mechanical engineering students to utilize biological inspiration during product development. Bioinspiration Biomimetics 2:S198–S209

Bryant Arnold CR, Stone RB, McAdams DA (2008) MEMIC: an interactive morphological matrix tool for automated concept generation. In: Mason JFaS (ed) Industrial engineering research conference

Bryant C, Bohm M, McAdams D, Stone R (2007) An interactive morphological matrix computational design tool: a hybrid of two methods. Paper presented at the ASME 2007 IDETC/CIE, Las Vegas, NV

Campbell NA, Reece JB (2003) Biology. Pearson Benjamin Cummings, San Francisco

Chakrabarti A, Sarkar P, Leelavathamma B, Nataraju BS (2005) A functional representation for aiding biomimetic and artificial inspiration of new ideas. Artif Intell Eng Des Anal Manuf 19:113–132

Cheong H, Shu LH, Stone RB, McAdams DA (2008) Translating terms of the functional basis into biologically meaningful words. In: 2008 ASME IDETC/CIE, New York City, NY

Chiu I, Shu LH (2007a) Biomimetic design through natural language analysis to facilitate cross-domain information retrieval. AIEDAM 21(1):45–59

Chiu I, Shu LH (2007b) Using language as related stimuli for concept generation. AIEDAM 21(2):103–121

Cross N (2008) Engineering design methods: strategies for product design. Wiley, Chichester

de Mestral G (1955) Velvet type fabric and method of producing same. USA Patent

Design & Intelligence Laboratory (2010) Biologue. http://home.cc.gatech.edu/dil/336. Accessed 10 Dec 2012

Dieter GE, Schmidt LC (2009) Engineering design, 4th edn. McGraw-Hill, New York

Dym CL, Little P (2004) Engineering design: a project-based introduction. Wiley, New York

Eguchi E, Tominaga Y (1999) Atlas of arthropod sensory receptors: dynamic morphology in relation to function. Springer, Tokyo

Forbes P (2006) The gecko's foot: bio-inspiration: engineering new materials from nature. W. W. Norton & Co, New York

Gero JS, Kannengiesser U (2002) The situated function—behaviour—structure framework. Artif Intell Des, pp 89–104

Goel AK, Chandrasekaran B (1992) Case-based design: a task analysis. Artif Intell Approaches Eng Des 2:165–184

Goel AK, Rugaber S, Vattam S (2009) Structure, behavior, and function of complex systems: the structure, behavior, and function modeling language. Artif Intell Eng Des Anal Manuf 23(1):23–35

Helms M, Vattam SS, Goel AK (2009) Biologically inspired design: products and processes. Des Stud 30(5):606–622

Hey J, Linsey J, Agogino AM, Wood KL (2008) Analogies and metaphors in creative design. Int J Eng Educ 24(2):283–294

Hill B (1995) Bionic—element for fixing the aim and finding the solution in the technical problem solving process. In: IDATER 1995 conference, Loughborough University, Loughborough. http://hdl.handle.net/2134/1509

Hirtz J, Stone R, McAdams D, Szykman S, Wood K (2002) A functional basis for engineering design: reconciling and evolving previous efforts. Res Eng Design 13(2):65–82

Hundal M (1990) A systematic method for developing function structures, solutions and concept variants. Mech Mach Theory 25(3):243–256

Hyman B (1998) Engineering design. Prentice-Hall, New Jersey

Klowden MJ (2008) Physiological systems in insects. Academic Press, Oxford

Lawrence E, Holmes S (1989) Henderson's dictionary of biological terms. Wiley, New York

Lindemann U, Gramann J (2004) Engineering design using biological principles. In: International design conference—DESIGN 2004, Dubrovnik

Linsey F (2008a) Biological flow correspondent term check meeting. Stroble JK. 20 June 2008, Rolla

Linsey F (2008b) Biological flow correspondent term check and validation meeting. Stroble JK. 10 July 2008, Rolla

Little A, Wood K, McAdams D (1997) Functional analysis: a fundamental empirical study for reverse engineering, benchmarking and redesign. In: 1997 ASME IDETC/CIE, Sacramento, CA

Lopez-Huertas MJ (1997) Thesaurus structure design: a conceptual approach for improved interaction. J Documentation 53(2):139–177

Mak TW, Shu LH (2004) Abstraction of biological analogies for design. CIRP Ann 531(1):117–120

Mak TW, Shu LH (2008) Using descriptions of biological phenomena for idea generation. Res Eng Design 19(1):21–28

Matrin E, Hine RS (2000) Oxford dictionary of biology. Oxford University Press, Oxford

McKean E (2005) The new Oxford American dictionary. Oxford University Press, New York

Mitchell BK (2003) Chemoreception. Encyclopedia of insects. Academic Press, Amsterdam, pp 169–174

Møller AR (2003) Sensory systems: anatomy and physiology. Academic Press, Amsterdam

Nachtigall W (1989) Konstructionen: Biologie und Technik. VDI, Dùsseldorf

Nachtigall W (2000) The big book of bionics: new technologies, following the example of nature. German Verlags-Anstalt

Nachtigall W (2002) Bionics: principles and examples for engineers and scientists. 2nd edn. Springer

Nagel JKS, Stone RB (2012) A computational approach to biologically-inspired design. AIEDAM 26(2):0

Nagel R, Tinsley A, Midha P, McAdams D, Stone R, Shu L (2008) Exploring the use of functional models in biomimetic design. J Mech Des 130(12):11–23

Nagel JKS, Stone RB, McAdams DA (2010a) Exploring the use of category and scale to scope a biological functional model. In: 2010 ASME IDETC/CIE, Montreal, Quebec, Canada

Nagel JKS, Nagel RL, Stone RB, McAdams DA (2010b) Function-based, biologically inspired concept generation. AIEDAM 24(4):521–535

Nagel JKS, Stone RB, McAdams DA (2010c) An engineering-to-biology thesaurus for engineering design. In: ASME IDETC/CIE 2010 DTM-28233, Montreal, Quebec, Canada

Nagel JKS, Stone RB, McAdams DA (2010d) Exploring the use of category and scale to scope a biological functional model. In: ASME IDETC/CIE 2010, DTM-28873, Montreal, Quebec, Canada

Nagel JKS, Nagel RL, Stone RB (2011) Abstracting biology in engineering design. Int J Des Eng 4(1):23–40

Otto KN, Wood KL (2001) product design: techniques in reverse engineering and new product development. Prentice-Hall, Upper Saddle River

Pahl G, Beitz W, Feldhusen J, Grote KH (2007) Engineering design: a systematic approach, 3rd edn. Springer

Sarkar P, Phaneendra S, Chakrabarti A (2008) Developing engineering products using inspiration from nature. J Comput Inf Sci Eng 8(3):1–9

Shu LH, Stone RB, McAdams DA, Greer JL (2007) Integrating function-based and biomimetic design for automatic concept generation. In: International conference on engineering design, Paris, France

Srinivasan V, Chakrabarti A (2009) SAPPhIRE—an approach to analysis and synthesis. In: International conference on engineering design, Stanford, USA

Srinivasan V, Chakrabarti A (2009) SAPPhIRE—an approach to analysis and synthesis. In: 3rd Symposium on research in product design, CPDM, IISc, Bangalore, India. doi:2-417

Stone R, Wood K (2000) Development of a functional basis for design. J Mech Des 122(4):359–370

Stroble JK, Stone RB, Watkins SE (2009a) An overview of biomimetic sensor technology. Sens Rev 28(2):112–119

Stroble JK, Stone RB, McAdams DA (2009b) Conceptualization of biomimetic sensors through functional representation of natural sensing solutions. In: International conference of engineering design, Stanford, California

Stroble JK, Stone RB, McAdams DA, Watkins SE (2009c) An engineering-to-biology thesaurus to promote better collaboration, creativity and discovery. In: CIRP design conference 2009, Cranfield, Bedfordshire, UK, pp 353–368

Stroble JK, Stone RB, McAdams DA, Goeke MS, Watkins SE (2009d) Automated retrieval of non-engineering domain solutions to engineering problems. In: CIRP design conference 2009, Cranfield, Bedfordshire, UK, pp 78–85

Szykman S, Racz J, Sriram R (1999) The representation of function in computer-based design. In: Proceedings of the ASME design theory and methodology conference, Las Vegas, NV

The Biomimicry Institute (2009) Biomimicry: a tool for innovation. http://www.biomimicryinstitute.org/about-us/biomimicry-a-tool-for-innovation.html. Accessed 10 Oct 2009

Toko K (2000) Biomimetic sensor technology. Cambridge University Press, Cambridge

Tsujimoto K, Miura S, Tsumaya A, Nagai Y, Chakrabarti A, Taura T (2008) A method for creative behavioral design based on analogy and blending from natural things. In: 2008 ASME IDETC/CIE, New York, USA. DETC2008-49389

Ullman DG (2009) The mechanical design process, 4th edn. McGraw-Hill, Inc., New York

Ulrich KT, Eppinger SD (2004) Product design and development. McGraw-Hill/Irwin, Boston

Umeda Y, Takeda H, Tomiyama T, Yoshikawa H (1990) Function, behaviour, and structure. AIENG'90 Applications of AI in Engineering, pp 177–193

Vakili V, Shu LH (2001) Towards biomimetic concept generation. In: 2001 ASME IDETC/CIE, Pittsburgh, Pennsylvania

Vakili V, Shu LH (2007) Including functional models of biological phenomena as design stimuli. In: 2007 ASME IDETC/CIE, Las Vegas, NV

Vattam S, Helms M, Goel A (2010a) A content account of creative analogies in biologically inspired design. AIEDAM 24:467–481

Vattam S, Wiltgen B, Helms M, Goel A, Yen J (2010b) DANE: fostering creativity in and through biologically inspired design. In: Proceedings of first international conference on design creativity, Kobe, Japan, pp 127–132

Vincent JFV, Bogatyreva OA, Bogatyrev NR, Bowyer A, Pahl A-K (2006) Biomimetics: its practice and theory. J R Soc Interface 3:471–482

Voland G (2004) Engineering by design, 2nd edn. Pearson Prentice Hall, Upper Saddle River

Wilson JO, Rosen D (2007) Systematic reverse engineering of biological systems. In: 2007 ASME IDETC/CIE, Las Vegas, Nevada

Wilson J, Chang P, Yim S, Rosen D (2009) Developing a bio-inspired design repository using ontologies. In: 2009 ASME IDETC/CIE, California, USA

Chapter 5
Function-Based Biologically Inspired Design

Jacquelyn K. S. Nagel, Robert B. Stone and Daniel A. McAdams

Abstract A "big picture" approach to a systematic, function-based (drawing from a Pahl and Beitz approach) biologically inspired design is presented in this chapter. The approach supports two different starting, or perhaps motivating, points: a customer need motivated product design and a biological system motivated product opportunity. Both approaches rely on a designer's ability to create a functional model that either captures customer needs or represents the biological system of interest. This methodology relies directly on the designer's ability to make connections between dissimilar domain information. Following presentation of the methodology are two validation approaches. One examines current biologically inspired products either in production or presented in the literature to demonstrate that the systematic design methodology for biologically inspired design can reproduce the existing design. The second validation exercise investigates three needs–based design problems that lead to plausible biologically inspired solutions.

Keywords Design methodology · Framework · Function-based design · Systematic design · Design tools · Functional modeling · Connections · Analogies · Sensor design · Problem-driven approach · Biology-driven approach · Concept development

---

J. K. S. Nagel (✉)
Department of Engineering, James Madison University, Harrisonburg, VA, USA
e-mail: nageljk@jmu.edu

R. B. Stone
School of Mechanical Industrial and Manufacturing Engineering, Oregon State University, Corvallis, OR, USA
e-mail: robert.stone@oregon.state.edu

D. A. McAdams
Department of Mechanical Engineering, Texas A&M University, College Station, TX, USA
e-mail: dmcadams@tamu.edu

A. K. Goel et al. (eds.), *Biologically Inspired Design*,
DOI: 10.1007/978-1-4471-5248-4_5,

5.1 Introduction

Biological organisms, phenomena, and strategies, herein referred to as biological systems, provide a rich set of examples that can be used to inspire engineering innovation. Bioinspired, or biomimetic, designs are publicly viewed as creative and novel solutions to human problems. Moreover, some biomimetic designs have become so commonplace that it is hard to imagine life without them (e.g., Velcro, airplanes). Although the bioinspired solutions are innovative and useful, the majority of inspiration taken from nature has happened by chance observation or dedicated study of a specific biological entity (e.g., gecko). This reveals a fundamental problem of working across the engineering and biological domains. The effort and time required to become a competent engineering designer creates significant obstacles to becoming sufficiently knowledgeable about biological systems (the converse can also be said). This research aims to go beyond the element of chance, reduce the amount of time and effort required to develop bioinspired solutions, and bridge the seemingly immense disconnect between the engineering and biological domains.

This chapter presents a comprehensive design approach, including a methodology and supporting tools, that integrates with Pahl- and Beitz-based (Pahl et al. 2007) or Otto- and Wood-based (Otto and Wood 2001) functional design techniques to facilitate biologically inspired design. As a function-based method, it offers several advantages: archival and transmittal of design information; reduces fixation on esthetic features or a particular physical solution; allows one to define the scope or boundary of the design problem as broad or narrow as necessary; and encourages one to draw upon experience and knowledge stored in a database or through creative methods during concept generation. Function, as used in systematic design, is recognized as a way to connect nature and engineering through a commonality. This approach addresses the knowledge requirement problem of working across the biology and engineering domains and, through functional modeling, offers the aforementioned advantages.

A design methodology that facilitates systematic biologically inspired design from a problem-driven or biology-driven perspective is given. Following a traditional design approach, the problem-driven perspective begins from customer needs or a given problem to solve. Following curiosity, the biology-driven approach encourages a designer to explore a biological system of interest to learn more about it or develop an innovative solution that could potentially be applied to an existing problem. The methodology also prescribes the tools of the framework to use while guiding the designer through the steps. Although systematic, the design methodology is also versatile by providing a designer multiple avenues that lead to a biologically inspired design.

Biologically inspired design is a young field within engineering design. Consequently, validation of the design methodology is challenging. Thus, two approaches to validation are pursued. One examines current biologically inspired products either in production or presented in the literature to demonstrate that the systematic design methodology for biologically inspired design can reproduce the existing design. The second validation exercise investigates three needs–based design case studies that lead to plausible biologically inspired solutions. The needs-based design case studies demonstrate the systematic design approach to biologically inspired design from the problem-driven perspective. A reference to a case study from the biology-driven perspective is also provided. The thought processes, model iterations, and connections that are the leaps that enable the ingenuity of the nature to be discovered and adapted for use in engineered systems are also presented.

Using functional representation and abstraction to describe biological systems presents the natural designs in an engineering context and allows designers to make connections between biological and engineered systems. Thus, the biological information is accessible to engineering designers with varying biological knowledge, but a common understanding of engineering design methodologies. By creating a bridge between the two domains through the perspective of function, engineers can leverage the elegant designs found in the world around them. This work demonstrates the feasibility of using systematic design for the discovery of innovative engineering designs without requiring expert-level knowledge, but rather broad knowledge of many fields.

5.2 Framework to Support Method

Various tools and techniques combine or interact to support the overall systematic biologically inspired design methodology. The engineering-to-biology thesaurus (see Chap. 4) is the backbone of this framework, as it assists in modeling biological systems and searching for inspiration or solutions. Consequently, the thesaurus also assists in concept generation, both directly and indirectly. Indirect assistance is through the modeling method and organized search tool, and direct assistance is through designer knowledge of a biological process (e.g., the conversion of sunlight to sugars) that could solve a set of design needs of a product. The following section makes this indirect assistance more explicit.

A framework, by definition, is an arrangement of parts that provides a system or concept a basic form (McKean 2005). In the framework developed here, the parts are identify, translate, represent, and conceptualize. The system or concept being supported is the systematic biologically inspired design methodology. What makes this framework particularly useful for design is the flexibility a designer is afforded when working toward a biologically inspired solution. Each tool and technique can be used individually and in multiple combinations. Prior work has shown how the engineering-to-biology thesaurus integrates with and improves organized search tools (Nagel et al. 2010; Nagel et al. 2010; Nagel et al. 2010), supports biological functional modeling method (Nagel et al. 2010; Nagel et al. 2010; Nagel et al. 2010), and integrates with concept generation approaches (Nagel 2010). Specific interaction benefits are pointed out in the Venn diagram of Fig. 5.1. Using all four parts results in systematic biologically inspired design.

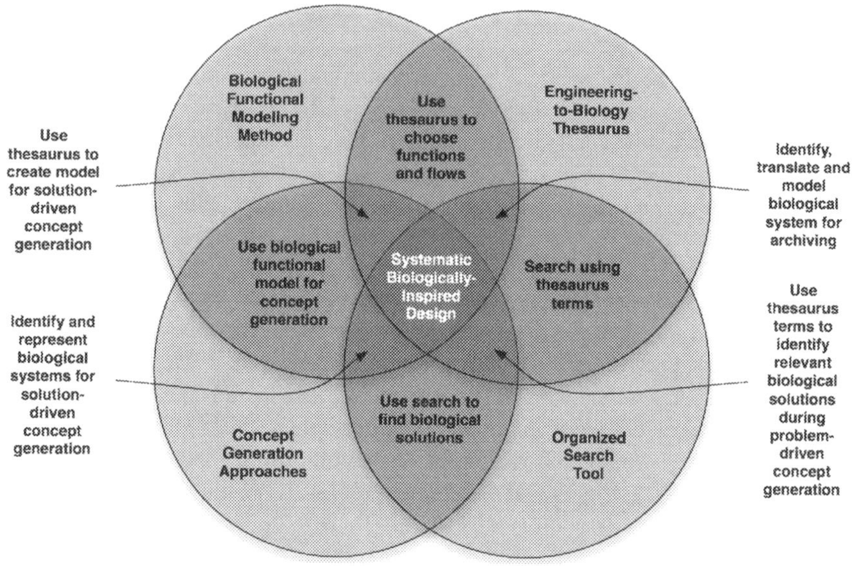

Fig. 5.1 Framework details

Table 5.1 Framework integration with existing design tools

Framework tool or technique	Existing tool leveraged by framework
Engineering-to-biology thesaurus	Functional basis
Biological functional modeling method	Functional basis
Organized search tool	Functional basis, design repository
Concept generation approaches	MEMIC, automated morphological matrix, design repository

Knowledge requirements for working between domains can be alleviated with design methods that integrate the existing engineering design tools. This framework alleviates knowledge requirements by integrating the function-based design methodologies, in particular, the systematic design process defined by Pahl et al. (2007). The framework also leverages existing design tools to further integrate with function-based design methodologies, as shown in Table 5.1. The existing design tools include a functional modeling lexicon, a repository of design information, and automated concept generation methods.

5.3 Biologically Inspired Design Methodology

In this section, the overall design methodology is given. In support of the systematic biologically inspired design methodology, several separate techniques coalesce to guide the designer from initial curiosity or customer needs to complete concept (Nagel and Stone 2011). The framework for making connections between biology and engineering is presented first followed by the design methodology.

Two holistic approaches by the Biomimicry Institute (2010) and Helms et al. (2009) offer a process for designers to follow, from a biology perspective and an engineering perspective, respectively, and a database of biological information that supports biologically inspired design. Additionally, the work by Helms et al. (2009) and Vattam et al. (2008) occurred in parallel with this work unbeknownst to the authors at the time. Both bodies of research made several similar observations that guided their work along distinct, but related paths. For example, Helms et al. (2009) first formally identified the two approaches of problem-based and solution-based bioinspired design that emerged from an earlier study of design students utilizing biological inspiration to solve engineering design problems (Vattam et al. 2008). The authors of this chapter made a similar observation while exploring how to use functional models as a means of finding related biological inspiration for design (Nagel et al. 2008) and identified two approaches to handle both problem-based and solution-based starting points for biologically inspired design. These approaches, however, do not offer a complete framework of design tools that integrate with the functional model-driven design process to support the designer throughout biologically inspired design. This chapter addresses that gap by offering a systematic methodology for both problem- and solution-based entry points along with a framework of supporting design tools that integrate with established functional model-based, engineering design methodologies.

The aim of the systematic design methodology of this chapter is to provide enough structure to assist in the bioinspired design process without hindering the creativity and inventiveness of the designer. This method is intended to foster and guide the abilities of the designer and encourage objective evaluation of the results. The systematic approaches of this method are further intended to render designing based on biological inspiration comprehensible and steer the efforts of designers down purposeful paths.

5.3.1 Making Connections

This research relies on the designer's ability to identify and formulate connections between biological and engineering domains. Just as there are different learning styles, there are multiple ways to make connections. Analogies (Casakin 2006; Gentner 1988; Gentner 1983; Gick and Holyoak 1980; Goel 1997; Hofstadter 1995; Tsujimoto et al. 2008; Linsey et al. 2008; Balazs and Brown 2001; Mak and Shu 2004; Bhatta and Goel 1997; Smith 1998; Nagai and Taura 2006) are the most widely used and have multiple forms. Direct, indirect, and compound analogies have all been used to connect a biological system to an engineering solution. A direct analogy mimics the biological system one to one. An indirect analogy uses the biological system to spur analogies for inspiration but does not mimic every aspect of the biological system. A compound analogy is the combination of multiple biological system attributes that lead to analogous engineered systems. The level of difficulty in accessing and transferring an analogy is largely dependent on how remote or close the distance between the domains is Johnson-Laird (1989). Because much of nature exhibits functionality and behavior in a comparable context to engineering, analogies with engineering are possible. To exemplify the connection-making process for analogies, consider a few textual examples.

A microflow detection sensor directly mimics the physiology and morphology of hair cells that make up the lateral line system in a fish. The connection is through the principle of a bending moment that is created from perpendicular flow against vertical hair cells. Direct analogy mimicry is achieved through fabrication of "hair-like" vertical structures on the end of a horizontal cantilever beam (Fan et al. 2002; Motamed and Yan 2005).

An indirect analogy bioinspired design was created between the common strain gage and the physiology of the campaniform sensillum or flexible exocuticle that many insects possess. An elliptical opening in the insect's cuticle, which is covered by a thin membrane layer, senses deformation because of the stress concentration (Gnatzy et al. 1987; Grunert and Gnatzy 1987). The connection for this system is that the opening causes mechanical coupling and global amplification to occur. Mimicry is achieved by optically measuring the stress concentration at a circular or elliptical hole in a rigid material when pressure is applied, resulting in a novel sensor that can sense strain in all directions (360°) (Wicaksono et al. 2004).

In the case of designing an electronic display that can be viewed in bright sunlight, a bioinspired compound analogy was used to solve the problem. For the display problem, hummingbird feather and morpho-butterfly wing attributes were combined to develop a solution (Vattam et al. 2008). Hummingbird feathers contain a series of alternating layers of thin films with different thicknesses instead of the intricate christmas-tree-like structures within an air gap that butterfly wings possess. The connections here are the air gap and "thin-film-like" structures, which are readily used in electronics processing today. Adding an air gap between thin films of varying thicknesses provided the right inspiration to develop the BrightView project (Vattam et al. 2008).

A designer must also be aware of analogies that hurt the design or ones that are overly complicated. Consider the biological phenomenon of abscission. When a leaf of a plant is damaged, it stops the flow of auxin and allows abscisic acid to dominate, thus forming a seal around the base of the leaf stem and over a period of time the leaf falls off (Campbell and Reece 2003). This biological system was used to inspire a solution to the problem of tiny parts sticking to a robot gripper in a microassembly process (Shu et al. 2006).

Considering indirect analogy for this case allowed the researchers to develop a sacrificial tool assembly. To highlight the analogy, consider the gripper as analogous to the plant, the sacrificial part of the tool as analogous to the abscission zone, and the tiny screw as analogous to the leaf that is released. Separation is achieved through the breakdown of the sacrificial part of the tool (abscission zone). Notice, liquids that are analogous to auxin and abscisic acid are not present in the final design.

Developing a direct analogy of abscission would require a flowing chemical that secures and releases the tiny screw. Release of the tiny screw would occur some time after the chemical flow stops and a chemical reaction takes place to loosen the part from the gripper. The indirect analogy that disregards liquids is the preferred solution concept. Therefore, another analogy form should be considered if the results lead to a bad design.

Two other approaches to formulating connections are through first principles (Hubka and Eder 1984; Lindemann and Gramann 2004; Otto and Wood 2001; Vincent and Mann 2002) and metaphors (Casakin 2007, 2006; Hey et al. 2008; Forty 1989). Analysis of physiology, structure, or behavior can lead to a connection made through first principles. Physical laws and concepts, such as the conservation of energy, that govern science as we know it also apply to natural systems.

Identifying a first principle shared by both domains leads to a connection and possibly innovation. Consider how ducks and other birds regulate their temperature during the winter to stay alive. The principle of heat exchange between the body and the legs is carried out to reduce the amount of heat lost through blood that is circulated through the legs (Lindemann and Gramann 2004). To date, metaphors for biologically inspired design have only been documented for architectural structures (Dollens 2009). The multiple approaches to formulating connections allow a designer to discover and become inspired in a manner that best suits him or her.

5.3.2 The Methodology

A pictorial representation of the method is given in Fig. 5.2. The flower is used to show that the methodology is an organic process that has systematic design roots. Each of the steps is discussed in greater detail below. The majority of, if not all, design processes are iterative, and this methodology follows the same convention. Cues for when to iterate are provided. Furthermore, the design methodology here should not be viewed as a rigid sequence in which one must follow each minute detailed step. Rather, it should be viewed as a starting point or a set of guidelines that aim to arrive at a biologically inspired design.

Figure 5.3 summarizes the avenues of the problem-driven approach that closely follows traditional systematic design, and the avenues of the problem-driven approach that starts from a known biological solution are summarized in Fig. 5.4. Figure 5.5 is a flowchart of the biology-driven approach that starts from curiosity. Also, the parts of the framework and existing design tools used in each step of the method are made explicit in Table 5.2.

Step 1: Needs or Curiosity

Initially, a designer can choose to start from a traditional set of customer needs or explore a curiosity. These two approaches are identified as problem-driven and

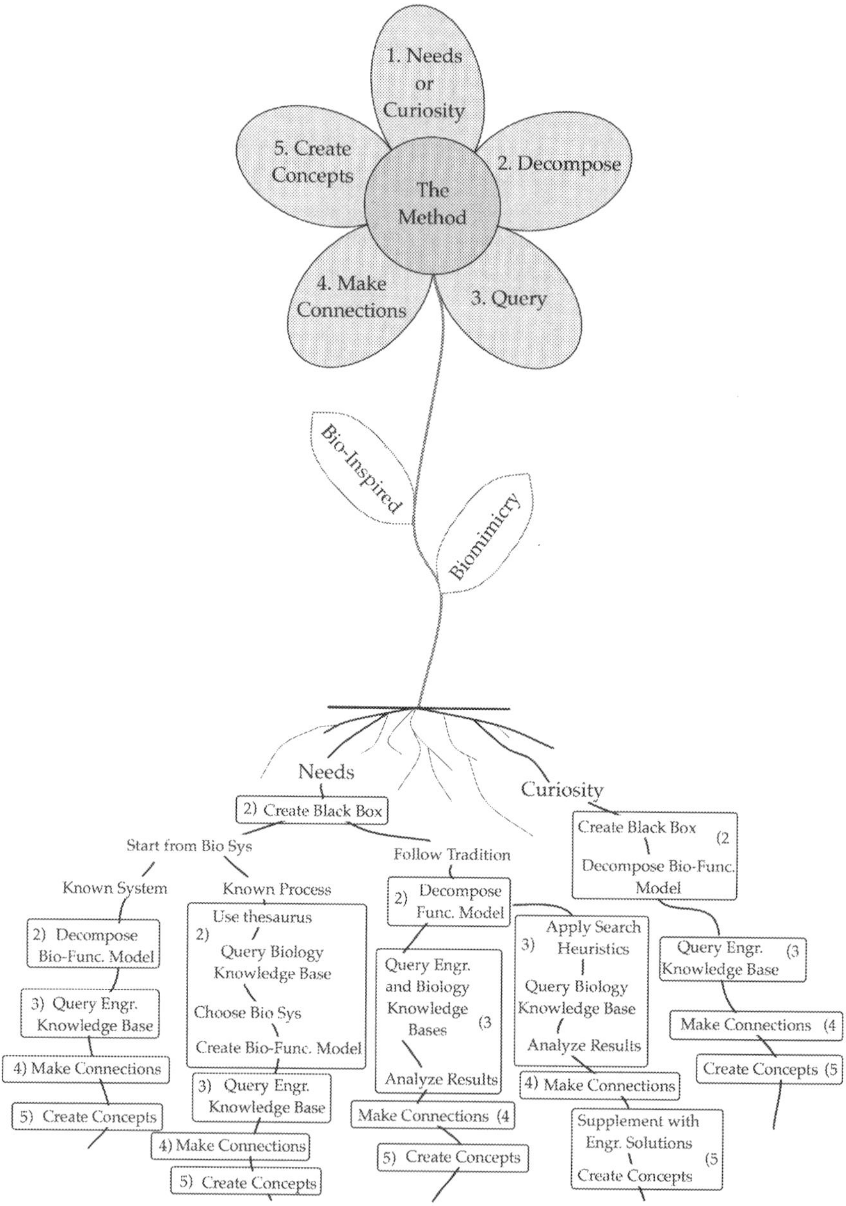

Fig. 5.2 An overview of the five steps of the systematic biologically inspired design methodology

biology-driven, respectively. Taking the problem-driven approach means the designer must gather a set of needs, requirements, and constrains. Many sources exist to aid the designer with proper needs gathering (Otto and Wood 2001;

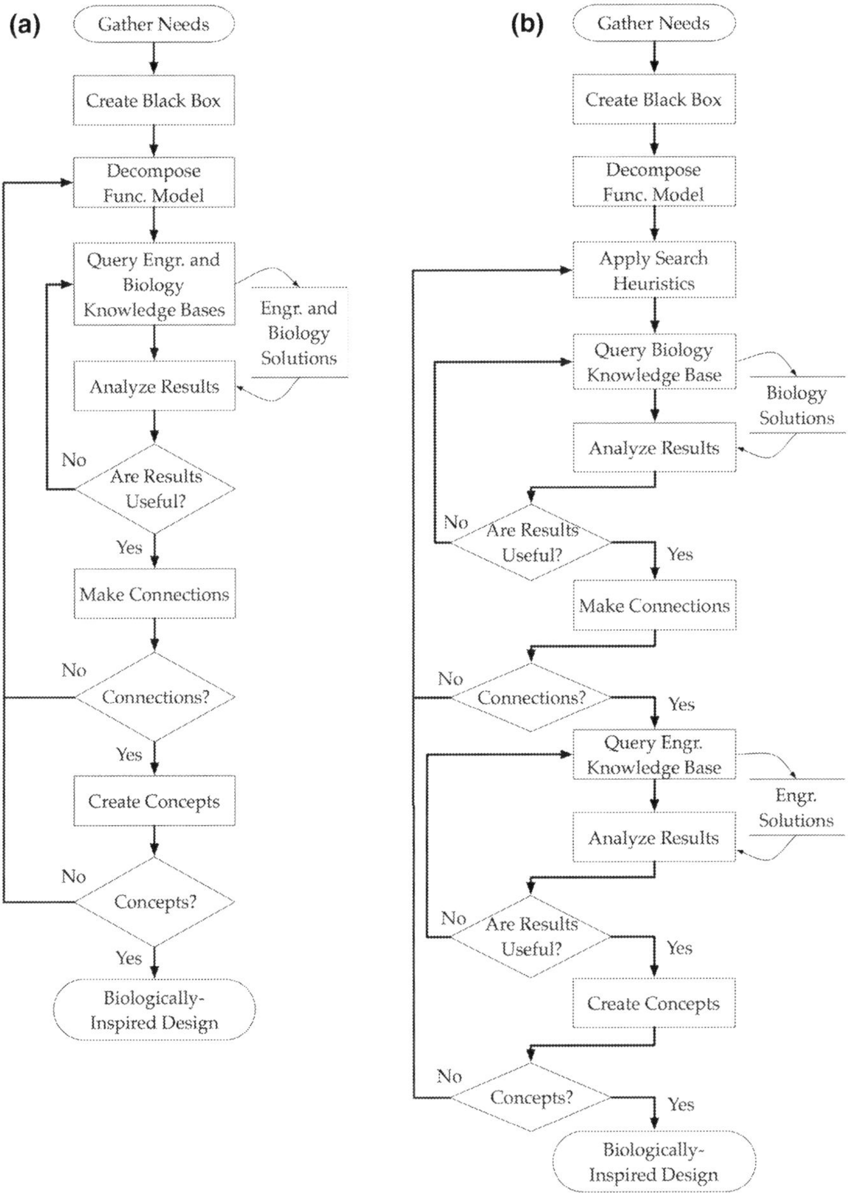

Fig. 5.3 Flowchart of the problem-driven approach that closely follows traditional systematic design (**a**) basic systematic approach to biologically inspired design using a function-based framework (**b**) systematic approach to biologically inspired design using a function-based framework with biological system identification guided by search heuristics

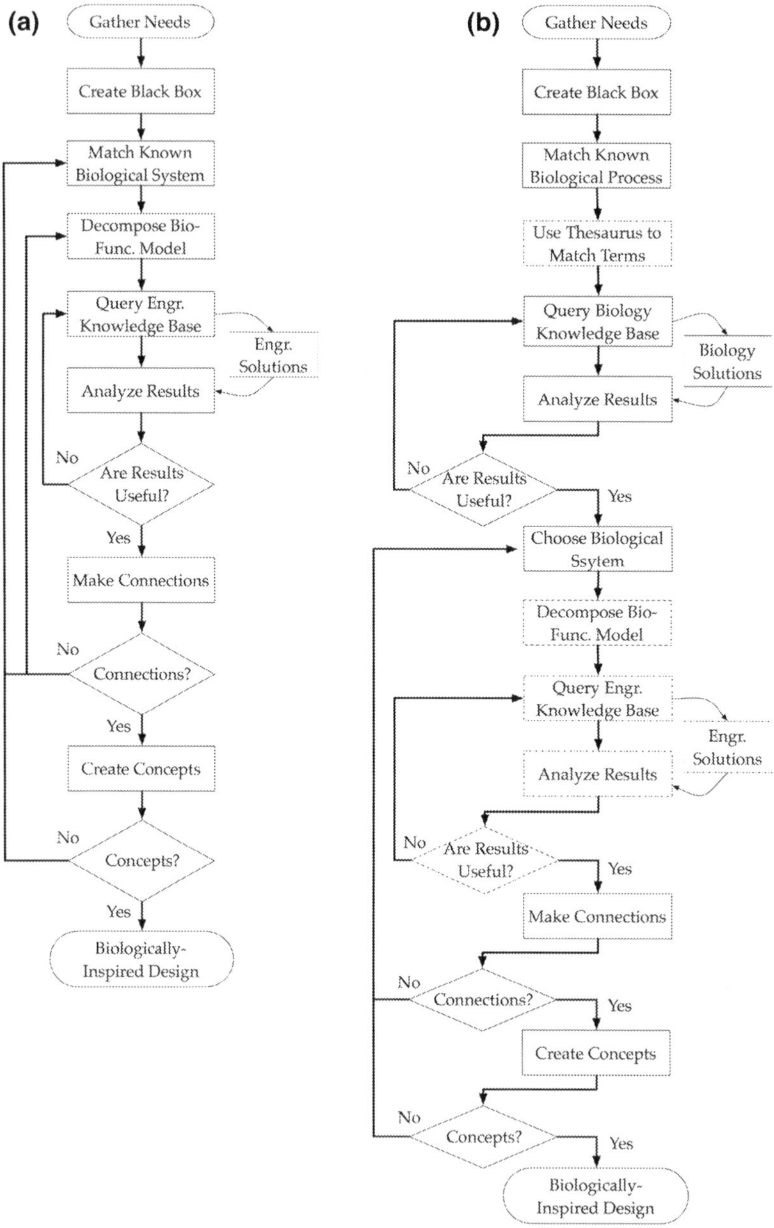

Fig. 5.4 Flowchart of the problem-driven approach that starts from a known biological solution (**a**) basic systematic approach to biologically inspired design using a function-based framework when an inspiring biological system is already identified (**b**) systematic approach to biologically inspired design using a function-based framework when translation of a biological process into functional language is needed

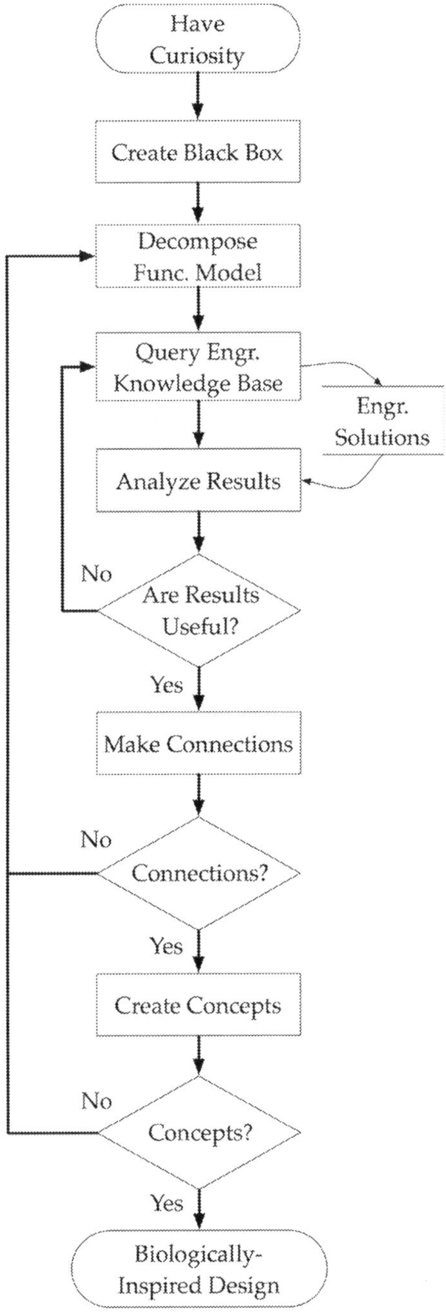

Fig. 5.5 Flowchart of the biology-driven approach

Table 5.2 Framework parts and existing design tools used in each step of the methodology

Tools	Steps	Needs or curiosity	Decompose	Query	Make connections	Create concepts
Framework	E2B thesaurus	X	X	X	X	
	Search tool	X		X		
	Biological functional model		X	X	X	X
	Biological concept generation			X		X
Existing	Design repository			X	X	X
	Functional basis	X	X	X		
	MEMIC			X		
	Automated morphological matrix			X		

Ullman 2009; Ulrich and Eppinger 2004; Voland 2004; Hyman 1998; Dym and Little 2004). Identifying customer needs is the most critical part of the design process as they form the basis for device functionality and specifications. Taking the biology-driven approach means a designer already knows of an interesting biological system in which he or she would like to investigate. Thus, the focus is on finding an application for the principles and concepts exhibited in that biological system rather than on finding principles to solve an engineering need. However, the organized search tool can be used to find a biological system to investigate initially.

Step 2: Decompose

The second step involves decomposing the needs or interesting biological system into, first, a black box model and, second, a functional model. All models created with this method use the Functional Basis modeling lexicon. With regard to traditional systematic design, the black box aims to abstract the overall function of the device that is to be designed. Whereas for a biological system, the black box model describes an interesting function, structure, behavior, or strategy of the system. Next, the input and output flows to the black box are determined. These flows are prompted by the customer needs from the first step or the needs/attributes of the biological system needed to achieve the black box functionality. The next task is to create the functional model. Decompose the black box description into subfunctions connected by flows of energy, material, or signal (Pahl et al. 2007; Otto and Wood 2001; Stone 1997; Stone and Wood 2000; Nagel 2010). Functional model creation is often an iterative task. Before moving on to the next step, check to see that all customer needs have been met by identifying the flows and subfunction chains that address them (Pahl et al. 2007; Otto and Wood 2001; Stone 1997; Stone and Wood 2000; Nagel 2010). When following the biology-driven approach, the designer can refer to the general biological modeling methodology presented in (Nagel et al. 2010) for assistance with creating a biological functional model.

Step 3: Query

Step 3 involves querying a knowledge base to identify solutions to each function/flow pair of the functional model. Two knowledge bases are required: one containing successful engineered systems and the other containing biological systems. To integrate with this method, both are indexed by engineering function and flow. The Design Repository[1] containing descriptive product information serves as the engineered systems' body of knowledge. It also includes product information such as functionality, component parameters, manufacturing processes, failure, and component connectivity. The Design Repository now contains design knowledge of over 113 consumer products and 30 biological systems. Instead of creating a large knowledge base containing functionally decomposed biological systems, similar to the design repository, an introductory biology textbook serves as the biological systems' body of knowledge. Although it is not indexed by engineering function, the engineering-to-biology thesaurus provides a starting point to find inspiration with engineering function and flow.

The tasks that comprise Step 3 begin with using the MEMIC software, or automated morphological matrix search tool, to query the Design Repository and the organized search software to query the biological corpus.

Based on the number of results returned for engineered and biological solutions, the search may need to be repeated. For engineered solutions, it can be helpful to abstract terms to the next level in the hierarchy. For example, if *transport,* a tertiary-level term, does not return any repository entries, then the secondary-level term *transfer* should be used. The same approach applies to flows.

Step 4: Make Connections

Step 4 involves making connections. Connections through analogies, metaphors, and first principles assist in bridging the biology and engineering domains. Section 5.3.1 contains a thorough discussion of how connections are formulated.

Step 5: Concept Generation

The fifth step involves performing concept generation and creating biologically inspired conceptual solutions. Concept synthesis involves analysis, reflection, and synthesis. Analysis is on the returned engineered and biological solutions from Step 3. Reflection is on the connections to the engineering domain formulated in Step 4. Synthesis is of the existing engineering solutions, engineering solutions inspired by biology and inventive solutions inspired by biology to derive a new idea. Once synthesis takes place, the result will be at least one concept. Depending on the number of solutions returned during Step 3 and the connections made during Step 4, multiple concepts may result. Evaluation of concepts follows systematic design; Pugh charts or decision matrices are used to rank concepts and narrow the selection to the superior few concepts (Otto and Wood 2001; Ullman 2009; Ulrich and Eppinger 2004; Voland 2004; Hyman 1998; Dym and Little 2004).

[1] www.designengineeringlab.org

Once a final concept has been reached, the next phase of systematic design, Detailed Design (Pahl et al. 2007), can initiate.

5.3.3 Comparison of Methodology Approaches

Notice that the major differences between the avenues to biologically inspired design as shown in Figs. 5.3 and 5.4 are within Step 2. When a designer follows traditional systematic design, flowchart A of Fig. 5.3, and decomposes a functional model from a black box model, the resultant model is referred to as a conceptual functional model. The conceptual functional model is used to query the engineering and biology knowledge bases. In the event that no connections can be formalized, then the designer should return to the query step and try different levels of functions and flows. The same holds true for when no concepts are synthesized.

Following flowchart B of Fig. 5.3 instructs the designer to use the organized search tool heuristics for the initial query. Once biological solutions are gathered, then the Design Repository is queried to supplement the biological solutions with engineering solutions. One difference here is that the results of the search tool should be screened first before moving on. In the event that no connections can be formalized, then the designer should return to the query step and try a different heuristic. The same holds true for when no concepts are synthesized.

If a designer knows of a biological system that can solve the black box functionality, then a biological functional model of this system can be created to drive the methodology as shown in flowchart A of Fig. 5.4. In the event that no connections can be formalized, then the designer should return to the decompose step and either modify the biological functional model or choose a different biological system for exploration. The same holds true for when no concepts are synthesized.

Consider the scenario when a designer can describe a known biological process that solves the black box functionality, as shown in flowchart B of Fig. 5.4. Using the biological process as a starting point, the designer can search for biological systems that perform that process using the organized search tool. If the search results are too narrow or uninspiring, the designer can use the engineering-to-biology thesaurus to identify new query terms. Once a biological system is chosen, then the designer can define the biological system with a functional model and use it to discover engineered solutions that perform the same functions to make connections back to the engineering domain, unless connections are readily facilitated. In the event that no connections can be formalized, then the designer should return to Step 2 and either choose a different biological system for exploration or, if no model was created, develop a biological functional. The same holds true for when no concepts are synthesized. In both process flows of Fig. 5.4, if the Design Repository results are limited, then the designer should try different levels of functions and flows to broaden the search.

The fifth approach under this methodology, termed the biology-driven approach, is driven by curiosity and the possibility of creating an innovative solution based on the inspiring biological system that may or may not be fit to a problem. The main use for this approach, however, is the creation of archivable biological knowledge that is in an engineering context. This type of knowledge can be reused in future design activities when stored in a knowledge base. The main tasks of this approach, as shown in Fig. 5.5, follow the steps of the systematic biologically inspired design method. In the event that no connections can be formalized, then the designer should return to the decompose step and modify the biological functional. The same holds true for when no concepts are synthesized. Another option is to return to Step 1, choose a different biological system for exploration, and start over. Also, if the Design Repository results are meager or non-existent, then the designer should try different levels of functions and flows to broaden the search.

5.4 Validation of Method

Validation of the systematic biologically inspired design methodology is achieved through the application of the methodology to (1) check whether it reproduces existing biomimetic products and (2) identify a closely related development version of a concept through the literature review. Analysis and reproduction of existing biomimetic products through primary function allow the verification of the methodology as the result is known. Validating the methodology for non-existing biomimetic products requires a review of literature to quantify whether the concept variants are realistic in some near-term form. Finding a closely related development version of the biologically inspired conceptual design indicates that the concept is feasible. Similarity is based on functionality and components chosen to achieve functionality.

Six existing biomimetic products are analyzed through the application of the method to demonstrate that the method can reproduce what is known. To further demonstrate the validity of this method, three needs–based design problems are presented. The methodology is considered successful when a designer can analyze a biological system and identify connections between biology and engineering through function that lead to inspiration of a concept. Further, more detailed validation cases of how the methodology can result in innovative solutions following the problem-driven approach are given in (Nagel 2010), and how the methodology can result in an innovative solution following the biology-driven approach is given in Nagel et al. (2010), Nagel (2010).

5.4.1 Proof Through Existing Biomimetic Technology

The first validation exercise is to analyze existing biomimetic products, apply the systematic design methodology, and verify that the biological system used to

Table 5.3 Analysis of existing biomimetic products to validate methodology

Existing biomimetic products	Mimicked biological system	Primary function/flow pair(s)	Source	Can method reproduce design?
Walking stick for visually impaired that uses sonar	Echolocation of bats	Detect solid	Biological corpus	Yes
Passive heating and cooling buildings	Termite mounds	Regulate thermal energy, distribute thermal energy, distribute gas, remove gas	Asknature.org	Yes
Self-cleaning surfaces	Lotus	Inhibit solid, inhibit liquid, decrease solid	Asknature.org	Yes
Motion detector	Compound vision	Detect solid, Sense solid	Biological corpus	Yes
Color changing material without harmful chemicals	Morpho-butterfly	Change visual signal	Asknature.org	Yes
Microassembly with sacrificial gripper	Abscission of plants	Separate solid	Biological corpus	Yes

inspire the original design is used in the results in such a way that could lead to a reproduction.

Table 5.3 lists six existing biomimetic products. These technologies represent the fields of electrical, civil, and mechanical engineering and material science. The validation studies here relied on technical descriptions of these products found in the literature. From the descriptions, primary function/flow pairs were identified and represented with Functional Basis terminology. The primary function/flow pairs are then used to query the biological knowledge base.

Both representation and querying utilize the engineering-to-biology thesaurus. If the mimicked biological system is within the query results and described in a way that makes a connection and results in a similar concept to the existing biomimetic technology, then it follows that the method can reproduce the design.

During this validation exercise, AskNature was added to the biological corpus knowledge base to augment topical limitations in the corpus. AskNature[2] is an online database that biologists, engineers, designers, chemists, etc., can contribute focused on biologically inspired design.

Following the five steps of the methodology, all six existing biomimetic technologies were reproduced. The three found within the organized search tool required the substitution of biological function and flow terms of the thesaurus, while the three found within the AskNature database needed substitution of only

[2] www.asknature.org

the flow term. Additionally, other biological systems were identified that also solve the function/flow pair, which, if a redesign was undertaken, could result in compound analogical design.

5.4.2 Smart Flooring Case Study

Validation through exploration of non-existing biomimetic products is performed here on a smart flooring application. Consider the following scenario. A customer wants to create a security/surveillance product that looks like ordinary carpet, mats, rugs, etc., to detect intruders, a presence of something in a room or movement. Requirements for the smart flooring include a detection mechanism unseen by human eye, durability, composition of common materials, and a quick detection response. Also, the system needs to be autonomous.

The new design should offer advantages over current systems. Current tagged-based systems require the user to carry a badge or other device to be tracked or monitored, and simply removing the trackable item can defeat the system. Radar or similar systems require calibration and an area map to be created. Each time the area layout is changed, the map needs to be updated. Video surveillance and heat signature systems can be very expensive and often require a person to watch the real-time video feed.

These needs and constraints are mapped to flows as shown in Table 5.4 to complete the first step of the methodology. Next, the black box model is created to guide the decomposition of the flows into a functional model. Figure 5.6 provides the black box model. Figure 5.7 provides the conceptual functional model. The model of Fig. 5.7 was created with a boundary of the flooring in place, electrical energy is supplied to the detection mechanism, and when an object or human interacts with the flooring, a signal is generated.

Table 5.4 Needs of smart flooring device mapped to flows

Needs/Constraints	Functional basis flow
Object/human to detect	Solid material
Quick detection response	Status signal
System power	Electrical energy

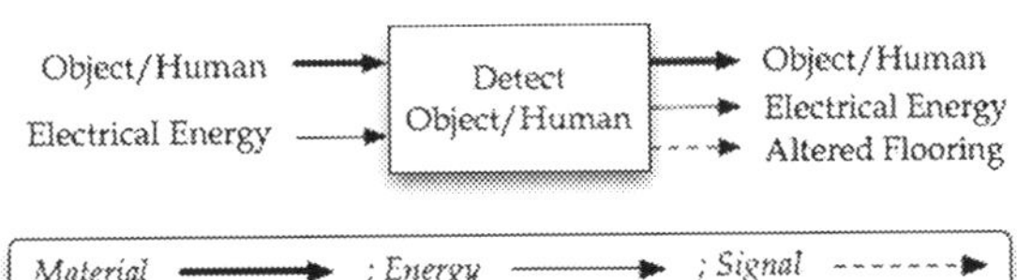

Fig. 5.6 Smart flooring black box model

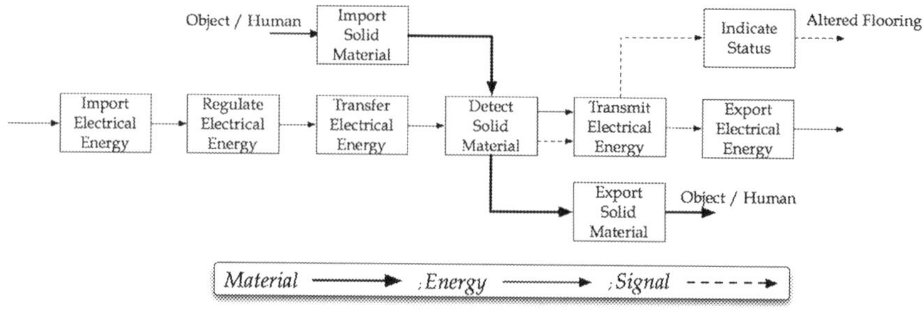

Fig. 5.7 Smart flooring conceptual functional model

From here the search heuristics were applied, following flowchart B of Fig. 5.3. Using the general inspiration heuristic, several interesting biological systems were found to perform the function of *detect*. An organized search of a biological corpus resulted in the following query results (Purves et al. 2001):

1. The hair cell
2. Electroreceptors found in electric fish
3. Epithelial cells
4. Genes that mark recombinant DNA
5. DNA
6. Why birds flock in large groups
7. Echolocation
8. Carotid and aortic stretch receptors
9. Membrane receptor proteins
10. Graded action potentials.

Next, following Step 4 of the methodology, the biological systems returned for the function of *detect* are analyzed to formalize connections.

Hair cells are analogous to cantilevers and would detect a presence when disturbed. In a similar manner, the carotid and aortic stretch receptors are analogous to flexible materials such as polymers. A polymer would detect a disturbance when pressure is applied.

Echolocation is analogous to radar. Radar is already used to detect objects; however, it is not a distributed system, as would be needed for a smart flooring concept. The final connection made from the above list is with the electroreceptor of fish. Electroreceptors generate an electric field for navigation of the environment, to locate objects, which is also analogous to radar. Echolocation uses sound waves where electrolocation uses electric waves.

Now that the query results have been analyzed and reflected upon to get to establish connections between biology and engineering, and the next task, following flowchart B of Fig. 5.3, is to query the engineering knowledge base and supplement the biologically inspired solutions with engineering solutions to complete the design. The returned engineering solutions, shown in Table 5.5, are

Table 5.5 Engineering solutions mapped to function/flow pairs

Function/Flow	Engineering solution
Import/electrical energy	Battery, circuit board, electric motor, electric wire, electric switch
Regulate/electrical energy	Actuation lever, capacitor, circuit board, automobile distributor, electric switch, heating element, transistor, transformer, thermostat, regulator, volume knob
Transfer/electrical energy	Battery, circuit board, electric wire, electric motor, electric socket, electric plate, electric switch, heating element, USB cable, light fixture, speaker
Transmit/electrical energy	Electrical wire, battery contacts, motor controller
Detect/solid material	Read head, line guide
Indicate/status signal	Light, tube, displacement gauge, LCD screen
Export/electrical energy	Circuit board, electric wire, electric switch

then synthesized. Functional representation enables a thorough understanding of the requirements while decreasing the tendency of designers to fixate on some particular physical solution for a problem. Therefore, this approach is form independent and relies on the designer's ability to develop a reasonable form for the concept. With the components of Table 5.5 and the established analogies, the final step of concept generation can begin.

The biological system of the hair cell prompts two concept variants. Both concept variants use wires and circuit boards to perform the functions involving electrical energy and would likely be powered by an external, low-voltage power supply. Regardless of the concept, a flexible circuitry layer and buffer layer would need to be underneath the visible flooring layer to connect the array to a processing unit and to protect the underlying circuitry, respectively. These are represented in the concept sketch close-up views as the orange layer between two black layers.

Wanting to remotely monitor a space means that the solution to indicate status signal must also be remotely located to the smart flooring. For both concept variants, the LCD screen is a viable option as the detection mechanism will produce a measurable result, which can be displayed on the LCD screen. The concept variants below focus on the biologically inspired detection mechanism; therefore, the power supply and LCD screen are not shown.

Recall that the critical need is unseen by the human eye. Considering the tactile response of the hair cell as a cantilever and flooring shaped as individual tiles, each tile could act as one cantilever to detect a load. The array of cantilever load sensors would then sense a pressure differential as a person walks across the smart flooring. A spring is added to the design to keep the tile from collapsing while allowing the cantilever to bend. A concept sketch is shown in Fig. 5.8. Detail design would need to be completed to determine the cantilever material that would best respond and last in a high-traffic environment.

The second concept variant simultaneously exploits the tactile response of the hair cell and the carotid and aortic stretch receptors and the change in electric

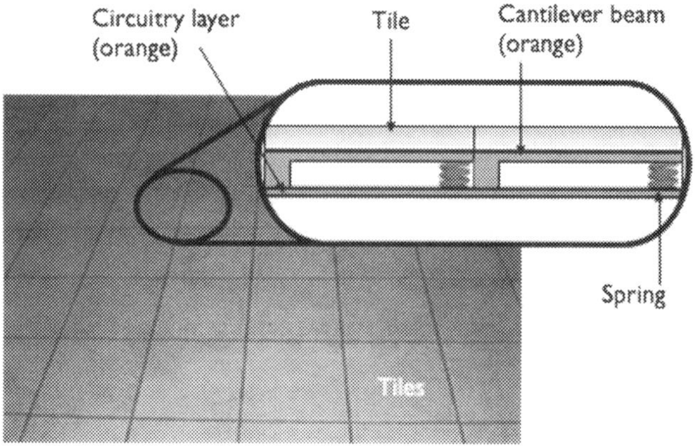

Fig. 5.8 Concept variant one for smart flooring

potential from the sensory cells. Hair cells are vertical structures, while stretch receptors are found in multiple orientations. Taking inspiration from the hair cell and stretch receptor morphology leads to a detector design that is comprised of a vertical structure that can be stretched in multiple orientations.

Offering flexibility and ruggedness for repeated deflection, this detector design could work for woven flooring such as carpet. Taking inspiration from the sensory cells, flexion of the carpet fibers would result in a change in resistivity, similar to a strain gage, or generate a voltage by the principle of piezoelectricity. Polyamide is a high-performance synthetic polymer and is commonly used in textiles. Fabricating polyamide tubes with a conductive gel or paste that can be woven into carpet to form an array would achieve the biologically inspired design. Materials research would need to be completed to determine whether the polyamide and conductive gel or paste would last in a high-traffic environment.

Alternatively, conductive thread, another detector solution that can be stretched in multiple orientations, could easily be woven into carpet and offer a change in resistivity. Conductive thread exists and is used in garments and accessories that merge technology into clothing. Materials research would need to be completed to

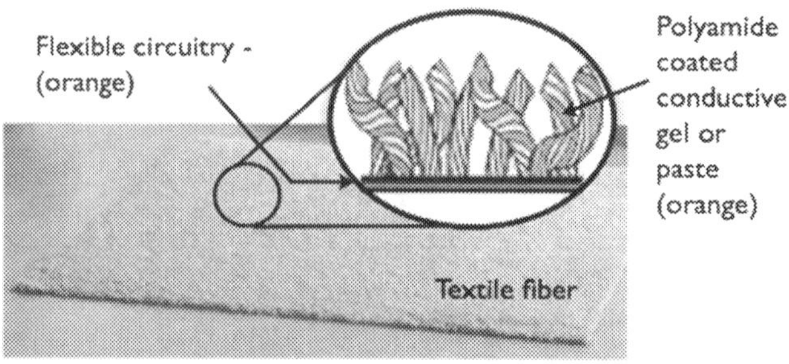

Fig. 5.9 Concept variant two for smart flooring

determine whether the conductive thread would provide a significant change in resistively or conductivity once woven into a carpet. A concept sketch is shown in Fig. 5.9. Shaded strands in the close-up view of the carpet fibers in Fig. 5.9 represent durable feedback elements woven into the carpet, providing hidden sensing capabilities.

Looking into the literature for a similar surveillance device uncovered a handful of attempts to create a "smart" floor or flooring. The first concept described above, individual tiles that detect a load placed in an array, has been done. Richardson et al. have developed hexagonal, puzzle-like pieces called nodes that are placed in an array to detect pressure (Richardson et al. 2004). Their tiles contain force-sensitive resistors and interlock to form a self-organizing network that passes data to the tile with an external data connection. An earlier approach to the smart floor involved only one measuring tile made of load cells, a steel plate, and data acquisition hardware and was not intended to be hidden (Orr and Abowd 2000). Rather, it was created as an alternative to biometric identification by recognizing a person's unique footstep profile.

Two other approaches that utilize load cells and layered flooring to conceal the sensors are nearly identical to the first concept variant. Liau et al. place a sensor in the center of every 60×60 cm wood-covered tile (Liau et al. 2008), where Addlesee et al. place a sensor at each intersection of four carpet-covered tiles (Addlesee et al. 1997). Load cells are similar in principle to cantilever beams in that deflection is transduced into an electrical signal that can be interpreted. Here, a literature review revealed that the first biologically inspired concept is feasible and has been attempted.

Investigating the second concept variant for smart flooring revealed only one existing design that is similar. Researchers at Infineon Technologies have woven conductive fibers into carpet and attached them to tiny sensor modules inlaid into the fabric to build a mesh network (IEE-Institution of Electrical Engineers 2003). The flooring can report where a person is located, which way they are moving, and if a sensor module has failed. Each conductor in the design is a copper wire coated with silver to prevent corrosion and then covered with polyester (IEE-Institution of Electrical Engineers 2003). A German textile company, Vorwerk, has teamed up with Infineon to develop the smart carpet (Vorwerk and TGCK 2004; Crane 2005). The Vorwerk/Infineon product is similar in structure to the second concept variant created here in that a conductor is concealed and woven into a textile product. Again, a literature review revealed that the biologically inspired concept is feasible.

This case study demonstrated that it is possible to systematically design using the search heuristics and take inspiration from biology in the process. By analyzing the biological system and making connections, a designer can become inspired. It is through these correlations that the designer can recognize existing designs that are similar or develop an innovative design.

Table 5.6 Chemical sensing needs mapped to conceptual flows

Customer need/Constraint	Functional basis flow
Selectivity	Status signal
Response time	Electrical energy
Reusable	Sensing layer (material)
Indirect sensing mechanism	Sensing layer (material)/Chemical stimulus (energy)

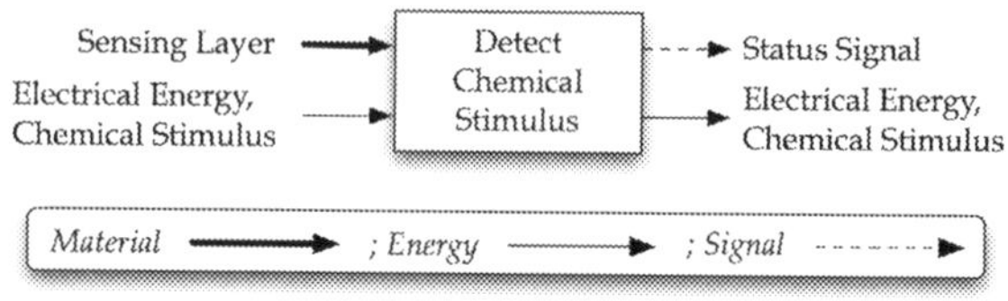

Fig. 5.10 Chemical sensor black box model

5.4.3 Chemical Sensor Case Study

The systematic biologically inspired design method presented here is able to inspire more than mechanical and electromechanical devices. This case study presents the design of a chemical sensor following the approach closest to traditional systematic design (flowchart A of Fig. 5.3).

For this study, the needs and constraints of the chemical sensing device are derived from the Handbook of Modern Sensors (Fraden 2004): selectivity (only senses the desired chemical in the presence of other species), quick response time, reusable, and utilizes an indirect sensing mechanism. These needs and constraints are mapped to flows as shown in Table 5.6 to complete the first step of the methodology. Next, the black box model is created to guide the decomposition of the flows into a functional model. Figure 5.10 provides the black box model. Figure 5.11 provides the conceptual functional model to complete Step 2 of the methodology.

The chemical sensing device black box and conceptual functional models show the generalized form of a chemical stimulus (i.e., chemical energy). This allows

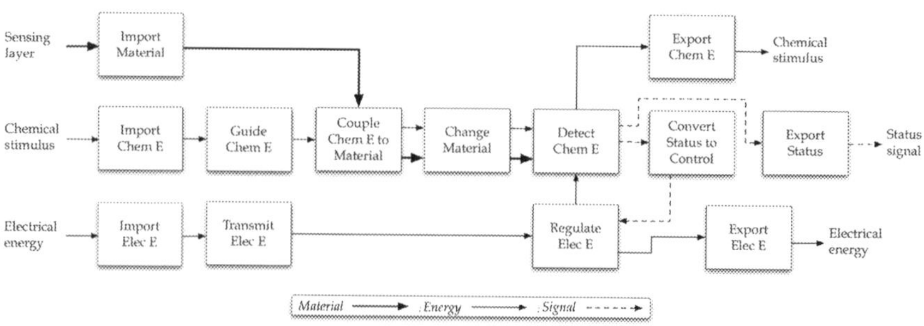

Fig. 5.11 Chemical sensor conceptual functional model

Table 5.7 Design repository solutions mapped to function/flow pairs

Function/Flow	Solutions from design repository
Import/material	Housing, reservoir, spring
Import/chemical energy	Container, nozzle
Guide/chemical energy	Tube
Couple/chemical energy to material	Basket, container, iron, nozzle, carburetor, burner, housing
Import/electrical energy	Battery, wire, circuit board, motor, cord, switch
Transmit/electrical energy	Wire, battery contacts, circuit board, compound eye
Change/material	Blade, impeller, heating element, punch, filter, staple plate, popcorn popper
Detect/chemical energy	Fly chemoreceptor, protein, Animalia chemoreception, Plantae chemoreception
Regulate/electrical energy	Circuit board, actuator, heating element, switch, resistor, diode
Convert/status to control signal	Circuit board
Export/chemical energy	Nozzle, bowl, tube, exhaust, bucket
Export/electrical energy	Wire, circuit board, cord, switch
Export/status signal	LCD screen, circuit board, wire, cord, level, speaker

the designer to query all possible forms of a chemical stimulus. The device substrate is also generalized as material to include all possible forms of material in the knowledge base. Figure 5.11 demonstrates the indirect sensing mechanism with *couple* and *change*, and the sensing element or transducer with *detect*. Electrical energy is utilized to power the sensor and transfer the detection status signal to the device capable of interpreting such signals, such as a computer. The boundary of the conceptual functional model includes the sensing layer and powered sensing element.

Following Step 3, the Design Repository was queried for engineering and biological entries. The components for each function/flow pair returned are shown in Table 5.7. For 10 of the 13 sensor functions, the component list was short and easy to choose from. The functions of *change*, *detect*, and *export signal* returned many possible components.

Considering the conceptual device as a whole and how one would use the device is an advantageous thought process for determining suitable components from a list. With regard to changing the material as the first step in the indirect sensing mechanism, the impeller, blade, and punch require mechanical movement to change a material, whereas the staple plate, filter, and heating element could change the chemical stimulus without mechanical movement. The conceptual design is not fully determined at this point and could be influenced by the component(s) chosen from biological inspiration. Therefore, the functions of *change* and *detect* will be considered together.

Following Step 4 of the methodology, connections need to be established to assist concept generation. Bacteria employ the two-component regulatory system for detection of extracellular signals, and the signaling pathways consist of modular units called transmitters and receivers, both of which are proteins (Stock et al. 2000). Fly antennae contain chemically sensitive cells (chemoreceptors) hidden deep within pores, which allow the insect to experience olfaction (i.e., sense of smell) (Mitchell 2003). The Animalia and Plantae mechanisms of chemoreception are not descriptive and are meant to guide a designer in a direction of research. Analysis of the biological components leads to the choice of fly chemoreceptor for the detect chemical energy function block.

Further exploration of the fly antennae reveals that the insect cuticle (a chitin-protein outer cover) has elaborations in the form of trichoids (hairs), pegs, pegs in pits and flat surfaces, all of which provide multiple pores for chemicals to travel through (Mitchell 2003). Within the pore is a fluid-protein pathway to the dendritic (sensory) cell membrane. Once the chemical molecule reaches the fluid surrounding the dendrite membrane, it bonds to an odorant-binding protein and is carried to one of the receptor sites of the membrane (Mitchell 2003). When the two make contact in the cation-concentrated fluid, a signal occurs as a voltage potential change across the membrane, which is the signal to be transduced. The sensing principles of fly antennae are complex and offer the designer inspirations for the function of detect and, as expected, for the function of change (Nagel et al. 2010).

A filter is analogous to the porous cuticle, which would narrow down the selection of chemicals or allow only one stimulus to interact with the sensing layer. The heating element is analogous to the odorant-binding proteins and cell membrane surface with receptor sites, in that an electrified element is capable of attracting polarized molecules (disregarding the heating aspect). An impeller could be used to steer the desired chemical stimulus to the sensing element, which is analogous to the odorant-binding proteins that shuttle stimuli to receptors.

Biological inspiration considering morphology leads to a sensing element that has specifically shaped cavities or is uniformly porous, and is a good conductor. Any material that can be patterned by photolithography can achieve the desired surface. Morphology of the fly antennae itself offers inspiration for a "stick-like" sensing element. Another connection exists between the engineering component filter and permeable or ion-selective membranes, which are used in current sensor technology. Therefore, permeable or ion-selective membranes are analogous to the porous cuticle. Further analogies exist between the heating element and electrical energy traveling through a conductor, which could be a copper wire, semiconductor, conducting polymer, etc. Semiconductor electrodes that allow absorption and desorption of chemical species are analogous to binding and removal of odorants from the cell membrane receptor sites within the fly antennae, which could also be used in the final concept.

Performing Step 5, concept generation leads to two concept variants. The first device supports a housing containing an electrified element (not for the production of heat) acting as a barrier to the transducer that chemical energy is guided to from the container, or space. The subsequent chemical energy is attracted to the electrified element, and once bonding occurs, the electrical properties of the electrified element change and generate a signal to be transduced as shown in Fig. 5.12. The electrical property change in the electrified element fulfills the requirement of an

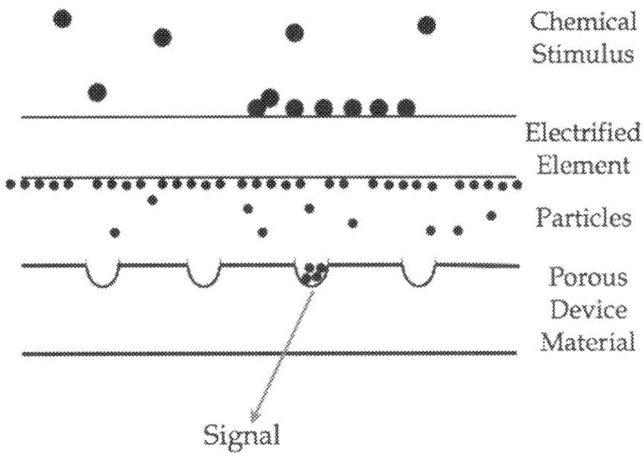

Fig. 5.12 Concept variant one for chemical sensor

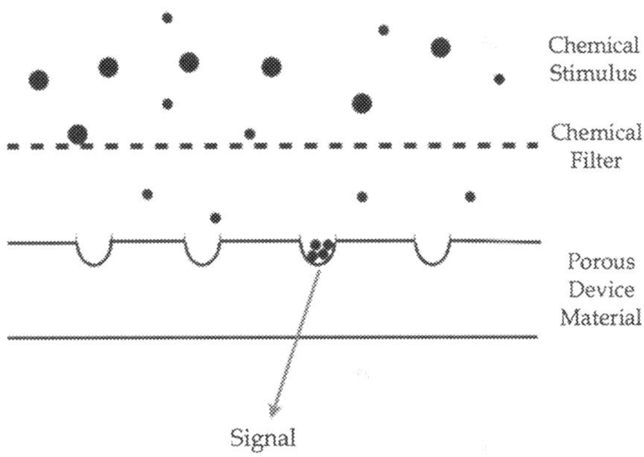

Fig. 5.13 Concept variant two for chemical sensor

indirect sensing mechanism, which also supports reusability as the absorbed particles could be removed by heating the element. An electronic circuit powers the transducer, decodes the sensor signal, and produces an electrical signal analogous to the input. The second device supports a housing containing a filter covering the sensing layer, which rests on the sensing element. Only the chemical species that pass through the filter interact with the sensing element. This interaction generates a signal to be transduced as shown in Fig. 5.13, but does not fulfill the requirement of an indirect sensing mechanism. The chemical stimuli, although filtered, still are allowed to interact with the detection layer. An electronic circuit powers the transducer, decodes the sensor signal, and produces an electrical signal analogous to the input. Further material research is needed to accurately define the sensing layers for both concept variants.

When researching whether a sensor with an electrified element on a transducer exists, the first concept described above, conducting polymer sensors that are typically used for detection of gases, was found. The conducting polymer is deposited atop interdigital electrodes, which make it the dielectric material for the electrodes. When the gas interacts with the polymer, the capacitance between the electrodes changes and the output signal decreases (György 2008; Bai and Shi 2007). These sensors are sometimes referred to as chemiresistors. Functionally, a chemiresistor is identical to the first concept.

The second concept is similar to an ion-selective electrode (ISE) or ion-selective field effect transistor (ISFET) used to measure pH. An ISE has high specificity to single-charged ions and is made of a doped glass. The glass allows only single-charged ions, such as hydrogen, to pass through to an internal solution of neutral pH monitored by an electrode. Concentration of hydrogen ions is correlated with a pH value by taking the logarithm of the concentration (Grundler 2007; Eggins 2002). An ISFET is the microelectronic version of an ISE and is similar in structure to an MOSFET (Grundler 2007; Eggins 2002; Liao et al. 1999). Instead of glass, the membrane is made as a thin film over an insulation layer of metal oxide or nitrate on a p-type doped substrate. Two n-type doped regions are added to the substrate for connection between the source and drain, while the gate is connected to the sensing layer beneath the membrane. Here, the sensing layer is a liquid or insulation material and the sensing element is an electrode. Functionally, an ISE is identical to the second concept.

This case study demonstrated that it is possible to systematically design a sensor and take inspiration from biology in the process. By analyzing the biological system and making connections, a designer can become inspired. It is through these correlations that the designer can recognize existing designs that are similar or develop an innovative design.

5.4.4 Heat Exchanger Case Study

This case study presents the design of a heat exchanging component for use in future hydrogen vehicles following the approach of a known process before querying is performed. Needs for the heat exchanger design are set by an external research group that wanted to explore biologically inspired concepts for distributing hydrogen on the microscale. The external group had exhausted all their avenues and expressed a need for more ideas, which satisfies Step 1 of the methodology. Through discussion, a common understanding of distributing hydrogen on the microscale and the role heat exchange would play in that process was reached. From here, the black box model, as shown in Fig. 5.14, was created. A known biological process that distributes material is circulation of blood by the vascular system of the human body. This is a known biological process that is used

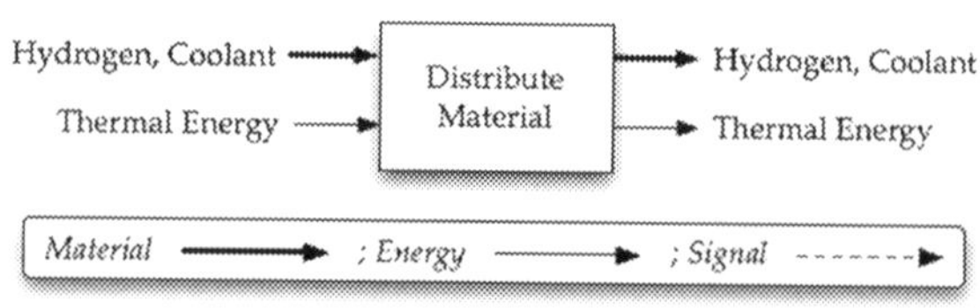

Fig. 5.14 Black box model of a heat exchanger

to begin the biologically inspired design process. When querying the biological knowledge base for circulation, many results lead to descriptions of the heart, but also mention gas exchange organs. This preliminary search did not result in a wide variety of inspiring biological systems. Therefore, to widen the inspiration space, the biological terms that correspond to the engineering function of *distribute* were retrieved from the engineering-to-biology thesaurus.

In the engineering-to-biology thesaurus, the following seven biological terms correspond to the engineering term of distribute: circulate, diffusion, exchange, disperse, scatter, spread, and spray. Circulate resulted in a narrow set of results. The list does reveal that the term *exchange* is used in biological literature. With a new keyword chosen, Step 3 of the methodology can be performed. The biological knowledge base was queried to identify biological systems that perform the function of *exchange*. The resulting query results from Purves et al. (2001) are as follows:

1. skin
2. gas exchange surfaces of animals
3. fish gills
4. lungs of a bird
5. respiratory system of insects
6. countercurrent heat exchanger in fish
7. stomata
8. abscisic acid.

It is important to note that the listed results are not the only results provided by the query. These eight were the most interesting and provided the most information for biological inspiration. Plants, fish, birds, mammals, and insects all offer inspiration for this case study. Notice that not all the biological solutions deal with gas exchange. Exploring other implementations of exchange increases the design space and creative solution potential.

As this design is for another research group, a meeting was held to discuss the biological systems that were identified to choose a biological system for further investigation as well as work through the remaining steps of the methodology. A connection established during the meeting is the similarity of engineering mass transfer concepts and the diffusion of respiratory gases into an insect. Although a direct relationship between the domains was recognized, the external research group did not see value in pursuing the relationship further. While reviewing a detailed description of the avian respiratory system, the discussion led to a connection that revealed the solution of a competing research group in the area of heat exchangers for hydrogen vehicles. During the first inhale/exhale cycle of a bird, the air travels to the trachea and into posterior air sacs and then is moved to the parabronchi area where the exchange takes place (Campbell and Reece 2003). During the second inhale/exhale cycle, the air moves forward to the cranial air sacs and then exits the bird (Vattam et al. 2008). Realizing that the air within the bird's lungs essentially travels in a circle sparked discussion of a spiral-shaped surface

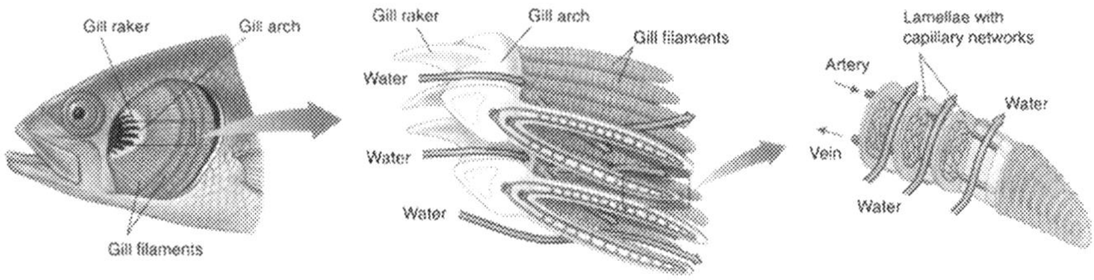

Fig. 5.15 Fish gills (Raven and Johnson 2002)

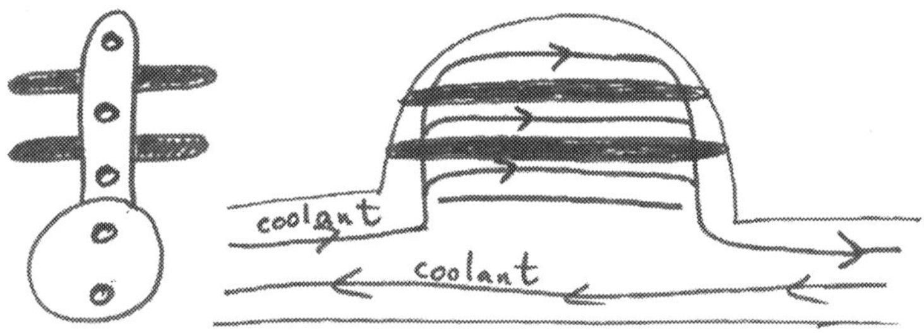

Fig. 5.16 Concept sketch for heat exchanger

that would effectively distribute hydrogen. That is, in fact, what a competing research group did (Venere 2010).

While reviewing a detailed description of gills along with Fig. 5.15, however, a researcher immediately made a connection between the gills of fish and hydrogen distribution. Considering the fish blood as the coolant and the water surrounding the fish as hydrogen, a distribution scheme inspired by the morphology and physiology of fish gills was discussed. The concept sketch is shown in Fig. 5.16. In this case, the biological functional model was not required, nor the engineering knowledge base query. The concept generation step began during the meeting, and the final concept was finalized by the external research group.

This case study demonstrated that it is possible to systematically design a portion of a larger problem with only a small amount of information as well as take inspiration from biology in the process. By analyzing the biological system and making connections, a designer can become inspired. It is through these correlations that the designer can recognize existing designs that are similar or develop an innovative design.

5.5 Summary

The design methodology presented here represents a function-based design approach devoted to developing biologically inspired solutions. It challenges and guides a designer to make connections between the engineering and biology domains to facilitate innovative design. Although systematic, the methodology provides a designer multiple avenues toward achieving a biologically inspired design. It is also envisioned as an inventive and iterative process, in terms of developing connections between systems at multiple levels of fidelity. As one level, or scale, of the biological system becomes understood, it leads to a deeper understanding and a greater curiosity to explore further, thus leading to multiple innovative designs. The four parts of the function-based framework, which can be used as stand-alone design tools or methods, coalesce in the systematic biologically inspired design method as demonstrated by the smart flooring, chemical sensor, and heat exchanger cases studies. Validation of the design methodology was proven through matching each of the generated concepts in the case studies with existing solutions or technologies. It was shown how the connections made between biological and engineered systems were key to arriving at the biologically inspired designs.

References

Addlesee MD, Jones A, Livesey F, Samaria F (1997) The ORL active floor. IEEE Pers Commun 4(5):35–41

Bai H, Shi G (2007) Gas sensors based on conducting polymers. Sensors 7:267–307

Balazs M, Brown D (2002) Design simplification by analogical reasoning. In: Cugini U, Wozny M (eds) From knowledge intensive cad to knowledge intensive engineering, vol 79. Springer, US, pp 29–44

Bhatta S, Goel A (1997) An analogical theory of creativity in design. In: Leake D, Plaza E (eds) Case-based reasoning research and development, vol 1266. Springer, Berlin Heidelberg, pp 565–574

Biomimicry Institute (2010) Biomimicry: a tool for innovation. http://www.biomimicry institute.org/about-us/biomimicry-a-tool-for-innovation.html

Campbell NA, Reece JB (2003) Biology. Pearson Benjamin Cummings, San Francisco

Casakin H (2006) Visual analogy as a cognitive strategy in the design process: expert versus novice performance. J Des Res 4(2), doi:10.1504/JDR.2004.009846

Casakin H (2006b) Assessing the use of metaphors in the design process. Environ Planning B: Planning Des 33(2):253–268

Casakin H (2007) Metaphors in design problem solving: implications for creativity. Int J Des 1(2):21–33

Crane D (2005) New high-tech sensor-Laiden smart carpet may revolutionize building security. http://www.defensereview.com/new-high-tech-sensor-laiden-smart-carpet-may-revolutionize-building-security/

Dollens D (2009) BioDigital architecture uses metaphor to design living systems. http://sensingarchitecture.com/3832/biodigital-architecture-uses-metaphor-to-design-living-systems-dennis-dollens-video/

Dym CL, Little P (2004) Engineering design: a project-based introduction. Wiley, New York

Eggins BR (2002) Chemical sensors and biosensors. Analytical techniques in the sciences. Wiley, Chichester

Fan Z, Chen J, Zou J, Bullen D, Liu C, Delcomyn F (2002) Design and fabrication of artificial lateral line flow sensors. J Micromech Microeng 12:655–661

Forty A (1989) Of Cars, Clothes and Carpets: design metaphors in architectural thought. J Des Hist 2(1):1–14. doi:10.1093/jdh/2.1.1

Fraden J (2004) Handbook of modern sensors : physics, designs, and applications. Springer, New York

Gentner D (1983) Structure-mapping: a theoretical framework for analogy. Cogn Sci 7:155–170

Gentner D (1988) Analogical inference and access, vol Analogica. Lecture notes in artificial intelligence. Morgan Kaufmann Publishers, Los Altos

Gick M, Holyoak K (1980) Analogical problem-solving. Cogn Psychol 12:306–355

Gnatzy W, Grunert U, Bender M (1987) Campaniform sensilla of Calliphora *vicina* (Insecta, Diptera) I. Topography. Zoomorphology 160:312–319

Goel A (1997) Design, analogy and creativity. IEEE Expert Intell Syst Appl 12(3):62–70

Grundler P (2007) Chemical sensors an introduction for scientists and engineers. Springer, Berlin

Grunert U, Gnatzy W (1987) Campaniform sensilla of Calliphora *vicina* (Insecta, Diptera) II. Typology. Zoomorphology 106:320–328

György I (2008) Conducting polymers: a new era in electrochemistry. Springer, Berlin

Helms M, Vattam SS, Goel AK (2009) Biologically inspired design: products and processes. Des Stud 30(5):606–622

Hey J, Linsey J, Agogino AM, Wood KL (2008) Analogies and metaphors in creative design. Int J Eng Educ 24(2):283–294

Hofstadter DR (1995) Fluid concepts & creative analogies: computer models of the fundamental mechanisms of thought. Basic Books, New York

Hubka V, Eder EW (1984) Theory of technical systems. Springer, Berlin

Hyman B (1998) Engineering design. Prentice-Hall, New Jersey

IEE-Institution of Electrical Engineers (2003) Research news—Walk this way for the smart floor. Electron Syst Softw 1:5–7

Johnson-Laird P (1989) Analogy and the exercise of creativity. In: Vosniadou S, Ortony A (eds) Similarity and analogical reasoning. Cambridge University Press, Cambridge, pp 313–331

Liao HK, Yang ES, Chou JC, Chung WY, Sun TP, Hsiung SK (1999) Temperature and optical characteristics of tin oxide membrane gate ISFET. IEEE Trans Electron Devices 46(12):2278–2281

Liau WH, Wu CL, Fu LC (2008) Inhabitants tracking system in a cluttered home environment via floor load sensors. IEEE Trans Autom Sci Eng 5(1):10–20

Lindemann U, Gramann J (2004) Engineering design using biological principles. In: International design conference—DESIGN 2004, Dubrovnik, 2004

Linsey J, Wood K, Markman A (2008) Modality and Representation in Analogy. AIEDAM 22(2):85–100

Mak TW, Shu LH (2004) Abstraction of biological analogies for design. CIRP Ann 531(1):117–120

McKean E (2005) The new Oxford American dictionary. Oxford University Press, New York

Mitchell BK (2003) Chemoreception. In: Vincent HR,Ring TC (eds) Encyclopedia of insects. Academic Press, Amsterdam, pp 169–174

Motamed M, Yan J (2005) A review of biological, biomimetic and miniature force sensing for microflight. In: Paper presented at the IEEE/RSJ international conference on intelligent robots and systems (IROS)

Nagai Y, Taura T (2006) Formal description of concept-synthesizing process for creative design. In: Gero JS (ed) Design Computing and Cognition'06. Springer, Dordrecht, pp 443–460

Nagel JKS, Stone RB (2012) A computational approach to biologically-inspired design. Artif Intell Eng Des, Anal Manuf, special issue DCC 2010 26(2):161–176

Nagel JK (2010) Systematic design of biologically-inspired engineering solutions. Doctoral Dissertation, Oregon State University, Corvallis

Nagel JKS, Stone RB, McAdams DA (2010) An engineering-to-biology thesaurus for engineering design. In: ASME IDETC/CIE 2010 DTM-28233, Montreal, Quebec, Canada

Nagel JKS, Stone RB, McAdams DA (2010) Exploring the use of category and scale to scope a biological functional model. In: ASME IDETC/CIE 2010, DTM-28873, Montreal, Quebec, Canada

Nagel JKS, Stone RB (2011) A systematic approach to biologically-inspired engineering design. Paper presented at the ASME IDETC/CIE 2011, DTM-47398, Washington, D.C., USA 2011

Nagel RL, Tinsley A, Midha PA, McAdams DA, Stone RB, Shu L (2008) Exploring the use of functional models in biomimetic conceptual design. J Mech Des 130(12):11–23

Nagel JKS, Stone RB, McAdams DA (2010) Function-Based Biology Inspired Concept Generation. In: Mukherjee A (ed) Biomimetics, Learning From Nature. In-Tech, Croatia

Nagel JKS, Nagel RL, Stone RB, McAdams DA (2010d) Function-based biologically-inspired concept generation. Artif Intell Eng Des Anal Manuf 24(4):521–535

Orr RJ, Abowd GD (2000) The smart floor: a mechanism for natural user identification and tracking. Paper presented at the Conference on Human Factors In Computing Systems (CHI), Hague, Netherlands

Otto KN, Wood KL (2001) Product design: techniques in reverse engineering and new product development. Prentice-Hall, Upper Saddle River

Pahl G, Beitz W, Feldhusen J, Grote KH (2007) Engineering design: a systematic approach, 3rd edn. Springer, Berlin

Purves WK, Sadava D, Orians GH, Heller HC (2001) Life, the science of biology, 6th edn. Sinauer Associates, Sunderland

Raven PH, Johnson GB (2002) Biology. McGraw-Hill, Boston

Richardson B, Leydon K, Fernström M, Paradiso JA (2004) Z-Tiles: building blocks for modular, pressure-sensing floorspaces. Paper presented at the Conference on Human Factors In Computing Systems (CHI), Vienna, Austria

Shu LH, Hansen HN, Gegeckaite A, Moon J, Chan C (2006) Case study in biomimetic design: handling and assembly of microparts. Paper presented at the ASME 2006 IDETC/CIE, Philadelphia, PA

Smith GF (1998) Idea generation techniques: a formulary of active ingredients. J Creative Behav 32(2):107–133

Stock AM, Robinson VL, Goudreau PN (2000) Two-component signal transduction. Annu Rev Biochem 69:183–215

Stone RB (1997) Towards a theory of modular design. University of Texas at Austin, Austin

Stone R, Wood K (2000) Development of a functional basis for design. J Mech Des 122(4):359–370

Tsujimoto K, Miura S, Tsumaya A, Nagai Y, Chakrabarti A, Taura T (2008) A method for creative behavioral design based on analogy and blending from natural things. In: 2008 ASME IDETC/CIE, New York, USA, 2008. DETC2008-49389

Ullman DG (2009) The mechanical design process, 4th edn. McGraw-Hill, Inc., New York

Ulrich KT, Eppinger SD (2004) Product design and development. McGraw-Hill/Irwin, Boston

Vattam S, Helms M, Goel A (2008) Compound analogical design: interaction between problem decomposition and analogical transfer in biologically inspired design. In: Third international conference on design computing and cognition, Atlanta, 2008. Springer, Berlin, 377–396

Venere E (2010) Engineers design, build major component for hydrogen cars. http://www.physorg.com/news186221598.html

Vincent JFV, Mann DL (2002) Systematic technology transfer from biology to engineering. Philos Trans Royal Soc Lond A 360:159–173

Voland G (2004) Engineering by design, 2nd edn. Pearson Prentice Hall, Upper Saddle River

Vorwerk & Co. TGCK (2004) Infineon thinking carpet. http://www.vorwerk-carpet.com/sc/vorwerk/bildmeldung_thinkCarpet_en.html

Wicaksono DHB, Pandraud G, Craciun G, Vincent JFV, French PJ (2004) Fabrication and initial characterisation results of a micromachined biomimetic strain sensor inspired from the campaniform sensillum of insects. In: IEEE Sensors 2004, 542–545

Chapter 6
Information-Processing Theories of Biologically Inspired Design

Ashok K. Goel, Swaroop Vattam, Bryan Wiltgen and Michael Helms

Abstract Starting from *in situ* studies, we develop an information-processing theory of biologically inspired design. We compare our theory with two popular theories of biologically inspired design: Biomimicry 3.8 Institute's Design Spiral and Vincent et al.'s BioTRIZ. While Design Spiral and BioTRIZ are normative and prescriptive, our information-processing theory provides a descriptive and explanatory account of the design paradigm. We examine if and how the process of biologically inspired design is different from that of other design paradigms beyond the differences between biological and technological systems. We posit that biologically inspired design appears to be a distinct design paradigm in part because it entails solution-based analogies in addition to the problem-driven analogies typical of other design paradigms.

Keywords Biologically inspired design · Biomimicry · Biomimetics · Bionics · Compound analogy · Creativity · Cross-domain analogy · Design · Innovation · Problem decomposition · Problem-driven analogy · Problem–solution coevolution · Solution-based analogy · Task analysis · Task model

A. K. Goel (✉) · S. Vattam · B. Wiltgen · M. Helms
School of Interactive Computing, Georgia Institute of Technology,
Atlanta, Georgia 30332, USA
e-mail: goel@cc.gatech.edu

S. Vattam
e-mail: svattam@cc.gatech.edu

B. Wiltgen
e-mail: bwiltgen@cc.gatech.edu

M. Helms
e-mail: mhelms3@cc.gatech.edu

A. K. Goel · M. Helms
Center for Biologically Inspired Design, Georgia Institute of Technology,
Atlanta, Georgia 30332, USA

A. K. Goel et al. (eds.), *Biologically Inspired Design*,
DOI: 10.1007/978-1-4471-5248-4_6,

6.1 Introduction

Biologically inspired design (Bar-Cohen 2011; Benyus 1997; Chakrabarti and Shu 2010; Shu et al. 2011; Vincent and Mann 2002; Yen and Weissburg 2007) perhaps is one of the most important movements in contemporary design of technological systems. The paradigm espouses use of analogies to biological systems in generating conceptual designs for new technological systems. This paradigm has inspired many designers in the history of design, such as Leonardo da Vinci and the Wright brothers. However, it is only over the last generation that the paradigm has become a movement, pulled in part by the growing need for environmentally sustainable design and pushed partly by the desire for creativity and innovation in design. The design of windmill turbine blades mimicking the design of tubercles on the pectoral flippers of humpback whales is one example of biologically inspired design (Biomimicry 3.8 Institute 2011a). The tubercles are large bumps on the leading edges of the flippers, which create even, fast-moving channels of water flowing over them (Fish and Battle 1995). The whales thus can move through the water at sharper angles and turn tighter corners than if their flippers were smooth. When applied to wind turbine blades, they improve lift and reduce drag, improving the energy efficiency of the turbine (Ashley 2004). The design of fog harvesting devices (Shanyengana et al. 2002) inspired by the arrangement of hydrophilic and hydrophobic surfaces on the back of the Namibian beetle (Hamilton and Seely 1976) is another example of biologically inspired design (Biomimicry 3.8 Institute 2011b).

Note that the designs of biologically inspired wind turbine blades and fog harvesting devices illustrate both sustainable design as well as creative design. However, although biologically inspired design is rapidly growing as a design movement, its practice at present is *ad hoc*, with little systemization of either biological knowledge from a design perspective or of the processes of biologically inspired design. Transformation of the promising paradigm of biologically inspired design into a principled methodology requires rigorous development of precise theories of biologically inspired design (e.g., Vincent and Mann 2002).

For the purposes of our discussion, it is useful to make three distinctions between different kinds and parts of design theories. Firstly, a design theory provides an account of both the *contents of knowledge* and the *processes of design* in some class of domains. Hubka and Eder (1988), for example, provide both a content account of technological systems and a process account of designing them. Similarly, in the context of biologically inspired design, Vincent et al. (2006) provide both a content account of biological systems in terms of energy, information, structure, substance, space, and time, and a process account of biologically inspired design called BioTRIZ. In previous work, we have described contents of knowledge of biological and technological systems in terms of the structure, behavior, and function of specific systems (Goel et al. 2012), the function and behavior of abstract design patterns (Goel et al. 2011a), and types of knowledge that are transferred from biological systems to technological systems in biologically inspired design (Vattam et al. 2010a). We also described partial process accounts of biologically inspired design such as compound analogies (Vattam et al. 2010a), problem-driven and solution-based design processes (Helms et al. 2009), and problem–solution coevolution (Helms and Goel 2012). Our goal in this chapter is to further develop the process account of biologically inspired design.

Secondly, process accounts of design can be *normative and prescriptive* or *descriptive and explanatory*. Pahl et al.'s (2007) theory of systematic design, for example, is normative and prescriptive; Simon's (1996) theory of design, on the other hand, is mostly descriptive and explanatory. Most information-processing theories of biologically inspired design are normative and prescriptive. This includes BioTRIZ (Vincent et al. 2006), Biomimicry 3.8 Institute's Design Spiral (Biomimicry 3.8 Institute 2009), Srinivasan and Chakrabarti's (2010) GEMS model, as well as Nagel et al.'s (2013) account. In contrast, our information-processing theory of biologically inspired design is descriptive and explanatory; it emerges out of a series of *in situ* studies of the practice of biologically inspired design (e.g., Helms and Goel 2012; Helms et al. 2009, 2010; Vattam et al. 2007, 2010a, b; Wiltgen and Goel 2011). Linsey et al. (2008) and Mak and Shu (2008) too have empirically investigated biologically inspired design from an information-processing perspective.

Thirdly, descriptive theories can focus on the microstructure or macrostructure of cognition. Newell and Simon (1972), for example, describe specific memory, inference, and attention processes for some kinds of human problem solving and decision making. Their work focuses on the microstructure of cognition. Similarly, Gentner (1983), Hofstadter (1996), Holyoak and Thagard (1996), and Kolodner (1993) provide alternative cognitive accounts of analogy. In contrast, we conduct a task analysis of the macrostructure of behavior in biologically inspired design. (See Crandall et al. (2006) and Shraagen et al. (2000) for the methodology of task analysis.) Our task analysis helps generate a task model of biologically inspired design. A task model describes the processes and the knowledge that result in the accomplishment of complex, extended, open-ended tasks such as design (e.g., Chandrasekaran 1990) as well as design methods such as design by plan instantiation (Brown and Chandrasekaran 1989) and case-based design (Goel and Chandrasekaran 1992).

To summarize, our main goal in this chapter is to develop a task model of cross-domain analogies from biology to engineering in the conceptual phase of biologically inspired engineering design. In general, an information-processing theory of biologically inspired design may describe the behaviors of an individual designer, the interactions among a team of designers, or the behaviors of a design team viewed as a unit. Although we are interested in all three levels of aggregation, in this work, we focus on interdisciplinary design teams of biologists and engineers viewed as the unit of analysis. We view this work as a step in a bigger, longer-term agenda of developing (1) a design methodology that promotes systematization of biologically inspired design, (2) pedagogical techniques that foster education and training in biologically inspired design, and (3) interactive technologies that facilitate the work of teams engaged in biologically inspired design. In this paper, first we briefly summarize the context and methodology of an empirical study of biologically inspired design. Then, we conduct a task analysis and develop a task model of creative analogies in biologically inspired design. Next, we compare our theory with the two best-known information-processing theories of biologically inspired design: Design Spiral and BioTRIZ. Finally, we examine if and how biologically inspired design is fundamentally different from other design paradigms.

6.2 Research Context and Methodology

In this section, we briefly describe the context of our observations and the methodology of our analysis of biologically inspired design practice. Since 2006, we have observed ME/ISyE/MSE/BME/BIOL 4740, a yearly, interdisciplinary, project-based course on biologically inspired design taught jointly by biology and engineering faculty at the Georgia Institute of Technology. The class is composed of mostly senior-level undergraduate students from biology, biomedical engineering, industrial design, industrial engineering, mechanical engineering, and a variety of other disciplines. Although it evolves a little every year, the course is consistently structured around lectures, found object exercises, journal entries, and one or more design projects. Some lectures focus on case studies of biologically inspired design. Some lectures pose problems for students to solve in small groups. Other lectures discuss design processes involved in biologically inspired design, such as reframing engineering problems in biological terms, functional decomposition of a design problem, use of analogy in design, and quantitative analysis of designs. Yen et al. (2010, 2011) provide a detailed account of the teaching and learning in the course.

We base this work on the 2006–2010 classes of the ME/ISyE/MSE/BME/BIOL 4740 course. The extended design projects in the classes were the focal points of our data collection. The projects involved identification of a design problem of interest to the team and conceptualization of a biologically inspired solution to the identified problem. Each design project grouped together an interdisciplinary team of typically 4–5 students. The design projects included both problem-driven and solution-based designs. Each team had at least one student with a biology background and a few from different engineering disciplines. Each team identified a problem that could be addressed by a biologically inspired solution, explored a number of solution alternatives, and developed a final solution design based on one or more biologically inspired designs.

In this chapter, we analyze two design projects by one team from one class of ME/ISyE/MSE/BME/BIOL 4740: the Shark Attack project and the Levee project. Note that we have changed the project names for the sake of anonymity. The same design team within the class conducted both projects. We report on this design team in particular because one of the authors (Wiltgen) participated in the team;

Wiltgen and Goel (2011) provide details of the participant observation. Wiltgen collected artifacts generated by the design team, including team presentations and reports. He also collected all his individual class assignments and made audio recordings of all team meetings. We then performed a task analysis informed by Wiltgen's observations of the design process. (Figures 6.1 and 6.2 are our reconstructions of parts of the design process).

We also collected design reports from other design teams in various classes of ME/ISyE/MSE/BME/BIOL 4740. Although we report in detail only on the two projects of Wiltgen's team, our task model is intended to capture a variety of design processes that we observed in our studies of biologically inspired design. For additional examples of design projects in the course, please see Helms and Goel (2012), Helms et al. (2009, 2010), Nelson et al. (2009), Wilson et al. (2010), and Vattam et al. (2007, 2010a, b).

6.2.1 The Shark Attack Project

The Shark Attack project lasted a little more than a month. In this project, each team in the class first selected a small set of biological organisms. Afterward, each team member was assigned to research a particular organism. Each team member was next assigned to find a human problem that he or she thought his or her researched organism could help solve and to investigate any existing solutions to

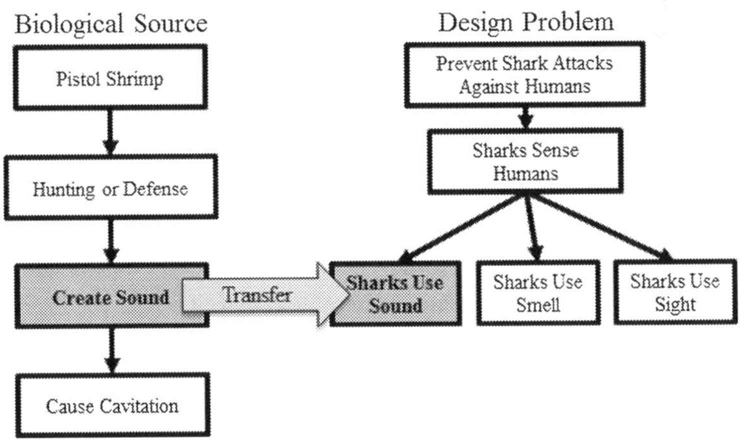

Fig. 6.1 The level of functional decomposition of the target design problem and the source biological system in the Shark Attack project at which analogical transfer from the biological source to the target problem occurred. The design team was inspired by the way the snapping shrimp (or pistol shrimp) creates sound. They used this inspiration to design a device to prevent shark attacks on humans by generating sounds using cavitation. (This figure was derived from the design artifacts of the Shark Attack project.)

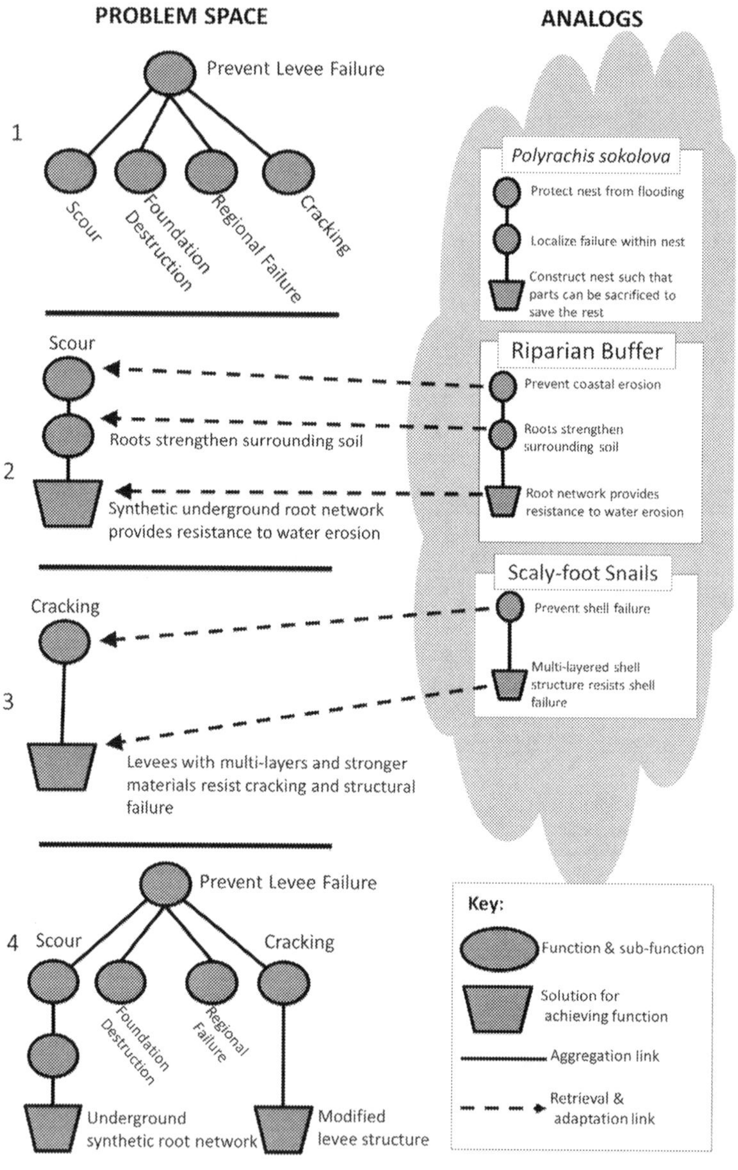

Fig. 6.2 The process of compound analogy in the Levee project during one iteration. Part 1 illustrates the decomposition of the target design problem into four sub-problems that the design team addressed. Parts 2 and 3 depict a high-level mapping between a sub-problem and a biological system. Part 4 shows the individual solutions to the sub-problems in relation to the original design problem. Note that only two analogies made by the design team are shown. (This figure was derived from the design artifacts of the Levee project.)

the problem. Each team member was then assigned to create a mapping between the researched organism and the design problem. Finally, the design team was assigned to pick one team member's work to pursue and to decide what organisms the team should keep or discard. The designing continued from there. For details of the assignments, please see Yen et al. (2011, 2013).

The Shark Attack project was directed at preventing shark attacks off of the coast of the United States without harming the sharks. The students designed an underwater sound-based shark repellant device inspired by the snapping shrimp (Ritzmann 1974), a small shrimp with the ability to create loud, underwater sound waves using one of its claws. The device worked by emitting sounds, generated by the same cavitation mechanism that the snapping shrimp uses to emit sound but at a frequency that sharks dislike. The design team envisioned that the device would repel sharks without harming them.

6.2.2 The Levee Project

Following the Shark Attack project, the Levee project lasted about two and a half months. Instead of starting with a biological system as in the Shark Attack project, the Levee project began by identifying a design problem. The Levee project occurred across two major iterations, with the first iteration being shorter in time than the second. For this project, each team member was first assigned to come up with a set of design problems. The team was then assigned to decide upon and investigate one of those problems as well as existing solutions to the problem. At the same time, each team member was assigned to find one natural system that solves all or part of the problem. The designing continued from there. (Again, see Yen et al. 2011, 2013 for details of the assignments).

The goal of the Levee project was to prevent another flood disaster like that caused by 2005's Hurricane Katrina by strengthening the levee system in New Orleans. The design team identified several modalities of failure for the levees in New Orleans, such as scouring, overtopping, and joint failure, and devised biologically inspired solutions for each except for overtopping. For example, the team drew inspiration from scaly-foot snails (Yao et al. 2010) based on which the levee design would be resistant to incoming water forces in part by being multilayered like the snails' shells.

6.3 Task Analysis

From our observations and analysis, we identified several tasks and processes of biologically inspired design. In this section, we analyze some of those tasks and processes.

6.3.1 Cross-Domain Analogies

By definition, biologically inspired design engages cross-domain analogies, for example, analogies from biology to engineering. Although we have observed that extended episodes of biologically inspired design involve both within domain and cross-domain analogies (Vattam et al. 2010a), it is the essentialness of cross-domain analogies that defines the paradigm of biologically inspired design. Both the Shark Attack and the Levee projects illustrate cross-domain analogies from biology to engineering.

6.3.2 Problem–Solution Coevolution

Conceptual design in biologically inspired design entails problem–solution coevolution (Helms and Goel 2012). That is, the design process iterates between defining and refining the problem and the solution, with both the problem and the solution influencing each other (Maher and Tang 2003; Dorst and Cross 2001). As a solution (S) is developed and evaluated for a given problem (P), it reveals additional issues, spawning a new conceptualization of the problem (P + 1). The process continues with the development of a new solution (S + 1) and will iterate until a final solution is decided upon.

For example, consider the Shark Attack project. The team originally pursued Wiltgen's individual idea, which was inspired by the snapping shrimp, to design a decoy-like device that would attract the sharks to a location away from human population using sound. Analysis of design artifacts suggests that the team discovered new sub-problems upon evaluation of the idea, such as durability of the decoy if sharks were going to attack it instead of humans. The overall problem then evolved to account for these newly identified issues, for example, that one must design a device that can handle getting attacked by a shark. The final design, however, did not incorporate a solution to this sub-problem. We hypothesize that this is because the team switched from a shark-attracting system to a shark-repelling system (a radical evolution!), as a result of which the system was no longer in danger of getting attacked by a shark, thus eliminating the sub-problem altogether; this illustrates problem–solution coevolution in biologically inspired design.

6.3.3 Problem Decomposition

Biologically inspired design engages decomposition of the target design problem as well as functional decomposition of the biological system that acts as a source analogue to the design problem (Vattam et al. 2007, 2010a). Decomposition of a large, complex problem into smaller, simpler problems is a familiar idea in design (e.g., Brown and Chandrasekaran 1989; Chandrasekaran 1990; Dym and Brown 2012; Hubka and Eder 1988; Pahl et al. 2007). Functional decomposition of a system into sub-systems too is a familiar idea in design (e.g., Dym and Brown 2012; Hubka and Eder 1988; Pahl et al. 2007; Simon 1996). Thus, it is not surprising that decomposition of target design problems and functional decomposition of source biological systems permeate biologically inspired design.

However, these decompositions appear to play a special role in biologically inspired design. The decomposition of the target design problem and the functional decomposition of the source biological system help identify the appropriate level for the analogical transfer from the biological system to the design problem. Figure 6.1 illustrates both functional decomposition of the design problem and functional decomposition of the biological system in the Shark Attack project. Here, both the design problem and the biological system are functionally decomposed until a level is reached where a function in the problem decomposition is similar enough to a function in the system decomposition for analogical transfer to occur. In this example, the ability of the snapping shrimp to defend itself and hunt using sound was an inspiration for the design of a device to prevent shark attacks using sound.

6.3.4 Problem Decomposition and Compound Analogies

Although problem decomposition is common in almost all kinds of design, it appears to play a second special role in biologically inspired design. In particular, we found that biologically inspired design often entails compound analogies in which a new design concept is generated by composing the results of multiple cross-domain analogies (Vattam et al. 2007, 2010a). This process of compound analogical design relies on an opportunistic interaction between the processes of memory and problem solving. In this interaction, the target design problem is decomposed functionally, solutions to different sub-functions in the functional decomposition are found through analogies to different biological systems retrieved from memory, and the overall solution is obtained by composing the solutions for achieving the different sub-functions. Thus, the sub-functions in the functional decomposition of the design problem act as probes into a functionally indexed memory of biological systems.

As an example, let us consider the Levee project. Figure 6.2 illustrates a part of the design process in the project. During the two design iterations in the Levee project, the design team identified six sub-problems that were related to levee failure: scour (the eroding of soil by water), regional failure (local levee failures causing whole regions to flood), cracking (water forces breaking and breaking through the levee), foundation destruction (water undermining the foundation of non-earthen levees), overtopping (water going over the levee), and joint failure (water forces breaking and breaking through the joints of concrete levees). Let us just consider the first design iteration here. Our analysis of the design artifacts produced by the design team in the first iteration suggests that the team identified the first five sub-problems and addressed the first four. The team did not identify a single biological source that would overcome all four of these sub-problems. Instead, the final design solution for the first iteration, which was more along the lines of a set of solutions rather than a tightly integrated device, was a compound solution; it took inspiration from (1) riparian buffers for the scour sub-problem, (2) riparian buffers and *Bacillus pasteurii* bacteria for the foundation destruction sub-problem, (3) scaly-foot snails for the cracking sub-problem, and (4) *Polyrachis sokolova* ant nests for the regional failure sub-problem. Figure 6.2 depicts the processes of analogy for the scour and cracking sub-problems; it also depicts a high-level breakdown of the *Polyrachis sokolova* biological analogue.

6.3.5 Problem-Driven and Solution-Based Analogies

We observed the existence of two high-level analogical processes for biologically inspired design based on two different starting points—*problem-driven analogy* and *solution-based analogy* (Helms et al. 2009; Vattam et al. 2007, 2010a; see also Wilson et al. 2010). As Fig. 6.3a illustrates, in a problem-driven analogical process, designers identify a problem that forms the starting point for subsequent problem solving. They usually formulate their problem in functional terms (e.g., stopping a bullet). In order to find biological sources for inspiration, designers "biologize" the given problem; that is, they abstract and reframe the function in more broadly applicable biological terms (e.g., what characteristics do organisms have that enable them to prevent, withstand, and heal damage due to impact?). Designers use a number of strategies for finding biological sources relevant to the design problem at hand based on the "biologized" question, and then they research the biological sources in greater detail. Important principles and mechanisms that are applicable to the target problem are then extracted to a solution-neutral abstraction and applied to arrive at a trial design solution.

The Levee project is an example of the problem-driven analogical process of biologically inspired design. The design team began with the problem of preventing levee failure and conducted a search for biological sources. The search returned such organisms as the scaly-foot snails and riparian buffers. Principles were extracted from these solutions (such as multilayered structures and dense underground networks, respectively), which were then applied to the design problem.

On the other hand, in the solution-based analogical process, as illustrated in Fig. 6.3b, designers begin with a biological source of interest. The designers understand (or research) their biological source to a sufficient depth to support the extraction of deep principles from it. Then, they find human problems to which the principle can be applied. Finally, they apply the principle to develop a design solution to the identified problem.

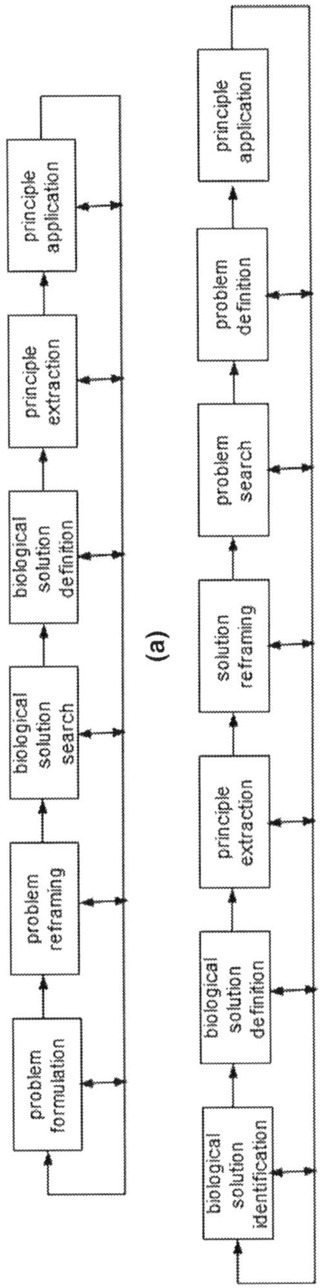

Fig. 6.3 The twin analogical processes of biologically inspired design: **a** problem-driven analogy and **b** solution-based analogy

The Shark Attack project is an example of the solution-based analogical process of biologically inspired design. Early in the design project, each designer in the ME/ISyE/MSE/BME/BIOL 4740 class was given an assignment to investigate a biological source, to determine a design problem for which the biological source might be useful, and to derive an analogical mapping between the two. Wiltgen selected the snapping shrimp. After doing some amount of research on the snapping shrimp, he picked the shark attack problem as a potentially good match and came up with the initial solution to the problem. The design team chose to pursue Wiltgen's design idea. During the course of the design project, the team developed an understanding of the snapping shrimp and the principle of cavitation as a means to create sound. Although other factors also likely influenced the design episode, evidence for the influence of the snapping shrimp analogue is apparent in the final design of the shark-repelling device that used cavitation as its sound generation mechanism.

We note that the problem-driven and solution-based analogical design processes have different characteristics. Compared to problem-driven analogical processes, solution-based analogical processes tend to exhibit not only design fixation but also a fixation on the structure of the biological design (Helms et al. 2009; Vattam et al. 2007). Again compared to problem-driven processes, solution-based design processes also tend to more often result in the generation of multi-functional designs, that is, where a single design principle meets multiple functional goals (Helms et al. 2009; Vattam et al. 2007).

It is important to note also that our findings about problem-driven and solution-based analogical processes of biologically inspired design are *not* limited to the design projects in the ME/ISyE/MSE/BME/BIOL 4740 class. Vattam et al. (2007) describe a narrative analysis of about seventy real-world case studies reported in the literature on biologically inspired design. We found that a majority of the successful case studies entailed solution-based analogy. Further, Vattam et al. (2007) found that some case studies of biologically inspired design described in the literature as based on problem-driven analogical design in fact appear to have been based on solution-based analogical design. Solution-based analogy also characterizes the two examples we briefly described in the introduction to this chapter: the design of wind turbine blades inspired by the design of pectoral flippers of humpback whales and the design of fog harvesting systems inspired by the design of hydrophilic surfaces on the back of the Namibian beetle. These findings have had a substantial impact on the teaching and learning in the ME/ISyE/MSE/BME/BIOL 4740 course. For example, until 2008, the course used to have design projects that emphasized only problem-based analogical design as is commonly prescribed by normative theories of biologically inspired design. As a result of our empirical findings and the descriptive account we developed to explain the findings, the instructors of the class introduced design projects oriented toward solution-based analogical design. Yen et al. (2013) describe the evolution of the class in detail.

It may be important to note also that our notions of problem-driven and solution-based analogical processes in biologically inspired design have little in common with similarly named processes described by Kruger and Cross (2006). Kruger and Cross are referring to the observation that once a designer has proposed a conceptual solution to a design problem, subsequent information processing often is driven by both the design problem and the proposed conceptual solution. In contrast, our problem-driven and solution-based analogical processes pertain to two different kinds of analogies entailed by biologically inspired design: one starts with the design problem and makes an analogy to a biological solution (problem-driven analogy), and the other starts with a biological solution and makes an analogy to a design problem (solution-based analogy).

6.4 A Task Model

In this section, we first summarize a generic task model of analogical design. We will then elaborate this generic task model to incorporate the findings presented in the previous section to obtain our task model of biologically inspired design.

Figure 6.4 illustrates a generic task model of analogical design based on our earlier work (Goel 1997; Goel and Bhatta 2004) that is consistent with information-processing theories of analogical reasoning in general (Gentner 1983; Gentner et al. 2001). The overall *task* is design (see Fig. 6.4). This is accomplished by using the *method* of analogical reasoning. The analogical reasoning method sets up *sub-tasks* like retrieval of a source analogue, mapping and transfer of relevant knowledge across source and target to obtain the new solution, and evaluation and storage of the new solution. Each sub-task (e.g., retrieval) might, in turn, be accomplished by one of several methods (e.g., feature-based similarity matching for retrieval). *Knowledge*, here, refers to the knowledge inputs and outputs associated with the processing of each task, sub-task, or a method. For example, the knowledge associated with the sub-task of transfer includes what may get transferred between the source and the target design situations; this can include, among

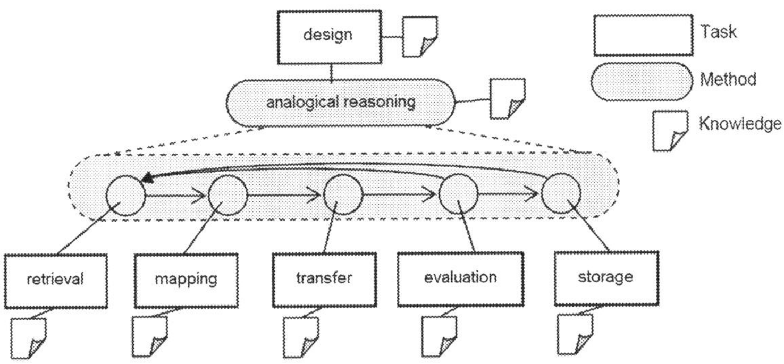

Fig. 6.4 A generic task model of analogical design

others, elements of a previous design like components and relationships between components.

Our task model of biologically inspired design is based on the above generic task model of analogical design, but will extend the generic task model to incorporate two key findings from our study described above: (1) problem-driven analogy and solution-based analogy and (2) compound analogy.

6.4.1 Incorporating Problem-Driven and Solution-Based Analogies

Earlier, we identified two processes followed by designers engaged in biologically inspired design, suggesting two *methods* for the *task* of biologically inspired design: the problem-driven analogy and the solution-based analogy. The problem-driven analogical process incorporates the design sub-tasks of problem formulation, problem reframing, biological solution search, biological solution definition, principle extraction, and principle application. Similarly, the solution-based analogical process incorporates the design sub-tasks of biological solution identification, biological solution definition, principle extraction, solution reframing, problem search, problem definition, and principle application.

As one might expect given that biologically inspired design is based on cross-domain analogies between biological and technological systems, there are correspondences between many of the sub-tasks in the generic task model of analogical design (Fig. 6.4) and the sub-tasks in the problem-driven and solution-based analogical processes of biologically inspired design (Fig. 6.3). Figure 6.5 illustrates some of these correspondences. For example, the "biological solution search" sub-task in the problem-driven analogical process and "problem search" sub-task in the solution-based analogical process correspond to the "retrieval" sub-task in the generic task model of analogical design. The aggregate of "biological solution definition," "principle extraction," and "principle application" sub-tasks in the problem-driven analogical process corresponds to the "mapping" and "transfer" sub-tasks in the generic task model of analogical design; similarly, the aggregate of "problem definition" and "principle application" in the solution-based analogical process corresponds to the "mapping" and "transfer" sub-tasks in the generic task model of analogical design.

On the other hand, there are also sub-tasks in the problem-driven and solution-based analogical processes of biologically inspired design (Fig. 6.3) that are not directly matched by sub-tasks in the generic task model of analogical design (Fig. 6.4). In particular, there are "problem abstraction" and "solution abstraction" sub-tasks in our task model of biologically inspired design that are preparatory to the sub-tasks of retrieval, mapping, and transfer that follow. Figure 6.6 illustrates the task model for biologically inspired design after incorporating problem-driven analogy and solution-based analogy.

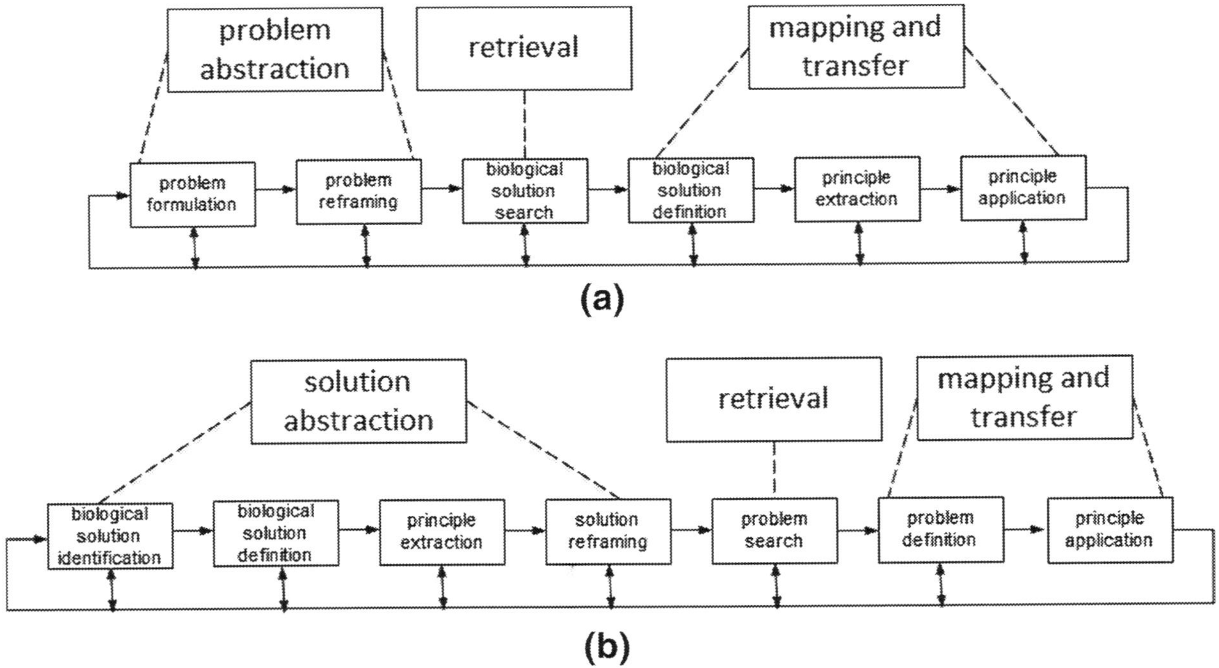

Fig. 6.5 Correspondences between the generic task model of analogical design and the processes of (**a**) problem-driven analogy and (**b**) solution-based analogy

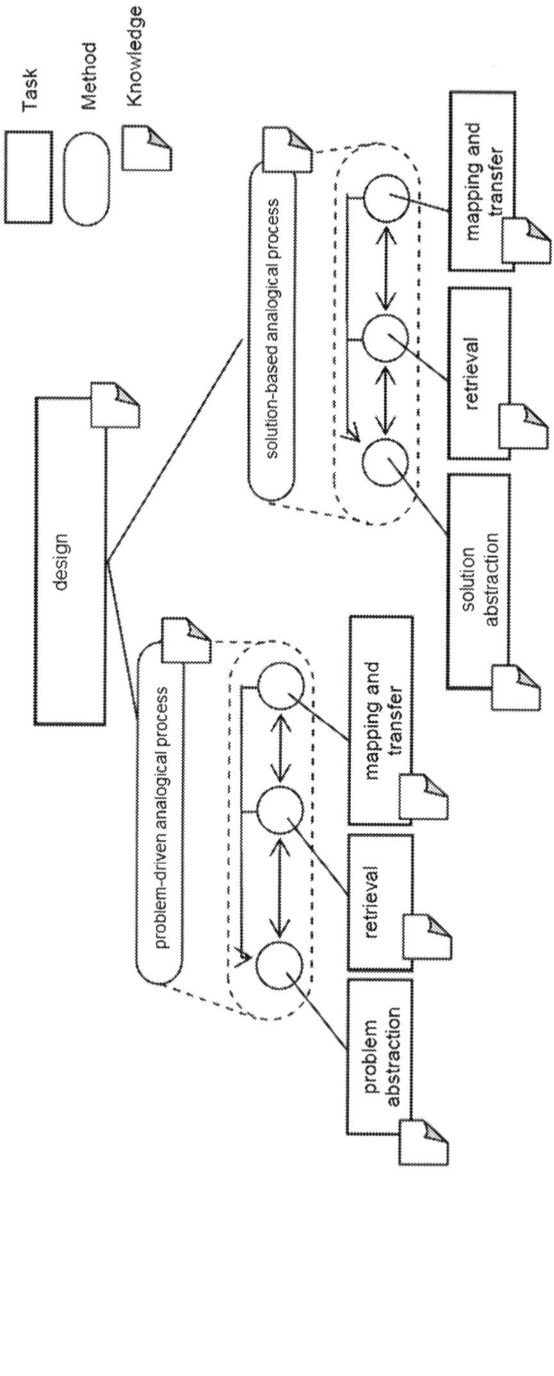

Fig. 6.6 A task model of biologically inspired design after incorporating the two processes of problem-driven analogy and solution-based analogy

6.4.2 Incorporating Compound Analogies

Our task model of biologically inspired design should also account for compound analogies. Figure 6.7 illustrates the task model of biologically inspired design after incorporating the processes of compound analogy, problem-driven analogy, and solution-based analogy. Here, S1 represents the initial solution obtained. We add a new sub-task "evaluate" to both problem-driven and solution-based methods. This sub-task evaluates the initial solution S1 generated by a method. If the evaluation of S1 indicates that S1 addresses only a part of the design problem, then a new design sub-problem is spawned to address the remaining part(s) of the problem. Addressing the new sub-problem may lead to another partial solution S2. The sub-task "compose" composes S1 and S2 to obtain a more complete solution to the original problem. For expediency, it is assumed here that sub-task execution for compound analogy is sequential, represented by one-way arrows between the circles denoting the evaluation, designing, and composition. The actual process may in fact involve much more complex interactions.

6.5 Comparative Analysis

In this section, we will compare our task model of biologically inspired design with existing information-processing theories along some of the dimensions that appear to characterize the design paradigm: (1) cross-domain analogies, (2) compound analogies, and (3) the twin processes of problem-driven analogy and solution-based analogy. We will not consider here the issues of problem–solution coevolution or problem decomposition because they appear to characterize most all design paradigms. In particular, we will compare our task model with the two best-known information-processing theories of biologically inspired design: Biomimicry 3.8 Institute's (2009) Design Spiral and Vincent et al.'s (2006) BioTRIZ.

We should note, however, that Design Spiral, BioTRIZ, and our own task model are all still under development and evolving. Deldin et al. (2013) describe the AskNature Webportal that provides a functionally indexed database of biological systems in support of Design Spiral. Similarly, Vincent (2013) provides an ontology for representing biological systems in support of BioTRIZ.

6.5.1 Design Spiral

Perhaps the most popular design process used for biologically inspired design is Biomimicry 3.8 Institute's (2009) Design Spiral illustrated in Fig. 6.8. In particular, Fig. 6.8a illustrates Biomimicry 3.8 Institute's Challenge to Biology Design Spiral and Fig. 6.8b illustrates Biology to Design Spiral.

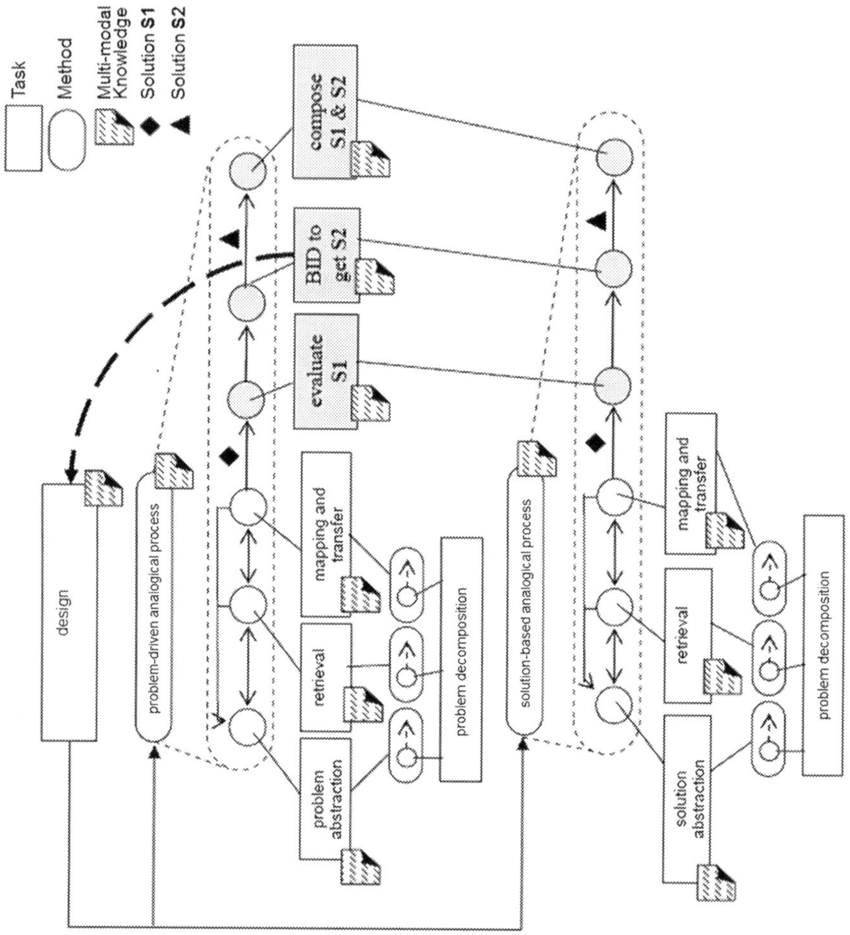

Fig. 6.7 Our task model of biologically inspired design after incorporating the processes of compound analogy, problem-driven analogy, and solution-based analogy

The basic idea of design spiral as a design process has been in the design literature for some time. For example, Evans (1959) describes design spiral for ship design and Boehm (1988) describes design spiral for software design and development. However, the Biomimicry 3.8 Institute pioneered the Design Spiral process in biologically inspired design and related it to its AskNature database (Biomimicry 3.8 Institute 2008).

A comparison between our task model and the Biomimicry 3.8 Institute's Design Spiral reveals the following similarities and differences:

- Both Design Spiral and our model address cross-domain analogies between biological and technological systems.

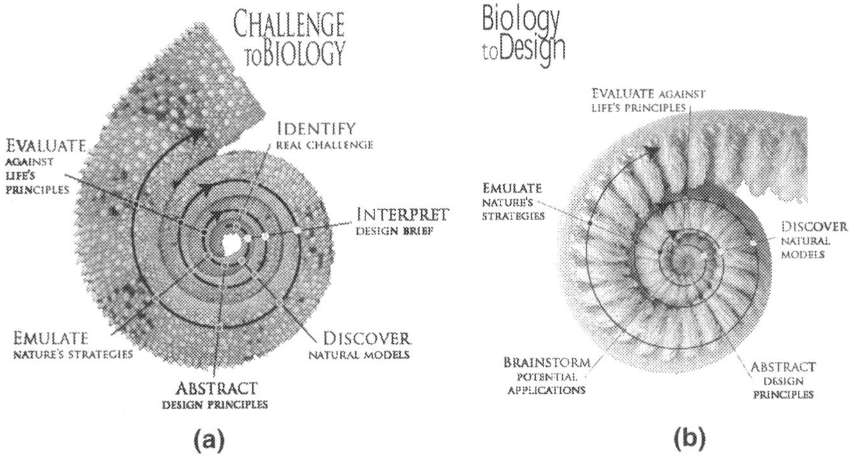

Fig. 6.8 **a** Biomimicry 3.8 Institute's Challenge to Biology Design Spiral (adapted from Biomimicry 3.8 Institute 2009). **b** Biomimicry 3.8 Institute's Biology to Design Spiral (adapted from Biomimicry 3.8 Institute 2009)

- Design Spiral appears to be a normative and prescriptive theory for biologically inspired design. In contrast, our task model is a descriptive and explanatory theory based on *in situ* observations of biologically inspired design in practice.
- Both Design Spiral and our model characterize biologically inspired design as an iterative process. Design Spiral also captures the notion that the first iteration is at a high level of abstraction, the next is more detailed, and so on.
- Although there are important differences, Challenge to Biology Design Spiral appears similar to our problem-driven analogical process, and Biology to Design Spiral seems similar to our solution-based analogical process.
- Design Spiral does not directly address compound analogies. However, the iterative nature of Design Spiral suggests that the process can support compound analogies. Our model explicitly accounts for compound analogies.
- The steps in Design Spiral are a little informal. We are trying to incrementally formalize the tasks in our problem-driven and solution-based analogical processes. For example, Wiltgen et al. (2011) describe methods of retrieval of biological systems relevant to a design problem from a digital library based on discrimination networks.

6.5.2 BioTRIZ

BioTRIZ (Vincent et al. 2006) is an information-processing theory of biologically inspired design derived from the earlier theory of engineering invention called TRIZ (Altshuller 1984). The TRIZ theory begins with a repository of design cases with known solutions, where each case is indexed by contradictions that arose in the original design situation. For example, consider a case in the repository that represents the design of an airplane wing. In this case, the designer faces the contradiction of obtaining a material that is both strong and lightweight and solves it using a solution, say S_1. This case is then indexed by the contradiction "strong yet lightweight material." Additionally, if the particular solution S_1 belongs to a more general way of resolving contradictions of a particular kind, it may be categorized as a generic abstraction, such as "use porous materials" (to resolve the contradiction of strong yet lightweight material). TRIZ posits the existence of forty such generic ways of resolving contradictions, called inventive principles. The inventive principles were extracted by dropping the specifics of a particular case and domain and retaining the essence of how a particular class of contradictions is solved, so we can imagine each principle pointing to numerous cases (potentially belonging to different domains) in which that principle was used to resolve a contradiction. The contradictions and the principles typically are organized in a contradiction matrix (TRIZ Matrix 2004).

When the designer is presented with a design problem and intends to use TRIZ, she reformulates the problem to identify certain key contradictions in the requirements of the design. For each contradiction, she is reminded of a inventive principle that is applicable for resolving that contradiction. In addition to suggesting the essence of a solution for resolving that contradiction, the inventive principle also points to a number of cases in which that inventive principle was instantiated (See Fig. 6.9). These cases can originate from domains different from the one in which the designer is currently working. TRIZ, however, does not address the issue of how transfer occurs.

Vincent et al. (2006) recently developed a modified version of TRIZ, called BioTRIZ, specifically for biologically inspired design. The primary difference between the two theories is a change in the features that compose the contradiction matrix. Whereas TRIZ defines 39 features with which to determine contradictions

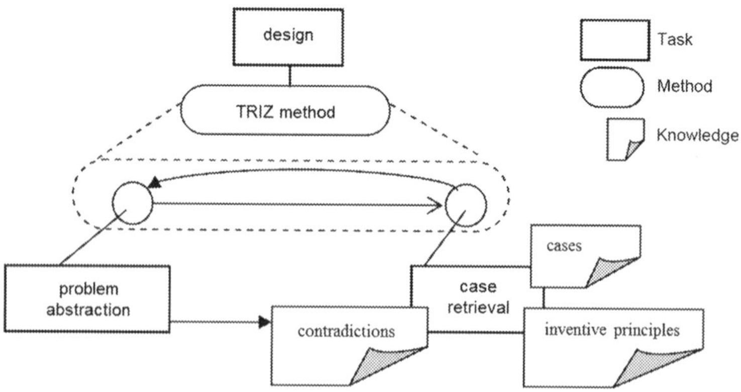

Fig. 6.9 The TRIZ information-processing model of invention

and index into inventive principles, the current version of BioTRIZ has six "operational fields": substance, structure, space, time, energy, and information.

A comparison of our task model and BioTRIZ reveals the following similarities and differences:

- Both BioTRIZ and our model address cross-domain analogies between biological and technological systems.
- BioTRIZ is a prescriptive theory of biologically inspired design, derived from best practices in mechanical engineering design. In contrast, our task model is a descriptive theory based on *in situ* observations of biologically inspired design.
- The processing in BioTRIZ is problem-driven. The processing in BioTRIZ always begins with a specification of a design problem. It does not directly address the solution-based analogical process. Our task model accounts for both problem-driven and solution-based analogies.
- BioTRIZ does not directly address compound analogy. However, since a design problem may contain multiple contradictions, and resolving those contradictions may require the invocation of multiple inventive principles, compound analogy appears to be feasible in BioTRIZ.
- BioTRIZ provides a detailed account of cross-domain analogical retrieval. Our information-processing theory too provides a detailed account of analogical retrieval (Wiltgen et al. 2011).

6.6 What Makes Biologically Inspired Design Different?

The above comparative analysis of theories of biologically inspired design brings us to questions we are often asked by other design theorists: is biologically inspired design different from other design paradigms? Or, put a little differently, what precisely makes biologically inspired design a novel design paradigm from an information-processing perspective? Note that the question here is not whether biological and technological systems are different. As Vincent et al. (2006) note, "biology and technology solve problems in design in rather different ways"; biological systems often use information for functions for which technological systems tend to use energy. French (1994) and Vogel (2000) make detailed analyses of the similarities and differences between biological and technological systems; biological systems in general tend to be more multifunctional than technological systems. Instead, the question here is the following: are the *processes* of biologically inspired design fundamentally different from that of other design paradigms?

As we noted above, Design Spiral and BioTRIZ are normative and prescriptive theories of biologically inspired design; they can be, and have been, applied to many kinds of design. In fact, TRIZ, from which BioTRIZ evolved, arose out of a study of mechanical engineering inventions, and it has been applied to many kinds of design, ranging from mechanical engineering to organizational design

(TRIZ Matrix 2004). Similarly, the design spiral process is extensively used in the design of artifacts ranging from ships to software.

However, our task analysis offers some insights into what may make biologically inspired design a different design paradigm from an information-processing perspective. Firstly, biologically inspired design by definition is based on cross-domain analogies. While many design processes in and out of biologically inspired design sometimes engage cross-domain analogies (e.g., Goel 1997; Goel and Bhatta 2004), and while biologically inspired design also frequently engages within domain analogies (Vattam et al. 2010a), insofar as we know there are not many other kinds of design that by definition are based on cross-domain analogies.

Secondly, biologically inspired design often entails compound analogies. For example, the target design problem is decomposed functionally, solutions to different sub-functions in the functional decomposition are found through analogy to different biological systems retrieved from a functionally indexed memory, and the overall design solution is obtained by composing the solutions for achieving the different sub-functions. While problem decomposition is common to almost all kinds of design (e.g., Brown and Chandrasekaran 1989; Chandrasekaran 1990) and while other kinds of design too sometimes engage in compound analogies (e.g., Goel and Chandrasekaran 1992), compound analogy appears to be a stronger characteristic of biologically inspired design compared to other kinds of design.

Thirdly, biologically inspired design engages two different analogical processes, namely problem-driven analogy and solution-based analogy. We first observed these two analogical processes in our *in situ* studies of biologically inspired design in practice. Subsequently, we conducted a narrative analysis of some seventy case studies of biologically inspired design and confirmed this observation. Insofar as we know, information-processing theories of analogy (e.g., Clement 2008; Dunbar 2001; Gentner 1983; Gentner et al. 2001; Gick and Holyoak 1983; Goel 1997; Holyoak and Thagard 1996; Keane 1988; Kolodner 1993) focus on and emphasize problem-driven analogy. Further, insofar as we know, information-processing theories of all other kinds of design focus on and emphasize problem-driven design (e.g., Boehm 1988; Brown and Chandrasekaran 1989; Chandrasekaran 1990; Dorst and Cross 2001; Dym and Brown 2012; French 1996; Goel and Bhatta 2004; Goel and Chandrasekaran 1992; Hubka and Eder 1988; Maher and Tang 2003; Pahl et al. 2007; Simon 1996). Therefore, that biologically inspired design entails both problem-driven and solution-based analogies appears to be another definitional characteristic of biologically inspired design.

Veros and Coelho (2011) take a different approach to comparing various methods of biologically inspired design. In particular, they apply Biomimicry 3.8 Institute's Challenge to Biology Design Spiral and our problem-driven and solution-based analogical design processes to a class of design problems. One of their key findings is that the solution-based analogical process tends to more often produce multifunctional designs, which confirms our own empirical observations (Helms et al. 2009; Vattam et al. 2007).

6.7 Conclusions

In this chapter, we first presented a task analysis of biologically inspired design based on our empirical observations of its practice in ME/ISyE/MSE/BME/BIOL 4740, Georgia Tech's introductory course on biologically inspired design. Our task analysis indicates some of the fundamental processes of biologically inspired design: (1) Biologically inspired design by definition engages cross-domain analogies. (2) Problems and solutions in biologically inspired design coevolve. (3) Problem decomposition is a fundamental process of biologically inspired design. (4) Biologically inspired design often involves compound analogy, entailing a complex interplay between the processes of problem decomposition and the processes of analogical retrieval from memory. (5) Biologically inspired design entails two distinct but related processes: problem-driven analogy and solution-based analogy.

We then presented a task model of biologically inspired design. Since problem–solution coevolution and problem decomposition are common to most design paradigms, our task model focused on two other results of our task analysis: that biologically inspired design often involves compound analogy and that biologically inspired design entails both problem-driven analogy and solution-based analogy. Our task model of biologically inspired design is a descriptive theory of biologically inspired design that has had significant influence on the teaching and learning in ME/ISyE/MSE/BME/BIOL 4740 (Yen et al. 2011, 2013).

Next, we compared our task model with the two best-known information-processing theories of biologically inspired design: Biomimicry 3.8 Institute's Design Spiral and Vincent et al.'s BioTRIZ. Both Design Spiral and BioTRIZ are normative and prescriptive theories of biologically inspired design. Both address cross-domain analogies and, at least implicitly, allow compound analogies. However, BioTRIZ appears to support only problem-driven analogies.

Finally, we analyzed what makes biologically inspired design a novel design paradigm and, in particular, how the processes of biologically inspired design differ from that of other design paradigms. Our task model suggests that biologically inspired design differs from other kinds of design in the use of cross-domain analogies, the use of compound analogies, and the use of both problem-driven and solution-based analogies. We are presently investigating the methodological, technological, and pedagogical implications of our information-processing theory of biologically inspired design.

Acknowledgments We are grateful to the instructors and students of the ME/ISyE/MSE/BME/BIOL 4740 class from 2006 through 2010, especially Professors Jeannette Yen, Craig Tovey, and Marc Weissburg, as well as the design team that developed the Shark Attack and Levee projects. We thank the US National Science Foundation for its support of this research through a CreativeIT Grant (#0855916) titled "Computational Tools for Enhancing Creativity in Biologically Inspired Engineering Design," and a TUES Grant (#1022778) titled "Biologically Inspired Design: A Novel Biology-Engineering Curriculum." This chapter augments and amplifies the information-processing analysis of biologically inspired design developed in Vattam et al. 2010b and Goel et al. 2011b. Figures 6.1 through 6.7 and Figure 6.9 have been adapted from Goel et al. 2011b. We thank the reviewers of the earlier papers and this chapter for their critiques.

References

Altshuller G (1984) Creativity as an exact science. Gordon and Branch Publishers, Luxembourg

Ashley S (2004) Bumpy flying Scalloped flippers of whales could reshape wings. Sci Am 291(2):18, 20

Bar-Cohen Y (ed) (2011) Biomimetics: nature-based innovation. CRC Press, USA

Benyus J (1997) Biomimicry: innovation inspired by nature. William Morrow, New York

Biomimicry 3.8 Institute (2008) AskNature. http://www.asknature.org/. Last retrieved on April 25, 2011

Biomimicry 3.8 Institute (2009) Biomimicry: a tool for innovation. <http://www.biomimicryinstitute.org/about-us/biomimicry-a-tool-for-innovation.html>. Last retrieved on April 25, 2011

Biomimicry 3.8 Institute (2011a) Energy. <http://www.biomimicryinstitute.org/case-studies/case-studies/energy.html>. Last retrieved on April 28, 2011

Biomimicry 3.8 Institute (2011b) Water vapor harvesting: namib desert beetle. <http://www.asknature.org/strategy/dc2127c6d0008a6c7748e4e4474e7aa1>. Last retrieved on April 28, 2011

Boehm BW (1988) A spiral model of software development and enhancement. IEEE Comput 21(5):61–72

Brown D, Chandrasekaran B (1989) Design problem solving: knowledge structures and control strategies. Morgan Kaufmann, San Mateo

Chakrabarti A, Shu L (2010) Biologically inspired design. AIEDAM 24:453–454

Chandrasekaran B (1990) Design problem solving: a task analysis. AI Mag 11(4):59–71

Clement J (2008) Creative model construction in scientists and students: the role of imagery, analogy, and mental simulation. Springer, Dordrecht

Crandall B, Klein G, Hoffman R (2006) Working minds: a practitioner's guide to cognitive task analysis. MIT Press, Cambridge

Deldin JM, Laux J, Shuknecht (2013) The AskNature database: enabling solutions in biomimetic design. In: Goel, McAdams, Stone (eds) Biologically inspired design: computational methods and tools (this volume)

Dorst K, Cross N (2001) Creativity in the design process: co-evolution of problem-solution. Des Stud 22:425–437

Dunbar K (2001) The analogical paradox. In: Gentner D, Holyoak KJ, Kokinov BN (eds) The analogical mind: perspectives from cognitive science. MIT Press, Boston

Dym C, Brown D (2012) Engineering design: representations and reasoning, 2nd edn. Cambridge University Press, Cambridge

Evans JH (1959) Basic design concepts. J Am Soc Naval Eng 71(4):671–678

Fish F, Battle J (1995) Hydrodynamic design of the humpback whale flipper. J Morphol 225:51–60

French M (1994) Invention and evolution: design in nature and engineering, 2nd edition. Cambridge University Press, Cambridge, 1988

French M (1996) Conceptual design for engineers, 3rd edn. Springer, London

Gentner D (1983) Structure mapping: a theoretical framework for analogy. Cogn Sci 7(2):155–170

Gentner D, Holyoak K, Kokinov B (eds) (2001) The analogical mind: perspectives from cognitive science. MIT Press, Cambridge

Gick M, Holyoak KJ (1983) Schema induction and analogical transfer. Cog Psych 15:1–38

Goel A (1997) Design, analogy, and creativity. IEEE Expert 12(3):62–70

Goel A, Bhatta S (2004) Use of design patterns in analogy-based design. Adv Eng Inform 18(2):85–94

Goel A, Bras B, Helms M, Rugaber S, Tovey C, Vattam S, Weissburg M, Wiltgen B, Yen J (2011a) Design patterns and cross-domain analogies in biologically inspired sustainable design. In Proceedings AAAI spring symposium on AI and sustainable design. Stanford University, Palo Alto, 45–51 Mar 2011

Goel A, Chandrasekaran B (1992) Case-based design: a task analysis. In: Tong C, Sriram D (eds) Artificial intelligence approaches to engineering design, vol II: innovative design. Academic Press, San Diego, pp 165–184

Goel A, Vattam S, Wiltgen B, Helms M (2011b) An information-processing account of creative analogies in biologically inspired design. In: Proceedings eighth ACM conference on creativity and cognition, Atlanta, 71–80 Nov 2011

Goel A, Vattam S, Wiltgen B, Helms M (2012) Cognitive, collaborative, conceptual and creative—Four characteristics of the next generation of knowledge-based CAD systems: a study in biologically inspired design. Comput Aided Des 44(10):879–900

Hamilton W, Seely M (1976) Fog basking by the Namib Desert beetle, Onymacris unguicularis. Nature 262(5566):284–285

Helms M, Goel A (2012) Analogical problem evolution in biologically inspired design. In: Proceedings 5th international conference on design computing and cognition, College station. Springer, Berlin

Helms M, Vattam S, Goel A (2009) Biologically inspired design: process and products. Des Stud 30(5):606–622

Helms M, Vattam S, Goel K (2010) The effect of functional modeling on understanding biological systems. In: Proceedings ASME 2010 international design engineering technical conferences & computers and information in engineering conference (IDETC/CIE 2010). Montreal, Quebec, Canada, 15–18 Aug 2010

Hofstadter D (ed) (1996) Fluid concepts & creative analogies: computer models of the fundamental mechanisms of thought. Harvester Wheatsheaf, London

Holyoak K, Thagard P (1996) Mental leaps: analogy in creative thought. MIT Press, Cambridge

Hubka V, Eder W (1988) Theory of technical systems: a total concept theory of engineering design. Springer, Dordrecht

Keane M (1988) Analogical problem solving. Ellis Horwood, Chichester

Kolodner J (1993) Case-based reasoning. Morgan Kauffman, San Francisco

Kruger C, Cross N (2006) Solution-driven vesus Problem-driven design: strategies and outcomes. Des Stud 27(5):527–548

Linsey J, Wood K, Markman A (2008) Modality and representation in analogy. AI Eng, Des, Manuf 22:85–100

Maher ML, Tang H (2003) Co-evolution as a computational and cognitive model of design. Res Eng Design 14(1):47–64

Mak T, Shu L (2008) Using descriptions of biological phenomena for idea generation. Res Eng Design 19(1):21–28

Nagel J, Stone R, McAdams D (2013) Function-based biologically inspired design. In: Goel, McAdams, Stone (eds) Biologically inspired design: computational methods and tools (this volume)

Nelson B, Wilson J, Yen J (2009) A study of biologically-inspired design as a context for enhancing student innovation. In: Proceedings 39th ASEE/IEEE frontiers in education conference, 2009

Newell A, Simon H (1972) Human problem solving. Prentice-Hall, Upper Saddle River

Pahl G, Beitz W, Feldhusen J, Grote K (2007) Engineering design: a systematic approach, 3rd edition. Translated and edited by Wallace K, Blessing L, Springer, London

Ritzmann R (1974) Mechanisms for the snapping behavior of two alpheid shrimp, Alpheus californiensis and Alpheus heterochelis. J Comp Physiol 95:217–236

Shanyengana E, Henschel J, Seely M, Sanderson R (2002) Exploring for as a supplementary water resource in Namibia. Atmos Res 64(1–4):251–259

Shraagen J, Chipman S, Shalin V (eds) (2000) Cognitive task analysis. Lawrence Erlbaum, USA
Shu L, Ueda K, Chiu I, Cheong H (2011) Biologically inspired design. CIRP Ann Manuf Technol 60:673–693
Simon H (1996) Sciences of the artificial, 3rd edn. MIT Press, Cambridge
Srinivasan V, Chakrabarti A (2010) An integrated model of designing. ASME JCISE 10(3). doi:10.1115/1.3467011 (from http://computingengineering.asmedigitalcollection.asme.org/article.aspx?articleid=1402109)
TRIZ Matrix (2004) Interactive TRIZ contradiction matrix. (http://www.triz40.com/)
Vattam S, Helms M, Goel A (2007) Biologically inspired innovation in engineering design: a cognitive study. Technical report GIT-GVU-07-07, graphics, visualization & usability center, Georgia Institute of Technology
Vattam S, Helms M, Goel A (2010a) A content account of creative analogies in biologically inspired design. AI Eng Des, Anal Manuf 24:467–481 Special issue on biologically inspired design
Vattam S, Helms M, Goel A (2010b) Biologically inspired design: a macrocognitive account. In: Proceedings ASME 2010 international design engineering technical conferences & computers and information in engineering conference (IDETC/CIE 2010). Montreal, Quebec, Canada, 15–18 Aug 2010
Veros C, Coelho D (2011) Biologically inspired design: methods and validation. In: Coelho D (ed) Industrial Design—New Frontiers. InTech open access books; Available in creative commons: http://www.intechopen.com/books/industrial-design-new-frontiers/biologically-inspired-design-methods-and-validation
Vincent J (2013) Finding the gaps in BID. In: Goel, McAdams, Stone (eds) Biologically inspired design: computational methods and tools (this volume)
Vincent J, Mann D (2002) Systematic transfer from biology to engineering. Philos Trans R Soc Lond 360:159–173
Vincent J, Bogatyreva O, Bogatyrev N, Bowyer A, Pahl A (2006) Biomimetics: its practice and theory. J R Soc Interface 3:471–482
Vogel S (2000) Cat's paws and catapults: mechanical worlds of nature and people. W.W Norton and Company, New York
Wilson J, Rosen D, Nelson B, Yen J (2010) The effects of biological examples in idea generation. Des Stud 31(2):169–186
Wiltgen B, Goel A (2011) Case-based reasoning all over the place: the multiple roles of CBR in biologically inspired design. In: Proceedings workshop on human-centered and cognitive approaches to case-based reasoning at the 19th international conference on case based reasoning. Greenwich, London, 2011. (An updated version is available at http://dilab.gatech.edu/publications/wiltgengoel-icccbr-mbr-2011.pdf.)
Wiltgen B, Goel A, Vattam S (2011) Representation, indexing, and retrieval of biological cases for biologically inspired design. In: Proceeding 19th international conference on case based reasoning. Greenwich, London, Sep 2011
Yao H, Dao M, Imholt T, Huang J, Wheeler K, Bonilla A, Suresh S, Ortiz C (2010) Protection mechanisms of the iron-plated armor of a deep-sea hydrothermal vent gastropod. Proc Natl Acad Sci (PNAS) 107(3):987–992
Yen J, Helms M, Goel A, Vattam S (2010) Evaluating biological systems for their potential in engineering design. Adv Nat Sci 3(2):27–40
Yen J, Helms M, Tovey C, Weissburg M, Goel A (2013) Adaptive evolution of a biologically inspired design course. In: Goel, McAdams, Stone (eds) Biologically inspired design: computational methods and tools (this volume)
Yen J, Weissburg M (2007) Perspectives on biologically inspired design: introduction to the collected contributions. J Bioinspiration Biomimetics 2. doi:10.1088/1748-3190/2/4/E01 (from http://iopscience.iop.org/1748-3190/2/4/E01)
Yen J, Weissburg M, Helms M, Goel A (2011) Biologically inspired design: a tool for interdisciplinary education. In: Bar-Cohen Y (ed) Biomimetics: nature-based innovation. Taylor & Francis, USA

Chapter 7
Adaptive Evolution of Teaching Practices in Biologically Inspired Design

Jeannette Yen, Michael Helms, Ashok Goel, Craig Tovey and Marc Weissburg

Abstract At Georgia Tech in 2005, we developed an interdisciplinary undergraduate semester-long course, biologically inspired design (BID), co-taught each year by faculty from biology and engineering. The objective of this chapter is to share our teaching experience with those interested in teaching such a course themselves. The specific curriculum of a BID course must depend on the student mix, the institutional context, and instructor goals. Therefore, rather than presenting a particular curriculum, we present key problems that we encountered in our 8 years of teaching and how we addressed them. We expect that any who try to teach such a course will face one or more of the same challenges, and we offer numerous pedagogical approaches that can be tailored to their specific circumstances. By describing our solutions, their consequences, and the extent to which they met our expectations, we also point out where tough student challenges still exist that are in need of attention from the community.

Keywords Teaching biologically inspired design · Learning biologically inspired design · Problem-driven design · Solution-based design · Analogical design · Cross-domain analogy · Design by analogy · Understanding biological systems ·

J. Yen (✉) · M. Weissburg
School of Biology, Georgia Institute of Technology, Atlanta, GA 30332, USA
e-mail: jeannette.yen@biology.gatech.edu

M. Weissburg
e-mail: marc.weissburg@biology.gatech.edu

M. Helms · A. Goel
School of Interactive Computing, Georgia Institute of Technology, Atlanta, GA 30332, USA
e-mail: mhelms3@cc.gatech.edu

A. Goel
e-mail: goel@cc.gatech.edu

C. Tovey
School of Industrial and Systems Engineering, Georgia Institute of Technology, Atlanta, GA 30332, USA
e-mail: ctovey@isye.gatech.edu

A. K. Goel et al. (eds.), *Biologically Inspired Design*,
DOI: 10.1007/978-1-4471-5248-4_7,

Functional decomposition · Structure-Behavior-Function · Design evaluation · Team design · Interdisciplinary design · Interdisciplinary education · Design creativity · Engineering design · Engineering creativity · Multi-disciplinarity · Team-based learning · Analogical reasoning

7.1 Introduction

Biologically Inspired design (BID) is highly interdisciplinary. The following four examples of BID illustrate the importance of contributions from different disciplines. For Velcro (see, for example, Simonton 2004), a close morphological examination of a plant burr provided the inspiration for a successful design of a fastener. For RHex (Altendorfer et al. 2001), a deep understanding of cockroach locomotion and stability of multi-legged organisms and robotic engineering was needed for success in an all terrain vehicle. The development of the Geckel wet adhesive (Lee et al. 2007) required a thorough understanding of the chemical and material properties of the biological solutions for gripping by geckos and sticking underwater by mussels. For the honeybee web-hosting algorithm, which dynamically allocates Web server resources to hosted services (Nakrani and Tovey 2007), the inventors needed mathematics to understand the bee behavior, the nectar output of flower patches, and the patterns of internet traffic. Thus, taken as a whole, BID spans science and engineering. Since few designers are likely to have deep enough knowledge in a wide range of fields required for any given design, BID often is collaborative. To teach a course well in this design paradigm likewise requires expertise in biology, engineering, and design.

At Georgia Tech in 2005, we developed an interdisciplinary undergraduate semester-long course [ME/ISyE/MSE/BME/BIOL 4740: BID] taught in the fall semester each year (Weissburg et al. 2010; Yen et al. 2010, 2011). The course recruits students from these majors: mechanical engineering, industrial and systems engineering, materials science engineering, biomedical engineering, and biology; in addition, we sometimes get majors from industrial design, architecture, chemistry, mathematics, or nuclear engineering. The course is co-taught by faculty from biology and engineering. Although it is not possible to teach deep knowledge in all these disciplines in a one-semester course, we can show the students that such knowledge can take them on the exciting path of BID. At Georgia Tech, we restrict this course to juniors and seniors who have established their majors, thus are able to bring specialized knowledge to the table. One of the goals of the course is to show the students how a deep understanding of their field and experience in working on an interdisciplinary team can enable them to be more inventive and creative. The goal is to motivate them to learn as much as they can in their field, then come to class to practice how to collaborate. In today's information-rich setting with easy access to knowledge resources, and in an increasingly interdisciplinary and collaborative design world, we turn our emphasis in this chapter on how to retrieve that knowledge, communicate it effectively across disciplines, and utilize it to solve problems in interdisciplinary teams. We think BID is a useful environment for learning and applying these skills.

In Yen et al. (2011), we defined the following five learning goals of the BID course: (1) Novel design techniques; (2) Interdisciplinary communication; (3) Science and Engineering knowledge outside core domain; (4) Interdisciplinary collaboration; and (5) Application of existing knowledge to a new field. To reach these goals, we presented the skills we taught, the exercises that we developed to enable the students to practice these skills, and the format of the class to indicate how we deployed these exercises. Our pedagogical techniques were grounded in the theory and practice of interdisciplinary research and education, recommended in the cognitive and learning sciences (e.g., Ausubel 2000; Bransford et al. 2000; Bybee 1997; Lave and Wenger 1991; Vygotsky 1978) as well as recommendations for teaching science (e.g., National Research Council 2011) and biology (e.g., National Research Council 2009). In Yen et al. (2011), we provided details on some of the more complex exercises such as biological design in natural evolution, problem decomposition, and analogical reasoning. We presented lessons learned from the first 3 years of teaching this course (2005–2007). To summarize, we found that creativity improved through the use of analogical reasoning to link biological functions to engineering challenges. We experienced clear differences in how biologists and engineers solve problems, which identified interdisciplinary communication gaps that were overcome, to some extent, by giving students practice in both domains in their interdisciplinary projects as well as by motivating them to apply their own knowledge to new problems and domains.

During the last 5 years (2008–2012), the BID course retained all the learning goals mentioned above. As this course and the field at large mature, additional challenges have arisen and become part of our learning objectives. In this chapter, we relay the triumphs and tribulations encountered in our ambitious plan to provide students with the opportunity to collaborate across disciplines through BID as well as learn about BID itself. The objective of this chapter is to share our teaching experience with those interested in teaching a BID course themselves. The curriculum of a BID course is flexible and depends on the student mix as well as the institutional context and instructor goals. Therefore, rather than presenting a specific curriculum, we present key problems that we encountered in the 8 years of teaching and how we addressed them. We expect that anyone who tries to teach such a course will face one or more of the same challenges, and we offer numerous pedagogical approaches that can be tailored to different circumstances. By describing our solutions, their consequences, and the extent to which they met our expectations, we also point out where tough student challenges still exist that are in need of attention from the community.

7.2 Key Challenges

While the inventory of skills required to generate successful BIDs is vast, we focus on the following key challenges that we see students struggle with year after year.

1. Searching for biological systems
2. Understanding biological systems
3. Identifying and understanding good design problems
4. Analogy mapping and transfer
5. Communicating across discipline boundaries
6. Communicating complex system knowledge
7. Teaming in an interdisciplinary environment
8. Maintaining equal engagement throughout the process
9. Evaluating designs

7.2.1 Searching for Biological Systems

In past studies (e.g., Vattam and Goel 2011), we have documented that up to 25% of out-of-class time can be spent simply searching for the right biological organism to solve a particular problem. While tremendous resources already are being devoted to solving this problem technologically (as this volume attests), currently students comb through volumes of textbooks, scientific databases, and the Web to find what they need. When they do find something promising: (a) it is often written in academic or technical language that is difficult for a non-expert to understand; (b) it takes time to determine whether it will be applicable to their design problem; and (c) it is usually not design oriented and requires translation before it is useful. These problems are magnified when the design teams do not have individuals with broad-based biological knowledge. Even when biologists are represented, they may not have the background that would be most desirable. For example, students with strong knowledge of basic organismal biology (e.g., comparative physiology, functional morphology, behavior, invertebrate and vertebrate biology) are the most well-equipped to search and identify appropriate systems for human-scale problems. Student designers typically (although not always) attempt to solve human-scale problems as this familiar scale is where humans have the most experience. For design problems at smaller or larger scales (e.g., at the molecular scale such as filtering pharmaceuticals from the water supply or at large scales such as city planning), different specializations may be helpful (from organic chemistry to ecosystem structure). Programs such as ours, where biologists are less well represented than engineers, require specific curricular elements to increase the ability of students to mine the biological literature.

7.2.2 Understanding Biological Systems

Our course has a mix of engineering, biology, architecture, and design students who must work as a team to understand the key mechanisms of the biological system so that they are capable of abstracting the mechanism and applying it to an engineering design problem. Whereas the biologists may have a deep understanding of a given biological phenomenon in its biological context, it is a challenge for them to communicate that understanding in such a way that the non-biologists understand it well enough to use in a design. This is exacerbated by the number and breadth of biological systems the students are asked to learn about, the limited time available for deep understanding, and the natural tendency of students to either focus on structural details, and/or use improper analogies to facilitate or communicate their understanding (e.g., the analogy: xylem in a tree acts like a straw in a drink—is not accurate at the mechanistic level since a pressure gradient is used to transport water by the straw while the molecular force of cohesion is used to transport water up to its leaves from its roots). Biologists who have experience examining biological systems in terms of function (e.g., biomechanics, physiology, and behavior) initially may be more able to communicate their understanding of biological systems in an appropriate manner. We find that architects and designers tend to focus on the structural elements of the system, at least initially, and require practice in thinking about function in biology. Engineers think about function, but generally lack the requisite biological knowledge.

7.2.3 Identifying and Understanding Good Design Problems

Throughout their scholastic careers, students are taught how to solve problems that are *given* to them. Less frequently faced in an academic context, this course presents a unique set of challenges when students are asked to identify and define a problem of their own choosing. Students in our BID class have (in early course iterations) devoted up to half the semester defining their design problem when challenged with a wide-open problem landscape. We have learned that BID may originate with the standard process of problem-driven design or may begin from a solution-based approach, where the unique mechanisms of a biological solution of interest determine which problems one may wish to explore (Helms et al. 2009). Thus, in solution-based design, problem identification is a critical aspect of BID. As instructors, we must balance the requirements of good problem identification and formulation against the needs of teaching a complete BID process.

7.2.4 Analogy Mapping and Transfer

Students often manifest cognitive limitations, biases, and errors (Helms et al. 2009). Whereas students naturally and effortlessly make analogies during the process of design, their analogies can be superficial. Students fall prey to a kind of confirmation bias, focusing on initial superficial alignment between analogue and problem, while ignoring deeper dissimilarities until they are forced to confront them late in the design process. For example, a student team in 2011 attempted to design a collapsible bicycle helmet inspired by the girdled lizard, which bends itself into a circle of spiked bands for protections against predators. The design failed because "protection" in the case of a bicycle helmet means dissipation and absorption of energy from a collision, whereas in the case of the lizard, it means resistance to penetration by claws or teeth.

7.2.5 Communicating Across Discipline Boundaries

Communication often is hampered by differences in the specialized terminology of different disciplines. For example, for biologists, "stress" represents extreme conditions such as heat, lack of water, or predators, to which organisms must respond using physiological, behavioral, genetic, developmental, or other mechanisms. For mechanical engineers, "stress" is the measure of force per unit area in a deformable body. Such differences occur even within the broad field of engineering, but become increasingly large as more disciplines participate.

7.2.6 Communicating Complex Systems Knowledge

In many disciplines, there are systems so complex that it seems impossible to draw possible analogies to another field without extensive research and teaching experience. We have found that decomposing a particular function of a system into subfunctions allows others to understand at least the interactions occurring at one level accurately even without gaining a full understanding of how all the functions in a complex system are integrated. If a subfunction still remains out of grasp of understanding, then it too must be decomposed further into its underlying mechanisms until a principle, common to both disciplines, is reached. This journey may take the designers several levels deep down in the hierarchical breakdown of the problem or the natural system, but success is more likely when the team reaches this common understanding.

7.2.7 Teaming in an Interdisciplinary Environment

Students taking classes within their field of study often work alone, or in teams with others in their field. Few if any entering students in our BID class have shared a course with someone outside their major. Hence, it may be difficult initially for students to recognize the value of knowledge and approaches outside their discipline. This can be abetted by the institutional persona that encourages divisions in the perceived utility of different fields of knowledge. For instance, engineers at a technology institute may think their expertise is more valuable than others. An appreciation for everyone's talent needs to be nurtured throughout the time the team is working toward a common goal.

7.2.8 *Maintaining Equal Engagement Throughout the Process*

The roles of each discipline may change throughout the course, depending on the stage in the design process and a design's specific requirements. Initially, emphasis is placed on biological knowledge since the teams have to select an organism and understand how the biological system works. Once the teams enter the design process, everyone is actively engaged because BID is unfamiliar to most students, with more weight placed on the biologists to find, understand, and explain solutions. The engineers are the most engaged when there is a required feasibility or performance assessment, work which biologists are not as experienced to perform. Under deadline pressure to complete a design, team members who cannot contribute directly can feel marginalized or devalued.

7.2.9 *Evaluating Designs*

A good design for our purposes must simultaneously satisfy the following criteria: functionality, potential market, manufacturability, novelty or competitive advantage, and reasonable cost. Although different ways of teaching BID may not emphasize all of these criteria, student designs become amorphous and speculative without a focus on functionality, novelty, and manufacturability, whereas failure to consider market and advantage results in designs that do not solve real problems or do so in a way little different from current designs. Challenges occur because: (a) students may not be familiar with using some of these criteria in their design analysis; (b) applying some of these criteria may require students to apply quantitative methods outside of their domain; and (c) students have trouble balancing conflicting criteria. These challenges are exacerbated by the profusion of possible quantitative analyses that could be performed. It is difficult for students to select the few analyses that are crucial.

7.3 Summary of Core Development Areas

In this work, we present our efforts to identify and solve problems in the teaching of BID, as embodied in the following five areas.

7.3.1 Content

In moving away from a lecture-based course to a problem-based collaborative learning environment, we need to balance between providing knowledge about biology, engineering, and design (content) and hands-on practice engaging in the BID process.

7.3.2 Representation and Tools

For students to find and learn about biological systems, to communicate that knowledge to people from different backgrounds, and to apply that learning to ill-defined engineering problems of their own making, we must equip them with tools that facilitate understanding and communication and focus attention on aspects of systems that are important for design.

7.3.3 Design Process

We have implemented several design process formalizations to contend with the special needs of a process focused on analogical design. As more experience is gained in teaching BID, we see similarities to and differences from more standardized design process approaches. One key difference is solution-based design: this process starts with solutions presented in natural biological systems and translates appropriate functions to solve design challenges in an inventive fashion.

7.3.4 Design Evaluation

Students produce conceptual designs in this course. Given the need to teach the process in 15 weeks, and the emphasis on student-identified design problems, building and testing a prototype to demonstrate feasibility is not possible. Nevertheless, even for conceptual designs, students must convince themselves, the class, and instructors that the design could work and would have some advantage over existing products. Throughout the semester, examples of quantitative analyses give students practice in addressing issues that often crop up in BID, such as scaling, materials selection, and environmental impact.

7.3.5 Interdisciplinary Teaming

BID can serve as a catalyst for innovation because of the mix of disciplines. But just throwing the students together would not lead to success. With different cultures, values, processes, and vocabularies, as well as different technical backgrounds, we have learned that a number of different teaming techniques are necessary to ensure proper communication, balance, and respect in a properly functioning team.

In the next sections, we document the challenges faced in each core development area.

7.4 Content

Given the multiple course objectives, balancing content is a difficult task. We must communicate a breadth of biology and engineering knowledge, accounts, and tools for design processes and facilitate interdisciplinary communication. Additionally, there must be sufficient time for the students to practice with the tools they have learned. We describe specific elements of content that we have identified to help meet those challenges and ways to maximize the effectiveness of this content to avoid overloading the students (Table 7.1).

7.4.1 BID Stories

To maintain student enthusiasm, we began every class with what we call a BID-wow story: an account about an exciting, innovative bioinspired design. These consist of examples such as: the whale fin inspired windmill blade (Miklosovic et al. 2004) which is more efficient, quieter, and able to capture wind energy at lower wind speeds; the slime mold that connected nutrient sources placed in a petri dish in the same pattern as major cities around Tokyo and grew a transport system as efficient as the Tokyo railway (Tero et al. 2010); the spacious, transparent cabins of the 2050 AirBus concept plane (http://www.airbus.com/innovation/future-by-airbus/) with a bionic structure mimicking bird bones to make planes lighter and stronger; the butterfly-inspired sensor that responds to different chemical vapors using the ordered arrays of iridescent scales to outperform existing nano-engineered photonic sensors (Potyrailo et al. 2007), or; the cat's eye retro reflector (Percy Shaw's patent No. 436,290 and 457,536) that reflects light back to its source with minimum scattering, similar to eye shine created by the tapetum of a cat's eye. These fascinating stories are told as soon as the class bell rings, encouraging the students to be in class on time, and keeping them focused

Table 7.1 Five course elements (down) pertaining to content that address the 9 key challenges (across)

Content	Searching for biological systems	Understanding biological systems	Identifying and understanding good design problems	Analogy mapping and transfer	Communicating across discipline boundaries	Communicating complex system knowledge	Teaming in an interdisciplinary environment	Maintaining equal engagement throughout the process	Evaluating designs
BID stories	X							X	
Case studies	X	X	X	X	X	X	X	X	X
Found objects	X	X			X	X		X	
Evolution	X	X							
Focused reading		X	X						

on the thrill of invention. While these news stories pop up frequently, they do not provide the details of the source of inspiration nor the process of transfer. For this, we turn to case studies.

7.4.2 Case Studies

Case studies presented by local experts provide information that can meet a variety of challenges from increasing subject knowledge to developing design skills and can be an integral part of any BID class. Locally, we have many to choose from, and the research described by familiar and respected teachers at one's home institute adds to the impact. Many of these bioinspired designs have taken years of research and development. As a result, the lectures given by the BID practitioners have a wealth of very detailed knowledge that students sometimes find difficult to absorb. Although the stories are all astounding and fascinating, what should/could a student get from this? One strategy is not to be concerned about content but to use these meetings to give the students the chance to meet and talk to the people behind the design, and we did that initially. We invited a parade of professors who used two of the 30 class periods (nearly 3 h in all), sharing their excitement about the process and product. This was great motivation, and students still rate these kinds of expert lectures as a favorite part of the class, but it did not teach the student "how to." Over the years, we reduced the number of lectures and the length (45 min plus time for discussion) and provided the following guidelines to the lecturer:

1. *Describe the key feature of the natural system that provided your inspiration.* In particular, we asked experts to focus on 3 things regarding their inspiration. What were the structures that come from the biological system? In this case, structure refers to the system components that perform the function of interest in the system. Why did this function help the organism survive? How did the organism achieve that function? This is the deep biological knowledge.
2. *Decompose the challenge you faced into its functions and describe the function that your design addressed.* What structures are needed for this function, what use is this function to humans, how do existing solutions achieve this function, and what are the limitations of existing solutions? This is the deep engineering knowledge.
3. *How did you translate the biology into the engineered design?* This is the design process. From this, we saw how analogical reasoning was a key element in this translation process.
4. *Provide the 3 best articles on your BID, one on the biological inspiration, one on the details of the specific biological mechanism of interest, and one on how the biological system were translated into an engineering design that worked.* This teaches the students how to read scientific literature.

This format is consistent with our emphasis on system components, interactions, and functions as tools to help students define effective analogies [see Sect. 7.6.6, Structure–Behavior–Function (SBF), and structured representation for BID (SR.BID)]. This narrative produces a balanced mix of both biological content and design process and leaves the students wanting to hear more about what the scientist–engineer did. There were always a handful of energetic students who asked a lot of good questions and we would have a lively discussion that spilled out into the hallway after class. One of the professors remarked: "I got more questions in this class than when I gave the same seminar to faculty and grad students in my discipline!" However, in our experience, these case studies (even when presented by an individual involved in the research) are not sufficient for the students to actually grasp the BID process, even when presented in the uniform way described above. Moreover, the journal readings and technical depth of the papers were clearly overloading the students with scientific publications that were difficult to read and understand. Despite this being a favorite activity of the students, we limit these lectures to 4–6 per semester, and instead, we focus on more hands-on learning strategies.

7.4.3 The Found Object Exercise

Observing, experimenting with, researching, and describing the functions of biological objects is a central curricular element that meets a variety of challenges, but is particularly well suited to increasing the ability to search biological systems and increasing interdisciplinary communication. Students are asked to go outside, find something in nature, play with it until something intriguing is noticed, then find an article that explains how the natural system works (Yen et al. 2011). Through this exercise, we want the students to reconnect with their natural surroundings, spend enough time interacting with nature to find something that is marvelous, and then deepen their knowledge by reading about it first in general biology/ecology/behavior texts that point to good articles. This develops a sense of connection to nature or biophilia (Wilson 1984). The objective of this part of the exercise is to figure out what search strategy to use to find the information needed for BID and how to get this information out of articles from the primary literature. We use their interaction with nature as the stimulus for deepening their biological knowledge base. In class, the team members share what they found and decide who has the best found object. That person tells the class about it using a succinct knowledge representation template (explained in the representations section). For a class of 40 with 8 interdisciplinary teams of 5, it takes the entire class period to do this. At the end, we review the 8 best objects and discuss whether enough was presented to understand how the system works. We found that when students became facile with the knowledge representation template, they could zero in on the key function of interest without getting caught up in all the other fascinating details inherent in complex biological entities. This helps the finders hone their analytical skills, focusing them on only the most salient features for design. This also helps the speakers to hone their communication skills, conveying the key principles that a biologist and an engineer need in order to see the value of the biological strategy of interest. Additionally, through active participation in the process, these exercises add a breadth of amazing local biological systems to each student's repertoire of biological knowledge, strengthening their appreciation for the local natural environment around them.

7.4.4 Perspective on Evolution

At the other end of the spectrum from the problem of providing sufficient depth, we have the problem of providing an overall perspective on BID. Engineering students can find the variety of biological organisms and functions to be bewildering. Biology students can have difficulty establishing and maintaining focus on biology in the context of design. We provide a lecture on evolution early in the course to help students gain a perspective on biology in the context of design. This lecture includes the concepts of common ancestry and convergent evolution, multi-function optimization, and local versus global optimization. This lecture has been given every year and receives consistently positive comments from students. It appears to help them understand differences between evolution as a design process and intentional design, which enables a more sophisticated view of how to search and evaluate potential biological solutions.

7.4.5 Focused Readings in BID

One of the nagging struggles of this design class is to provide the correct depth of information at the right time to students, and to do it without overburdening the students, or suppressing their motivation to continue reading throughout the course. Early iterations of the class asked students to find technical, academic papers for biological systems of interest. Considering that a student may do five found object assignments, five case study lectures, and must research up to ten biological systems, a requirement consisting of two documents per assignment results in a massive overload of technical documentation (up to 40 technical papers!). Adding requirements on top of the technical reading, such as formulating summaries or key questions for presenters only exacerbated the situation. The problem still remains: how do we ensure that students engaged in BID conduct deep explorations of a select few organisms, while getting a broad range of exposure to many and in such a way that given some biological system, they are capable of acquiring the knowledge on their own?

One means to address the breadth and depth issue was the use of a general purpose textbook, such as Vogel's (2000) Cat's Paws and Catapults, instead of technical papers prior to the case studies. The textbook is written to be broadly applicable, yet provides sufficient depth to highlight the key principles as well as the challenges of applying those principles. By aligning the themes in the textbook chapters with the themes of the case studies, we provide salient real-world examples to reinforce the reading.

The problem of finding and understanding deep technical references for a few systems, balanced against the need to understand, for example, found objects, remains an unsolved challenge in the class. Simply put, students give higher priority to generating exciting designs than to time-consuming deep reading.

7.5 Representations and Tools

As we have emphasized, BID requires students to find and learn about interesting biological systems, to communicate that knowledge to people with backgrounds different than their own, and to apply that learning to ill-defined engineering problems of their own choosing. In this fast-paced, novel context, students become easily overwhelmed and unable to identify clear learning objectives. Student presentations of biological systems found locally (found objects) in years 2006 and 2007 best exemplify these early struggles. When asked to summarize the most interesting aspects of found objects: (a) discussion is dominated by the structural details of biological objects; (b) students superficially associate a wide variety functions to the design; (c) though we emphasize mechanistic explanations, they are rarely provided; (d) if mechanistic explanations are offered, they are often provided by reference to a common sense analogy (often incorrectly); and (e) technical explanations employing terms from one domain are not understood by a majority of students from a different domain. In this context, what specific tools or representation strategies can help focus students on aspects of systems that are important for design? Table 7.2 below lists five that we have used in our course. In Sects. 7.4.1–7.4.4, we describe the first four. We postpone the description of the fifth, SR.BID, until Sect. 7.6.

7.5.1 Search Strategies

Since our initial classroom deployments, we realized that students are challenged with converting engineering-centric design problem language to the corresponding biological terms necessary to find biological systems in the external information environment. This is a particular problem when classes are dominated by engineers who have had little basic biology and are unfamiliar with potentially relevant biological systems. Starting in 2006, we offered three useful techniques for helping students to identify keywords that might lead to fruitful searches. These techniques are documented in (Yen et al. 2011), but in brief, we asked students to: (1) identify

Table 7.2 Four course elements (down) pertaining to representation and tools that address the 9 key challenges (across)

Representations and Tools	Searching for biological systems	Understanding biological systems	Identifying and understanding good design problems	Analogy mapping and transfer	Communicating across discipline boundaries	Communicating complex system knowledge	Teaming in an interdisciplinary environment	Maintaining equal engagement throughout the process	Evaluating designs
Search strategies	X								
SBF		X			X	X			
DANE and biologue	X	X		X		X			
Functional decomposition		X	X	X					

key functions of interest; (2), invert the function to reveal generic principles (e.g., heating and cooling both concern principles of heat transfer); and (3) identify extreme environments in which high-performing biological systems might be found. Each of these techniques provides one or more keywords that may be useful when browsing large collections of literature for biological systems. Regardless of the technique used, search remains challenging; students report that as much as a quarter of their design time is spent searching for information on biological organisms and functions (Vattam and Goel 2011).

7.5.2 Structure–Behavior–Function Analysis

As noted above, students have trouble articulating the properties and functions of biological systems in a way that facilitates abstraction and transfer of design principles. Our first iteration of this course in 2005 lacked any advice about representation, which diminished our ability to teach students how to transfer knowledge from one domain to the other. The basic problem is that both engineering and biological systems can be described in a variety of ways, which obscures the fundamental cognitive step of transferring biological mechanisms as solution principles.

In 2006, we introduced a single lecture on SBF analysis (Bhatta and Goel 1997; Goel et al. 2009), which is grounded in cognitive theories of systems thinking.

Structure–Behavior–Function

- Structure, behavior, and function form an abstraction hierarchy for systems thinking; behavior is an intermediate level of abstraction between structure and function.
- Structure specifies the components of the system as well as the connections among them. For example, the structure of the electrical circuit in an ordinary household flashlight comprises of an electrical battery, a light bulb, a switch, electrical connections among the battery, bulb and switch.
- Behaviors specify the causal processes occurring in the system. For example, the behavior of the flashlight is that when the switch on the flashlight is pressed, current flows from the battery to the bulb, and the bulb converts electrical into light energy.
- Functions specify the outcomes of the system. For example, the function of the flashlight is to produce light when the switch is pressed.
- Behaviors provide causal mechanistic explanations of how the structure of the system accomplishes its functions. For example, the behavior of the flashlight explains how its structure accomplishes its functions.
- A behavior of a system specifies the composition of the functions of its subsystems into the system functions. For example, the behavior of the flashlight composes the functions of its components—the battery, bulb, and switch—into the function of the flashlight.

- A subsystem or component of a complex system can itself comprise a system and thus have its own SBF model. Hence, SBF models of a system can have a hierarchical structure. For example, consider the system of the basilisk lizard, which is well known for its ability to run across water. If the function (F) of interest of the basilisk lizard is "run on top of water," one can consider the opposing limbs, tail, and wide flat feet as part of the structural (S). The way in which the feet move in opposition are counter-balanced by the tail, and how the feet slap the water generating lift, then extend down and back creating more lift, thrust and a pocket of air in the water, and are then withdraw up and out through the air pocket could be considered the behavior (B) that generates the "run on top of water" function. One could consider the muscular-skeletal system of the legs as a subsystem of this system used to create a subfunction "generate movement of legs" which causes the higher-level "run on top of water" function. Likewise, one can consider the form the foot takes throughout the process as another subfunction, "change foot surface area." In this way one can decompose the "run on top of water" function into a number of subfunctions, including "generate movement of legs" and "change foot surface area," each of which could entail another SBF model. Similarly, one can consider the function "run on top of water" to be part of the function "escape predator" showing that one can navigate both up and down the levels of functional abstraction in the SBF model hierarchy. This kind of hierarchy will be discussed further in Sect. 7.5.4.

The origin of SBF analysis lies in Chandrasekaran's functional representation scheme (Chandrasekaran 1994; Chandrasekaran et al. 1993). Other researchers have developed similar cognitively oriented approaches to thinking about complex systems, for example, Rasmussen (1985). Gero and Kannengeisser (2004) describe the design process itself in terms of function, behavior, and structure. Erden et al. (2008) provide a recent review of functional modeling. Note that in SBF analysis, functions are mental abstractions chosen by the modeler, and not intrinsic to the complex system. In the case of engineering systems, a functional abstraction corresponds to an intended output behavior of a system, subsystem, or component. However, since functions are mental abstractions, we can also use SBF modeling to model natural systems, including biological systems, such as the human heart, and ecological systems, such as forests. Even more so than engineered systems, natural systems exhibit layers of varied functionality at different scales, feedback loops, and other types of causal processes that characterize complex systems.

Recently, there has been considerable interest in the use of SBF modeling in science education. Goel et al. (1996) proposed the use of SBF models for explaining complex systems in science education. Ebert-May et al. (2010) and Speth et al. (2011) found that construction of SBF models in college-level courses helped expose students' misconceptions of ecological and biological systems, respectively. Chan et al. (2010) found that high-achieving students in a college-level course on biomedical engineering paid more attention to behavior and function than did low-achieving students, and that attention to behavior and function improved student performance. Helms, Vattam and Goel (2011) found that SBF models of biological systems enable complex inferences that were not readily enabled by textual or diagrammatic representations of the systems. Vattam et al. (2011) discovered that use of SBF modeling for learning about ecosystems in middle school science classes resulted in statistically significant improvement in students' understanding of the structures, behaviors, and functions of aquatic ecosystems. Silk and Schunn (2008) summarize some of the benefits of SBF analysis in science education.

In 2007, we introduced SBF analysis as a framework for organizing found object exercises. Students were asked as part of the found object homework assignments (Sect. 7.4) and in their discussions to (a) focus on a single *function* of the organism in question, (b) identify the *structures* relevant to accomplishing that function, and (c) provide a *behavioral* explanation for how those structures give rise to the function. Instructors facilitated these discussions as necessary to guide students. (In the SBF vocabulary, behavior is synonymous with causal mechanistic explanation.)

As expected, students discussed structure at length, although they were unable to limit themselves to the discussion of a single function. As noted earlier, SBF is a hierarchical representation and systems are naturally functionally hierarchical. As a result, it was difficult for students to maintain a single level of functional abstraction during their discussions. Often students travelled "up" the functional hierarchy attempting to explain *why* the organism performed the function in question such as reproduction, survival, and escape from predators. The result was discussions about many high-level functions that lacked in detail. Less frequently, students travelled "down" the functional hierarchy, explaining a small portion of *how* the organism performed a function. These discussions usually resulted in very detailed, technical low-level discussions that only a few students could follow. One must continually emphasize to the students that, while the number of levels in a decomposition is very large, functions expressed at much lower or higher levels than the original problem may not always be useful for the purpose at hand, because they introduce constraints (lower levels) or goals (higher levels) not present in the initial problem definition. In addition to traversing levels of abstraction, students frequently confused the different senses of the word "behavior." Students often associate behavior with higher-level actions at the organism level, for example, mating behavior, territory marking behavior, seeking shelter from the heat rather than addressing the causal mechanisms, as this word is used in the cognitive sciences (Gero and Kannengeisser 2004; Goel et al. 2009).

To simplify the vocabulary, in 2008, we changed the SBF vocabulary to a What-Why-How vocabulary, mapping "What" to "Structure," "Why" to "Function," and "How" to "Behavior." This was an attempt to both remove the ambiguous interpretation of "behavior" and to formalize the levels of functional abstraction. Functional abstraction was considered in terms of "why" moving up the hierarchy (more abstract, superfunctions), and "how" moving down (more detailed, subfunctions). Again, students were asked to describe all biological systems in these terms, both conversationally and in formal homework assignments and design reports. Despite removing potentially confusing SBF language, students continued to describe these systems in a way that enabled them to avoid providing a mechanistic explanatory account. Table 7.3 gives an example from a midterm presentation in 2008.

It is illuminating to characterize the failure in Table 7.3 as a traversal in hierarchical levels. A *hierarchical level* corresponds to the vertical location of a function in a problem decomposition. Going down one level means to think about the subfunctions of the function under focus. Going up one level is to consider to what the focal function directly contributes. For example, if irrigation is the focal function, then acquiring, transporting, and dripping or spraying water would be subfunctions one level below. Sustaining plant growth would be a function one level higher, perhaps at the same level as harvesting; feeding the hungry would be a function considerably higher in the hierarchy.

In the case shown in Table 7.3, we see "What" addressing a function, "Why" addressing a higher-level function, and "How" addressing a structure, in this case color patterns. However, ideally, the "What" would address the components of the solution (e.g., structural color patterns), the "Why" would address the functions of solar absorbance and energy capture, and the "How" would explain the mechanism by which structural color patterns cause solar absorbance and energy capture.

The example in Table 7.3 uses "Why" to capture a function several hierarchical levels higher than the one that is really being considered. That is, while maintaining body temperature may be the top-level function or goal, it is several levels displaced from energy absorbance. This suggests an overloading of the term "Why" as both "the function of interest" and "the reason for the function of interest."

In our experience, students need a large amount of practice with these representations to employ them correctly. As a rule of thumb, we have found that restricting the analysis to one to three hierarchical levels above or below the "What" function is useful to focus the student's attention on the right structures, functions, and mechanisms. Levels above this cutoff often take the students to the ultimate evolutionary objective of a given biological "solution," which may not match the engineering problem for which a given function may be useful. For example, the ultimate evolutionary objective, to survive, is so universal that it gives no additional guidance in the search for connections between biology and engineering. Going too many levels down may introduce constraints specific to the particular way the biological function is achieved, and which may not be relevant if the goal is to abstract the function rather than copy precisely the mechanism.

Table 7.3 Using What–Why–How vocabulary fails to generate a mechanistic explanatory account

Iridescent butterfly wings	
"What"	Solar absorbance and energy capture
"Why"	To maintain body temperature
"How"	Structural color patterns

Despite student problems with correctly identifying answers to the "why," "how," and "what" questions, we have found the SBF schema helpful because it helps students ask useful questions in trying to understand complex biological systems. Such hierarchical analysis is used to decompose complex concepts into manageable pieces of information. Thus, SBF has become part of the language of discourse in the BID class.

7.5.3 DANE and Biologue

We also have experimented with interactive tools that use SBF models to help enhance student understanding of biological and engineering systems. Given the importance of knowledge representations and interactive tools, it is not surprising that recently there has been enormous amount of work on devising representations and tools to support BID (Biomimicry 3.8 Institute 2008, 2009; Bruck et al. 2007; Chakrabarti et al. 2005; Chakrabarti and Shu 2010; Cheong et al. 2011; Chiu and Shu 2007a, b; Nagel et al. 2008, 2010; Sarkar and Chakrabarti 2008; Sarkar et al. 2008; Sartori et al. 2010; Shu 2010). These tools differ in their representations of biological and engineering designs, strategies for searching for a biological solution potentially relevant to a design problem, the (implied) process of BID, evaluation of design solutions, and so on. However, these representations and tools for BID are normative and prescriptive. We believe that it is important to situate the development of representations and tools in real-life contexts. The BID course has provided a motivation and a context for using, often in new ways, existing knowledge representations such as SBF, and also for developing and evaluating new representations such as SR.BID (see Sect. 7.6.6) and new interactive tools such as DANE (Goel et al. 2012; Vattam et al. 2010b) and Biologue (Vattam and Goel 2011) (see below).

DANE provides a digital library of SBF models of biological and engineering systems, as well as tools for constructing SBF models of new systems. We introduced DANE into the BID class in 2009. Some students in the BID class found DANE useful for making sense of complex biological systems and constructing a conceptual understanding of the systems. (DANE can be downloaded from <http://dilab.cc.gatech.edu/dane/>.)

Biologue enables students to annotate and share documents on biological systems as a team, to tag the documents with SBF models, and to search for additional documents based on SBF tags. We introduced Biologue into the BID class in 2012. Some students in the BID class found Biologue useful for online annotation and sharing of biology articles. In controlled experiments, we discovered that Biologue enables subjects to more easily and accurately locate relevant biology documents online (Vattam and Goel 2011). While initial results from DANE and Biologue are promising, identifying ways to search for interdisciplinary analogies remains an open research area (as this volume attests).

7.5.4 Functional Decomposition

In addition to the SBF analysis introduced in 2007, we began introducing decomposition diagrams in the class. Our early cognitive studies (Vattam et al. 2007) provided some evidence that analogical matching between problems and biological solutions was taking place functionally, but implicitly so. That is to say, we had no formal method of determining the quality of a match between the problem and the biological solution identified to help solve it. If we could formalize the decomposition of both problem and solution functionality, we would provide a more formal method for making and evaluating the connection between problem and solution.

In 2007, we introduced functional decomposition (Dym and Brown 2012; French 1996; Pahl et al. 2007; Simon 1996). An interactive lecture on problem decomposition was provided where the class participated in a group decomposition exercise for designing a search and rescue vehicle that could walk on uneven and shifting surfaces, such as sand. Figure 7.1 shows the decomposition that was created interactively with students during that lecture. Assuming functional matching was the primary index used to retrieve biological solutions from memory, a diagram such as this should provide a number of functions, each of which may lead to an array of biological solutions that could be applied to solve one or more of the subfunctions identified. Thus, multiple biological solutions could be used in solving a single problem. This phenomenon, termed compound analogical design, is well documented in class (Helms et al. 2009; Vattam et al. 2010a, b).

Students were asked to provide similar "solution-neutral" decompositions of their problem for all presentations and reports, as well as to justify their analogies by matching the functions provided by a biological solution and the function in the decomposition. Figure 7.2 is a typical example of a student's problem decomposition. We found students consistently tailor these decompositions to the biological, and sometimes technological, solutions that they are already considering. That is, the functions that appear in the functional decomposition are nearly always aligned directly with a system that the students are considering. It is unclear whether this is selective pruning of the decomposition so that there are no functions for which there is no solution, or whether the decomposition is a result of a bottom-up approach where students fit functions from solutions to the problem decomposition. This bottom-up composition suggests that functional decomposition of a problem is not only formed in a top-down manner, but also may be partially formulated based on the availability of solutions.

7.6 Design Process

Students in BID class are asked to invent their own design challenge and to generate a creative, biologically inspired conceptual design that solves that challenge.

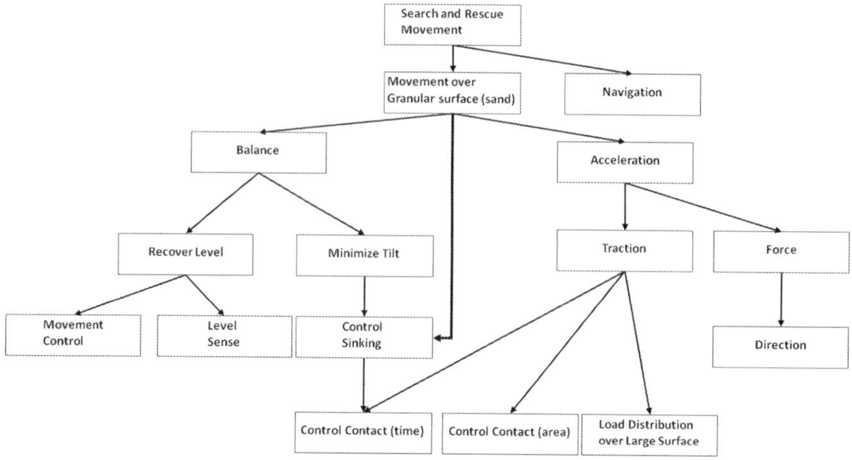

Fig. 7.1 Problem decomposition of a search and rescue vehicle to transport over uneven and unstable ground

Undergraduate students enrolling in a BID class may enter with little or no formal design process training. Even for engineering students, design is often in the context of a problem with very specific functional requirements, that is, the problem and evaluative criteria often are very clear. It is important to monitor the typical design experience of the student pool to determine how much to coach students through the process, particularly during problem definition.

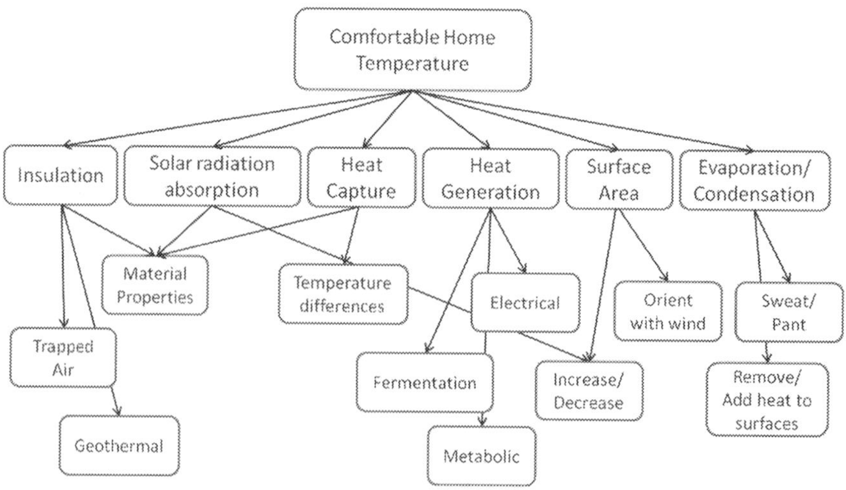

Fig. 7.2 Decomposition of the problem of maintaining a comfortable home temperature

Generating a new design problem, as well as learning a process for solving design problems adds an additional level of complexity to the design process. In early iterations of the class, we used a typical problem-driven design process, but we recognized that student designers spontaneously adopted a second design process, solution-based design. After close study of this process, we recognized that not only was the second process useful for students, but it also appeared to represent the more successful modern design approach outside of the classroom (Vattam et al. 2007). Beginning with the recognition that we needed to teach not one, but two design processes, we implemented a number of strategies for assisting students in structuring their design process (Table 7.4).

7.6.1 Problem-Driven and Solution-Based Design

Perhaps the most significant reworking of the class involved the organization of the class into sections representing these two dominant process modes for BID. Initially, we instructed students to find a problem, find a biological source, and apply the source to the problem to generate a solution. With regularity, half of the design teams would follow the problem-driven approach, and half of the design teams would instead fixate on an interesting biological solution and then find an appropriate problem to solve. Since each process seemed useful in different circumstances, we decided to formalize the different approaches and allow students to experience both.

On the very first day of class, students are now introduced to dozens of interesting biological systems in our "biology auction" exercise. The auction engages student's curiosity and imagination with a wide range of possible biological systems that can serve as design inspiration, either directly or indirectly. In addition, the immediate emphasis on biological systems reinforces the validity of biological knowledge and engages the biologists.

Over the next 6 weeks of the class, students identify one interesting biological system and figure out a means for using the interesting principles of that solution to solve a human-scale problem. We teach this process in class more formally as solution-based design and scaffold the process with exercises meant to help students (a) understand the mechanism of interest in their biological system, (b) abstract the mechanism used in their system, (c) identify a number of problems for which their system may provide a solution, and (d) formally analyze the analogy between their system and the problems they propose to solve in order to identify the best solution-problem match. Only in weeks five and six are students asked to begin producing conceptual designs.

We institute a more compressed problem-driven design cycle during weeks seven through ten. This begins with students: (a) defining a problem; (b) abstracting the problem; (c) finding biological solutions to the abstract problem; and (d) formally analyzing the analogy between their problem and the solutions they propose will solve their problem. Students craft a design solution in the last

Table 7.4 Six course elements (down) pertaining to the design process that address the 9 key challenges (across)

Design process	Searching for biological systems	Understanding biological systems	Identifying and understanding good design problems	Analogy mapping and transfer	Communicating across discipline boundaries	Communicating complex system knowledge	Teaming in an interdisciplinary environment	Maintaining equal engagement throughout the process	Design evaluation
Problem-driven and solution-based processes									
Source breadth	X								
Problem definition			X						
Problem focus		X	X		X				
Project format									
Analogical evaluation and SR.BID				X	X	X			X

week of this second cycle. This process of problem-driven design is an instantiation of the more basic cognitive process of analogical reasoning (Clement 2008; Dunbar 2001; Gentner 1983; Gick and Holyoak 1983; Goel 1997; Hofstadter 1996; Holyoak and Thagard 1995; Keane 1988; Kolodner 1993; Nersessian 2008). Solution-based design appears new and different from the perspective of design theory, all of which is problem-driven (e.g., Dym and Brown 2012; French 1996; Pahl and Beitz 1996). Thus, the BID course acts as a research laboratory for developing, identifying, and studying new BID constructs and processes.

With the remaining time, we allow students to continue to develop their ideas for either the first or second project. This allows students to experience both solution-based and problem-driven processes, while balancing the need to get deep into the design, permitting students to deal with the complex problems that come with more detailed design.

7.6.2 Source Breadth

The challenges associated with defining a problem, and finding and understanding biological systems, often result in students exploring few potential biological solutions (Wilson et al. 2010). This compounds the tendency of all novices to engage in design fixation. Students in early iterations of the course were required to explore only one problem and solve the problem with one or more biological solutions. Student designers in this environment investigated between two and ten biological solutions, while applying one or two biologically inspired mechanisms to solve the problem. However, in about two-thirds of the design projects, students fixated on the first biological system they encountered and only superficially explored other systems. Thus, in early classes student design teams shallowly investigated a handful of biological systems and came to deeply understand one or maybe two biological systems.

To counteract the effects of solution-fixation, student design teams are now required to report on at least thirty biological systems throughout the course. Each student examines a minimum of five biological sources before selecting one for their solution-based design, which means for a team of five, the entire team learns about 25 biological systems. These systems can be related to each other via convergent evolution, phylogeny, or exaptation. Furthermore, during problem-driven design, each student is required to report on five biological sources, or 25 natural systems for the team. In their final team design reports, a deep analysis is required of at least five of these systems. These systems may or may not overlap with systems discussed in their five found object exercises, again including up to twenty-five additional systems per team. Thus, student teams may explore and share knowledge about as many as *seventy-five biological systems* over the course of a semester. Furthermore, because students are trained in formally representing these systems using the function-based representation tools we provided (see Sect. 7.5), exploration of these systems is a structured process that (in theory) allows students to functionally index each system in their own memory for later retrieval during design episodes.

7.6.3 Problem Definition

In a semester-long design project, students (who again, are largely naive with respect to open-ended design problems) spend approximately half of the semester learning about and grappling with the complexity of their own design problems. Students seem to be motivated by tackling complex, often topical, issues such as oil-spill cleanup or eliminating traffic congestion, for which they may have little familiarity. In a class where students are expected to learn an incredible breadth of content and process knowledge, our task is motivating students to find interesting, challenging problems, without letting the definition of the problem itself become the core challenge.

Problem discovery and definition is usually the first step in the design cycle (Dym and Brown 2012; French 1996; Pahl et al. 2007). Even when the design cycle is solution-based, problem definition is quickly derived by working backward from a potential solution. Moreover, problem definition is inherently iterative. We have found, for example, that 70 % of the function requirements considered during the semester are discarded by the final design and as many as one-third of the final function requirements were identified during the final few weeks of a semester-long design.

We provide three scaffolds for students to help with problem definition. First, we give a lecture early in the semester that is inspired by Ron Bills, the CEO of Envirofit, entitled "What makes a problem a good problem?" this lecture provides an answer in terms of three W's: what stinks, who cares, and what are you going to do about it? This lecture sticks in the students' minds and keeps their focus on the practical. To reinforce this perspective, student design teams are required to answer the three W's during their preliminary design evaluation. Second, we provide the problem-structuring tool (SR.BID) described in Sect. 7.6.6. These tools provide a handle for students to gain traction on defining their problems. Third, we embed problem definition formally in many of the design assignments, forcing students to reflect, to explicitly represent their design problems, and come to a shared team understanding of them. Our observations in 2011 suggest that these interventions: (a) reduce the time students spend on problem development; (b) reduce the problem scope; and (c) enhance the level of detail, especially with respect to the number of performance criteria and specifications that are considered. We believe projects in 2011 were among the most practical designs produced since we began teaching this class, with no sacrifice in perceived creativity.

7.6.4 Problem Focus

The extent to which student problems should be defined by the instructor versus by the students involves some trade-offs that may strongly affect student performance. We have tried limiting problems to specific areas (for example, sustainable housing) and allowing students to choose their own problems. Constraining student problems to specific areas potentially allows student groups to share information and come to a deeper understanding of the problem as a result of their joint efforts. It also may ameliorate some of the difficulties associated with problem formulation, and identifying and understanding biological systems. However, many students express disappointment that they were not allowed to choose problems with which they were comfortable. Moreover, constraining the problem sometimes limited the opportunity of certain engineering disciplines to participate. We also learned that unless problems are highly constrained, student teams were able to find a wide variety of problems such that benefits of shared knowledge were weakened considerably. Ultimately, in our highly interdisciplinary environment, allowing students to self-identify problems leads to stronger application of engineering knowledge and student motivation and is preferred even though problem definition remains challenging. Constraining problems to certain domains may be more productive when students share a greater amount of disciplinary knowledge or attitudes.

7.6.5 Project Format

The project format can serve a variety of goals, some of which may be important, but are ancillary to BID practice. For instance, it is common practice at Georgia Tech for instructors to require capstone design projects to be sponsored by industry partners, where industry partners then can participate in the project as real-world customers. We initially framed the project format in an entrepreneurial context, that is, student design teams were expected to pitch their designs to appeal to a group of venture capitalists. While students found this quite motivating, they also spent a lot of time on branding, marketing, and selling their concept, rather than understanding and articulating the underlying principles. Still, the emphasis on design feasibility is important. Thus, we currently frame the design process in a more pragmatic sense. We ask students to prepare a presentation that validates the design concept in a way that would convince us (the instructors, and other guest evaluators) to invest in creating a prototype of their design.

7.6.6 Analogical Evaluation and SR.BID

Since BID is a cross-domain activity between biology and engineering, analogy making is a cornerstone process of BID but formalizing instruction for analogical mapping continues to challenge us. Students asked to find a biological analogy for a problem like "create a device for collecting water samples at a fixed depth underwater in a lake," can almost immediately produce answers like "puffer fish," "pelican," and "whale." Remarkably, students generate these analogies naturally, without instruction, and within minutes given a particular design challenge. When asked to describe why the analogy is a good fit, however, students require much more time and often are at a loss for a description that they themselves find satisfactory. When asked to evaluate these analogies for applicability or "goodness of fit" for generating a solution to the design problem, students can often categorize one analogy as better than another, but they are generally incapable of producing a consistent rationale for why one is better than the other. These observations have led us to seek better ways to focus student attention on appropriate analogical mapping and evaluation of goodness of fit.

Students seem to use more than functional similarity as memory probes for arriving at appropriate analogies. Similarity of structure, external environment, and performance characteristics too are often involved; for instance, the search technique of "finding extreme adapters" is used to find high-performing biological solutions by environmental similarity of problem–solution pairs. Capitalizing on these findings, we introduced in 2011, a new representation framework called Structured Representation for Biologically Inspired Design (SR.BID) that extends and expands the SBF representation. The core aim of SR.BID was to create a comprehensive representation linking biological solution descriptions and problem descriptions over the broader range of concepts used for analogical indexing, mapping, and evaluation. The SR.BID framework borrowed the structure (called specifications in SR.BID) and function concepts from SBF, and appended environment and performance concepts noted earlier. The "Four Box Method" uses the four concepts pictured in Figure 7.3 to organize student thinking about problems and solutions.

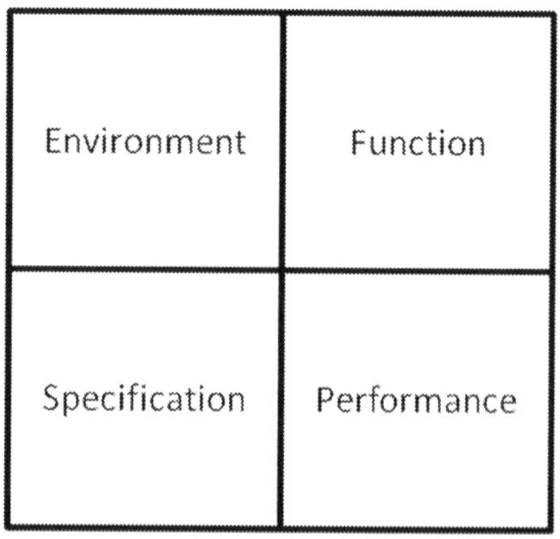

Fig. 7.3 Four Box Method of SR.BID

Table 7.5 Using the Four Box Method to describe the problem of building a better bike lock

Operating environment	Functions	Performance	Specifications
Outdoors and indoors	Protect bicycle from theft	Weigh less than 5 lbs.	Adjustable
Wind and precipitation	Prevent potential damage to bicycle caused by contact with lock	Withstand 4500 lbs. of force	Weatherproof
Daytime and nighttime		Withstand temperatures below 32° F and above 100° F	Flexible
Temperature variations		Waterproof, freeze proof, and shockproof	Strong materials
			Easy to use
			Deter from cutting
			Competitively priced

All assignments in the 2011 class were structured using the SR.BID framework. The following Table 7.5 provides an example of a four box diagram provided to describe the problem of building a better bicycle lock.

Furthermore, once given a problem specification and a specification of the biological solution in the same format, students were asked to do a side-by-side comparison where they identified whether elements were the same, similar, or different. In Table 7.6, we see a side-by-side comparison of the North American Elk antler with the bicycle lock problem. This side-by-side comparison forces students to consider not only where the problem and solution align, but also where solutions do not line up. Markman and Gentner (1993) have suggested that comparison of source and target problems in analogical transfer often is based on such alignment. We believe that in BID, highlighting these differences early in the analogical mapping stage helps students consider their sources more deeply, as well as identify potential transfer issues such as performance, size scaling, material composition, and manufacturing earlier in the design process.

From 2006 through 2011, we have experimented with a number of representations and tools to help students overcome common interdisciplinary design challenges. We have less experience with SR.BID than with other representations. However, student surveys indicate that the SR.BID organizational framework resulted in more pragmatic final projects and provided students with a more robust method for evaluating analogies, especially for identifying potential transfer failure points. On the other hand, students continued to provide shallow mechanistic explanations of biological systems. In 2011, we did not teach diagrammatic functional decomposition, and instructors felt as a result students lacked a deeper understanding of the connectedness of functions in both problem and solution descriptions.

Table 7.6 Side-by-side comparison of the biological solution (Elk antler) with the problem (bicycle lock) using SR.BID

Problem target		Biological source (Elk antler)
Operational environment		*Operational environment*
College students/adults	N/A	N/A
Global use in all habitats	Different	North America/Eastern Asia/forest habitat
Bike racks, poles, sign posts, fixed structures	Different	Elk head
Usable in all seasons/usable at all times of day	Different	Usable only during mating season
Temperature range: adaptable to outside temperature	Same	Temperature range: adaptable to outside temperature
Weather resistant	Same	Weather resistant
Functions		*Functions*
Prevent bike theft	N/A	Fight/protect against other male elks during mating season
Withstand applied stress	Same	Withstand applied stress
Maintain temperature	Same	Maintain temperature
Deter possible thieves	Similar	Deter predators
Specifications		*Specifications*
Stress withstanding materials	Same	Stress withstanding
Lightweight materials	Different	Strong materials
Inexpensive materials	Similar	Relatively low energy cost
Inert materials	Same	Inert materials
Detachable from bike	Similar	Ability to shed antlers after the end of mating season
Lifespan of over 5 years	Different	Lifespan equals the duration of mating season
Criteria		*Criteria*
Weight <5 pounds	Different	Weight of up to 40 pounds
Fits around average sized tree trunk	Different	Height of up to 3.9 feet
Fits on/around average sized bike frame	N/A	N/A

7.7 Evaluation

One of us once overheard an alumnus from our 3rd iteration of the course describe it to a prospective student. He said, "It's different from any other course. There are assignments like 'go outside and find something.' It's hard to know exactly what you're supposed to do." In most courses, it is clear to the students what specific information and skills they must master. Because BID education is process oriented, as opposed to content oriented, the students often have trouble gauging their own performance, particularly before their first projects are vetted. Evaluation by the faculty is necessary throughout, and most important quite early, to help students realize what it is they should be working on, how good their work is, and what mental activities are leading to productive outcomes. The first two course

Table 7.7 Six course elements (down) pertaining to evaluation that address the 9 key challenges (across)

Evaluation	Searching for biological systems	Understanding biological systems	Identifying and understanding good design problems	Analogy mapping and transfer	Communicating across discipline boundaries	Communicating complex system knowledge	Teaming in an interdisciplinary environment	Maintaining equal engagement throughout the process	Evaluating designs
Three W's			X		X		X	X	X
In-class feedback			X						X
Environmental impact assessment								X	X
Make-or-break QA									X
Materials assessment									X
Reports		X	X	X		X		X	X

elements in Table 7.7 below are the main methods we have found to get students on a productive track early in the course.

Ideas are commonplace; good ideas less so. Many of the ideas in the student journals were novel, but impossible to implement so as to achieve the desired functionality. Quantitative analysis is usually the key to assessing feasibility of a design. For example, how much must the surface area of a shoe expand to prevent sand from liquefying when an adult walks at normal speed? How much will serrations at the leading edge of a lawn mower blade reduce noise? Course elements 3–5 in Table 7.7 represent the quantitative assessments we require. In addition, we ask the students to perform some quantification of the three W's.

The last course element in Table 7.7 is the final report. This pulls together the biological sources, problem description, design description, analogical evaluation, and all of the quantitative analyses that are described in this and the previous sections. If the students can write a persuasive project summary and have correctly performed the underlying analysis, they can feel confident that they have delivered a good BID.

7.7.1 Three W's

The three W's, "What stinks," "Who cares," and "What are you going to do about it," were introduced in Sect. 7.6.3 as scaffolding for student problem definition. For the oral presentations of the first two projects, we require the students to state the three W's of their problem definition. This helps them select a worthwhile and well-defined problem. For the final reports, we also require quantitative justifications of each W. For example, the first W would ask how wasteful are lawn sprinklers compared with drip irrigation? The second W would ask how much clean water is wasted annually by lawn sprinklers and of what fraction of total clean water use does that consist? A more thorough answer to the second W would calculate the annual cost of the wasted water to a typical owner of a water sprinkler. If the annual cost is a few dollars, who is going to care, even if the overall cost is a hundred million? The third W would call for the quantitative analysis of the design to be sure it saves the amount of water claimed, and an estimate of the production cost.

7.7.2 In-class Feedback

During in-class work sessions, we circulate among groups, answering questions, critiquing designs, helping with analyses, and suggesting ideas. We have not kept records of these interactions, but we are sure that this feedback is indispensible to the students during the first and second design projects. We often have the ready knowledge to tell a group that an idea has already been tried, or that an organism's mechanism is not what they think it is, or that the basic nature of their problem is different from what they suppose. This kind of feedback helps eliminate dead ends early, before the team sinks much time into them.

The other kind of feedback that is very helpful during the early stages of work has to do with problem focus. Students frequently begin with too broad a problem and need to be advised to narrow their focus, often drastically. It has become much easier for us to provide this feedback now that we have taught the course for several years, because we have acquired some problem domain knowledge. For example, every year since 2006 at least one team has wanted to solve the problem of water. We have learned that worldwide water problems range from aquifer depletion, desertification, inefficient irrigation, and leaky toilets to collection, non-point-source pollution, filtration, and millions of children's deaths annually. Each of these differs by geographical region, culture, and other factors. We might suggest a focus on collecting potable water from the air for a hundred thousand refugees living in makeshift tents in Haiti, or on finding gray water alternatives to pure aquifer sources for farmers in the midwestern USA.

Occasionally, a group will have too narrow a problem focus. If their solution is good, it is usually enough to point out that there is not a sufficiently important "who cares," and urge the students to find a broader scope of application. Therefore, it is not usually vital to detect this flaw very early. Feedback during in-class presentations, discussed next, is sure to reveal such flaws that have not yet been detected.

We invite experts from various departments such as mechanical engineering, materials engineering, architecture, chemistry, psychology, and civil engineering, as well as local firms such as Perkins+Will, Interface, and David Oakey Designs and that are interested in sustainability, to attend the oral and poster presentations of the student projects. These presentations are typically given a week or two before the final project reports are due. Each team gets feedback from the visiting experts, the course instructors, and their fellow students. Surprisingly, we have found that some of the toughest questions come from other students. The visitors are the most likely to challenge fundamental assumptions or parameters of the entire project. We instructors, perhaps because we have been providing feedback all along, tend to ask the least unsettling questions. Instead, we usually probe to test whether or not the students have a deep understanding of the biologically inspired mechanism that is being transferred into the design.

Several times visitors have expressed regret that they had not been brought in earlier, because there is not enough time for the student team to act on their criticisms or ideas. On the other hand, these visitors are a scarce resource and we are leery of imposing too much on them. The best use of this resource seems to be during the presentations of the first and second designs, because the teams will choose one of those two to refine for their third design and therefore have several weeks to take an expert's comments into account.

7.7.3 Environmental Impact Assessment

Student responses to our course suggest that BID captures the imagination and attunes students to values of sustainability. In fact, many engineering students in our early courses reported they were more likely to consider sustainability and environmental impact of their designs as a consequence of learning BID, even though sustainability was not a design requirement. Subsequently, we added an environmental impact assessment (EIA) assignment to align student output with their greater sensitivity to environmental concerns.

The EIA assignment creates a number of challenges, given many engineering and biology curricula do not cover this kind of evaluation. We discovered it was necessary to familiarize students with the major environmental impact categories and their associated metrics (e.g., greenhouse gases in CO_2 kilogram equivalents and solid non-toxic waste in cubic feet). We identified some of the most common difficulties and created a quantitative homework assignment that forced the students to navigate them. The assignment was to compare the environmental impact of travel by air and travel by car. This forced students to understand the need to express the cost per function achieved (e.g., amount of CO_2 equivalents released per passenger miles travelled), and how to prioritize potential costs (e.g., the amount of clean water used per passenger mile is negligible compared to the impact of greenhouse gas emission). Afterward, when teams were working on their projects, we met with each group to discuss which impacts were important and how they were to be measured.

Several of the changes that we have described, namely identifying pitfalls and environmental impact categories in lectures, tailoring quantitative analysis assignments to these lectures, and discussing these issues with each team while they were developing their designs, had the net effect of changing quantitative assessment from a burdensome requirement of a final report to a key tool used during much of the design process.

7.7.4 Make-or-Break Quantitative Analysis

In the first three years of the course, we gave three quantitative homework assignments, each analysis tied to a specific reading or lecture. Our aim was to stimulate students to perform quantitative evaluations of their projects. These assignments were unpopular; many students, especially biologists, found them difficult. Starting in the fourth year, we changed these from individual to group assignments. To keep the biologists engaged, we offered extra credit to teams if a biologist presented the group's solution to the class. We observed that the quality of the student solutions improved, and that the student satisfaction with the assignments increased when the design team was jointly responsible for the exercise.

However, the degree to which all team members, in particular the biologists, learned how to do quantitative analysis is unknown.

Though we do not know whether everyone learned *how* to perform quantitative analysis, we do know that the students did not learn how to choose *what* quantitative analyses were worth doing. All final design reports were supposed to include a quantitative assessment that related to how well the design functioned. In the first few years, we were usually dissatisfied with their quality. Many assessments lacked depth or importance. Teams frequently analyzed aspects of the design that were not critical to its performance, choosing analyses with straightforward techniques as opposed to relevance. We elected to address this problem with a "make-or-break" lecture, in which we stress that usually there is a single quantitative issue *of function* that is critical to the success of the design. A bicycle helmet must be able to protect against a certain speed of collision; a condensation device for desert use must produce a certain amount of water per day; a levee must withstand a certain flood height. We told each team to figure out what would make or break their design. We then met with each team to discuss their choice. This discussion was important because otherwise a difficult time-consuming technical analysis could turn out to be irrelevant or a major design infeasibility could go undetected.

Our subsequent experience has led us to identify common issues that students confront in this analysis. One pitfall has to do with scaling. For example, a human-sized gecko could not climb walls easily because the mass increases as the cube of the length, but the surface area of the foot increases only as the square of the length. The adhesive force is proportional to the surface area, as a simple thought experiment will show. We created a new quantitative homework assignment for which scaling was the key. Since biological solutions often are scale-dependent, students often will have to deal with this issue, and some discussion of scaling seems key for successful analogical transfer of principles. The other common pitfall had to do with materials. This was so important that we made a materials assessment a separate requirement, as described in the next subsection.

7.7.5 Materials Assessment

Students in the first 3 years of the course would often base their design on a hypothesized material with certain physical properties, when no such material existed. When we reviewed the course after 3 years, we were a bit shocked to see that this single weakness rendered about one-third of all the designs infeasible! We began to warn students not to rely on imagined materials, encouraging them to use existing material or to design a hybrid material from known materials. Now, we require a materials analysis in the second or third week of the third project. The final reports typically incorporate this materials analysis.

Often the properties of a material have turned out to be the "make-or-break" quantitative question. In several cases, teams performed a computation-intensive finite-element analysis to answer the question. Usually, only one member of the team, a mechanical or materials engineer, knew how to perform such an analysis.

In our experience, therefore, a materials analysis seems necessary to prevent situations in which students produce unfeasible designs. In the most recent iteration of the course, fall 2012, only one of the eight final designs (a radically different toothbrush) depended on material of dubious manufacturability.

7.7.6 Reports

We have always required students to deliver both oral and written reports. In the first year, we tried different formats. For written reports, we asked for traditional write-ups of about 10 pages, posters, and pamphlets of 4–8 sides. For oral reports, we asked for either short poster presentations or PowerPoint presentations. We observed that students were highly motivated by poster presentations, and we have retained them. We found that written reports got much better if we specified a template in advance and tied all of the elements in the template to previous assignments. In this way, final reports served as a reflective synthesis of previous work and provided an opportunity for improvement. Report templates also provided students with focus. There are so many aspects of the process of BID that could be included in a report that students are at a loss for what to include or not include. In particular, in 2007, about 40 % of the final report documented the design process, while 60 % documented the actual final design. The template seemed to reinforce both the process learning goals for the students and the product/design goals. Creating good rubrics for a class is an extremely difficult problem, particularly in design. To grade the final designs, these 10 sections are awarded a specified portion (%) of the final grade as follows:

1. *Summary* (5 %). Specify the problem and the biological source; state the key analogy; describe the design solution and its value proposition as compared with existing solutions.
2. *Biological System Understanding* (10 %). For solution-based designs, convey a deep understanding of the primary natural system, with particular focus on explanation of the mechanism(s) of interest. For problem-based designs, provide a deep description of all mechanisms transferred to the design. In addition, describe at least briefly all natural systems that were considered, indicating why a system was or was not chosen for inspiration.
3. *Design Problem Understanding* (10 %). Motivate the problem, including what stinks, who cares, and what are we going to do about it. Also, give a detailed problem decomposition showing a logical analysis of the functions involved in the problem including function, operating environment, performance criteria, and constraints.
4. *Biological System to Design Problem Analogy and Comparison* (10 %). Describe similarities and differences between the biological systems and the

design problem. In addition, present arguments for and against the suitability of the biological systems to serve as a solution to the design problem.

5. *Visualization* (10 %). Supplement the written text with a variety of visual representations such as graphs, figures, drawings (CAD or freehand), and tables. Legends must be informative.
6. *Quantitative Analysis of Biological Mechanism* (20 %). Provide a succinct and quantitative analysis of the mechanics, material properties, or interacting processes of the biological system(s) that are transferred to the design.
7. *Quantitative Analysis of Design* (20 %). Provide a succinct and quantitative analysis of the key functions of the problem. Show how the new design integrates the principles derived from nature.
8. *Design Understanding* (10 %). Discuss the principal obstacles to achieving the design objectives that were encountered. Assess the value of the design (greater functionality, cost savings, increased sustainability, other potential applications).
9. *Cross-Domain Translation Creativity* ($\pm$10 %). This portion of the grade depends upon the creativity of the design based on its novelty with respect to current technology and previous BID designs, together with the potential usefulness of the proposed product.
10. *Literature* (5 %). This must contain key references from the primary literature (no Weblinks allowed) for the biological systems, existing solutions, similar problems, materials, and mechanics.

Item 9 in the list above requires clarification. The weights of the other items sum to 100 %. Item 9 permitted the project grade to go up or down by as much as a full level, for example, from B to A or C. We instructors did not fully agree as to how much the designs should be graded on the process rather than on the outcome. Item 9, being a highly subjective criterion, gave individual instructors leeway with respect to the rest of the grading rubric. To ensure fairness, we balanced the set of instructors assigned to each report. Note also that the weights assigned to these categories will vary, reflecting the instructors' course goals and institutional context.

In the most recent iteration of the course, we required a complete draft of the report a few weeks before the final version was due. We graded the drafts as carefully as we would have graded final versions. About half of the final reports were much improved over the drafts. (several were already excellent). This process required a lot of time from both faculty and students, but it significantly improved the outcome.

7.8 Interdisciplinary Training

Having the opportunity to work in interdisciplinary teams gives students that chance to examine a problem from a different viewpoint, share uncommon knowledge between disciplines, enable them to re-examine their own major, and in essence, seed their minds with new ideas. Teams in this class include at least one of each of these two disciplines: biologist, mechanical engineer, plus a mixture of these: systems engineer, materials scientist, designer (industrial designer, architect, and artist). Asking the students to show they are able to use each other's skills, starting from the biological inspiration, throughout the design process, to the final quantitative analysis of feasibility informs them of the importance of the interdisciplinary effort. These interactions expand their design space, promoting creative thinking and innovation in design (Table 7.8).

7.8.1 Faculty Engagement

How often does a biologist work on a team with a biomedical or mechanical engineer, a materials scientist, an industrial engineer or an architect or city planner? One key to effective bioinspired design is that it requires expertise in multiple fields. In our experience, there is no greater influence on the success of a final design than having a mentor to help facilitate the team design. Even one or two sessions with an expert can make a dramatic difference. For example, when a team of mechanical engineers, computer scientists, and biologists tried tackling the issue of desalination, they classified the problem as one of finding a way to generate water pressure for reverse osmosis. Having created a biology-based solution that required "no input energy," the team thought they had "solved" the problem. Five minutes with a faculty expert, and suddenly, now recast in terms of a thermodynamics problem, the team saw they had a major problem with their design (specifically that the system would quickly reach equilibrium after which no further desalination would occur). Over the years, we have identified those faculty who are open to interdisciplinary collaboration, can spare the time to facilitate a team over several one or 2 h sessions, and evaluate the output in such a way that makes sure the team correctly understands the principles of interest. Taking advantage of local expertise helps customize the course to the strengths of the institute.

7.8.2 Knowledge from Other Domains

As already mentioned, BID draws from many areas of scientific knowledge and cannot be accomplished without at least two or more disciplines working together. In our course, we emphasize the essential value of knowledge from other domains. In particular, it requires a sufficiently broad understanding of biology to facilitate search and selectively deep understanding once a particular biological source is targeted as a potential source for innovation. Whereas substitutes exist, there is still no resource quite as effective as *a good biologist*.

Table 7.8 Five course elements (down) pertaining to interdisciplinary training that address the 9 key challenges (across)

Interdisciplinary training	Searching for biological systems	Understanding biological systems	Identifying and understanding good design problems	Analogy mapping and transfer	Communicating Across discipline boundaries	Communicating complex system knowledge	Teaming in an interdisciplinary environment	Maintaining equal engagement throughout the process	Evaluating designs
Faculty engagement		X							
Knowledge from other domains	X	X	X	X	X	X	X	X	X
Nature auction		X			X			X	
ID teams					X		X		
Peer evaluations					X		X	X	

It is not easy for an engineer to identify keywords to search for a system with properties they seek: a BID thesaurus is useful, and biologists can act as a "translator." Although several groups are working on techniques, such as context-based searching, to help engineers bridge the knowledge gap without direct access to biological expertise, such an approach is neither optimal nor justified when there is easy access to biologists. Hence, we always place at least one biologist in a team of 5, although adding additional biologist team members, when possible, is sound practice. Just as for engineers, there are many kinds of biologists, so the particular ability of the biology teammate can affect strongly the choice of systems that can be examined by the team.

7.8.3 Nature Auction

It is important to emphasize the vital role of non-engineering disciplines in what is (ultimately) an engineering design exercise. One of our techniques is to throw the students into a fun but unfamiliar situation that establishes the importance of different types of knowledge. In our first class, we form temporary student teams, each containing at least one biologist. Then, we engage them in an extraordinary auction. *We are auctioning off nature.* The room is lined with often spectacular images of organisms (e.g., basilisk lizard) performing some uncanny feat (walk on water) with a caption that describes the behavior. The teams are given an equal number of "BID dollars" to select at least 3 (usually 4–5) biological systems to study for their first (solution-based) BID. They examine potential selections as a team and discuss the value of each natural system as the basis of their choice. Thus, begins the process of revaluing the role of nature, and the role of their fellow teammates! The BID auction not only provides them with a jumpstart on their investigation of interesting biological organisms, but also helps them learn about the knowledge, values, and perspectives of their teammates.

7.8.4 Interdisciplinary Teams

We continue to use the interdisciplinary team as a way to encourage the importance of acquiring and communicating knowledge outside of one's domain. In the final presentations, extra credit is given to the engineer who can explain the biological function and the biologist who can explain the engineering function. This embeds in each team the need for all participants to share their knowledge and is one of the pedagogical advantages of collaborative inquiry-based learning (Bransford et al. 2000; Bybee 1997). Eventually, everyone in the group understands the value of the biologist but usually, the biologist remains the source and search engine for interesting biological strategies. Similarly, not all the biologists succeed in becoming a materials engineer or mechanical engineer, but often understand the basic constraints and capabilities of these skills, and learn how to express themselves using concepts familiar to the engineers. All students come to the conclusion that they can address this complex problem more effectively by putting their skills together and learning how to apply their knowledge as a team to address the challenge.

7.8.5 *Peer Evaluations*

Team interactions can range from everyone working equally under strong leadership and team spirit to dysfunctional teams ruined by team members who do not or are unable to engage in the process. We ask each member of a team to evaluate themselves and their team, using a system based on a fictional reward. Students are given 1000 fictional dollars per team member, which they distribute between individuals (including themselves) based on the value to team. We ask each student to justify this distribution by commenting on the contribution of each team member (again including themselves). We alter the grades of students who average significantly higher or lower than 1,000. Our intentions are both to be fair and to motivate. Students know in advance that their grades may be lower than their team's grade if they do not contribute adequately. In some teams, everyone clearly valued the expertise offered by each discipline. However, in other teams, the engineers would not engage in the biological search and the biologists did not know how to engage in the quantitative assessments. More attention is needed to find ways to engage all the disciplines throughout the process.

7.9 Synthesis

This is an unusual course. We are not teaching biology or engineering, but we are asking the students to obtain a deep understanding of the specific biological system they intend to apply and translate into engineering design. A student's understanding of the biological system has to be deep enough that he or she can identify the biological knowledge that should be transferred to an engineering problem. Well-defined grading rubrics are useful so students know what constitute the traits of a BID expert. These should be directed at project evaluation, but also need to help students understand what sorts of mental process and activities help produce novel and useful (e.g., creative) designs.

Despite the unfamiliarity and challenges, students are eager to include this design process in their skill set because of the lure of invention, the novelty of BID, and its potential to lead to more sustainable practices. When given the freedom to work on a problem of their own choosing, motivation is not a significant problem. While engagement can wax and wane, depending especially on the ability of an individual to apply their domain skill set and feel useful, case studies, BIDwow, and auctioning off nature work well to maintain enthusiasm. Taking advantage of this enthusiasm by teaching bioinspired design allows us to reach quite a few learning goals, making it well worth the effort to identify some best practices. These recommendations are specifically for a course where we take inspiration from biology for design, going beyond copying or using nature.

Our experience has been that for effective BID, having biological expertise is necessary, either by having biologists on the team or available for consultation. When there is no option for student teams to include biologists, trying to find the right function from the right natural system is difficult. We recommend that experts be consulted for the best understood biological systems. Seasoned biology faculty that do research on biological systems and attend biology conferences have an edge in finding the natural systems that are rich in mechanistic details. Whereas using this expertise to give the students a good starting point may take away the chance to teach them how to search the biological literature, the students still will have many chances to hone their search strategies to find other exemplars of similar or inverted functions in extreme environments. Advanced students may even learn to find relevant examples based on phylogenetic relatedness, convergent evolution, or exaptations.

Focusing the found object exercise on some key biological functions (sensing, locomotion, and hierarchy) and comparing conjectured designs based on these solutions illustrates to the class the myriad of possibilities, teaches them about these key biological concepts and reduces design fixation. Skillful use of SBF, functional decompositions, and analogical reasoning to compare biological and engineering systems enable connections to be made between the biological functions and engineering needs. The tools for these cross-disciplinary interactions that we have developed work well. Students read the primary literature carefully and they can use SBF to focus and to keep them from getting lost in the inherent biological complexity. The newly developed SR.BID framework applied to the problem and biological solutions works well to identify the many functions to take across the divide and to evaluate analogies more deeply across functional, performance, and specification viewpoints.

When the student population is diverse, encouraging team-based problem solving is not only desirable, but necessary. Using bonus points given to the biologist who is able to explain the engineering principles or the engineer who is able to explain the biological principles tells the students that we think this ability to communicate across disciplines is useful in their training to be a practicing bioinspired designer. This cross-disciplinary practice serves as one way to keep disciplines engaged throughout the design process. Oral presentations where each team member speaks can help encourage all members to be active.

Design evaluation remains one of the more challenging aspects. Students need clear direction as to what does and what does not constitute a good problem, and to avoid common pitfalls in arriving at good designs (e.g., poor material selection, lack of appreciation of scaling, and appropriate EIA). Simple assignments pertinent to developing specific skills help students to incorporate these considerations in their final design, particularly when project grading rubrics indicate they are required.

Facilitator interaction is needed to make sure the biological mechanisms are understood correctly and the match to engineering functions makes sense, or teams may work diligently but unproductively. This is a particular problem since students consider the amount of time and energy devoted to a project as a sunk cost, which discourages them from unbiased evaluations of the potential for a given project. Timely and useful feedback is not a problem if the facilitator has a vested interest in the project. But having all these capabilities in a single instructor faced with any number of biological and engineering functions is rare, particularly when students are allowed to self-identify projects.

The final output of our course is a conceptual design that identifies the make-or-break criteria and theoretically testing the design's feasibility. Convincing tests of product success requires building and testing a prototype, which requires another semester of effort.

7.10 Conclusions and Next Steps

Problem areas that require additional attention are the search strategy for biological systems, a more complete method for teaching analogical mapping and evaluating good analogies, and evaluating good designs and good design problems. Searching and identifying useful biological systems with high potential for transfer to design still can be much improved when an expert biologist makes a suggestion or guides an exploration into the natural world. Capitalizing on evolutionary knowledge is key here, and computational methods, while promising, still cannot take the place of a skilled biologist. Although many connections can be made between biology and engineering, it is still difficult to figure out which analogical match is the best to pursue to solve the engineering challenge.

The BID class is also a research laboratory for studying BID artifacts and practices. On one hand, it has allowed us to apply, evaluate, and explore theories of creative design, analogical reasoning, and knowledge representation. On the other, in situ studies of BID already have led to the development of new descriptive theories of BID such as solution-based analogy, new knowledge representations such as SR.BID, new interactive tools such as DANE and Biologue, and new techniques such as the four box method for specifying design problems.

Our current course, focused on idea generation and conceptual design, does not include instruction in essential skills that can bring a creative idea to fruition. Translating a biological principle to a functioning device requires fundamental concept testing and experiments to build and test a BID prototype. Over and over, these new interdisciplinary designers realize that quantitative analyses can evaluate value and feasibility of the design but that implementation and testing provide the essential proof of success. Long (2012) found that making a physical prototype solves the problems of functionality, manufacturability, and improper quantitative analyses. A sequel to this course would be a fundamental concepts testing class to evaluate whether the make-or-break criterion is feasible. If it is feasible, then the

3rd class would be prototype building and testing. Some students in our classes have gone on to do this independently in other design classes. However, for certain projects, there may be some value in extremely limited prototyping within the context of our current 15 week course model. Rapid prototyping combined with hierarchical scaling have been implemented voluntarily by student groups in our BID class and this activity has gone far to evaluate designs. For certain classes of projects, incorporating this requirement would be feasible and useful.

On balance, BID provides a continuous and exciting growth process. Practicing this approach improves the facility with these new design skills, encouraging us to make friends outside our expertise. The novelty and allure of the BIDs provide motivation to go beyond the superficial and deeply understand the complexity of the problem and the complexity of the natural system.

Acknowledgments We are grateful to several colleagues who have contributed to this work over the last several years, including Inbal Flash-Gvili, Wendy Newstetter, Swaroop Vattam, and Bryan Wiltgen. We thank the US National Science Foundation for its support of this research through a TUES grant (#1022778) entitled "Biologically Inspired Design: A Novel Interdisciplinary Biology-Engineering Curriculum," and a CreativeIT Grant (#0855916) entitled "Computational Tools for Enhancing Creativity in Biologically Inspired Engineering Design."

References

Altendorfer A, Moore N, Komsuoglu H, Buehler M, Brown HB Jr, McMordie D, Saranli U, Full R, Koditschek DE (2001) Rhex: a biologically inspired hexapod runner. Auton Robots 11:207–213

Ausubel D (2000) The acquisition and retention of knowledge: a cognitive view. Kluwer, New York

Bhatta S, Goel A (1997) Learning generic mechanisms for innovative design adaptation. J Learn Sci 6(4):367–396

Biomimicry 3.8 Institute (2008) AskNature—the biomimicry webportal. http://www.asknature.org/. Retrieved 15 Apr 2012

Biomimicry 3.8 Institute (2009). Biomimicry: a tool for innovation. http://www.biomimicryinstitute.org/about-us/biomimicry-a-tool-for-innovation.html. Retrieved on 15 Apr 2012

Bransford J, Brown A, Cocking R (eds) (2000) How people learn: brain, mind, experience, and school. National Academy, Washington, DC

Bruck H, Gershon A, Golden I, Gupta S, Gyger L, Magrab E, Spranklin B (2007) Training mechanical engineering students to utilize biological inspiration during product development. Bioinsp and Biomim 2:S198–S209

Bybee R (1997) Achieving scientific literacy. Heinemann, Portsmouth

Chakrabarti A, Shu L (2010) Biologically inspired design. AIEDAM 23:453–454

Chakrabarti A, Sarkar P, Leelavathamma B, Nataraju B (2005) A functional representation for aiding biomimetic and artificial inspiration of new ideas. AIEDAM 19:113–132

Chan Y, Yu A, Chan C (2010) Assessing students' integrative learning in biomedical engineering from the perspectives of structure, behavior, and function. In: Proceedings of 40th ASEE/IEEE frontiers in education conference, 27–30 Oct 2010, Washington, DC

Chandrasekaran B (1994) Functional representation: a brief historical perspective. Appl Artif Intell 8:173–197

Chandrasekaran B, Goel A, Iwasaki Y (1993) Functional representation as a basis for design rationale. IEEE Comput 26:48–56

Cheong H, Chiu I, Shu, Stone R, McAdams D (2011) Biologically meaningful keywords for functional terms of the functional basis. J Mech Des 133: doi:10.1115/1.4003249

Chiu I, Shu L (2007a) Biomimetic design through natural language analysis to facilitate cross-domain information retrieval. AIEDAM 21:45–59

Chiu I, Shu L (2007b) Using language as related stimuli for concept generation. AIEDAM 21:103–121

Clement J (2008) Creative model construction in scientists and students: the role of imagery, analogy, and mental simulation. Springer, Dordrecht

Dunbar K (2001) The analogical paradox. In: Gentner D, Holyoak KJ, Kokinov BN (eds) The analogical mind: perspectives from cognitive science. MIT Press, Boston

Dym C, Brown D (2012) Engineering design: representation and reasoning. Cambridge University Press, New York

Ebert-May D, Momsen J, Long T, Wyse S (2010) Dearly departed Grandma Johnson: revealing student understanding of carbon cycling with structure-function-behavior models. In: Proceedings of 95th annual meeting of the Ecological Society of America

Erden M, Komoto H, van Beek T, D'Amelio V, Echavarria E, Tomiyama T (2008) A review of function modeling: approaches and applications. AIEDAM 22:147–169

French M (1996) Conceptual design for engineers, 3rd edn. Springer, London

Gentner D (1983) Structure-mapping: a theoretical framework for analogy. Cog Sci 7:155–170

Gero JS, Kannengiesser U (2004) The situated function–behaviour–structure framework. Des Stud 25:373–391

Gick M, Holyoak KJ (1983) Schema induction and analogical transfer. Cog Psych 15:1–38

Goel A (1997) Design, analogy and creativity. IEEE Intell Syst 12(3):62–70

Goel A, Gomez A, Grue N, Murdock W, Recker M, Govindaraj T (1996) Towards design learning environments—exploring how devices work. In: Proceedings of international conference on intelligent Tutoring systems, Montreal, Canada

Goel A, Rugaber S, Vattam S (2009) Structure, behavior and function of complex systems: the SBF modeling language. AIEDAM 23:23–35

Goel A, Vattam S, Wiltgen S, Helms M (2012) Cognitive, collaborative, conceptual and creative—four characteristics of the next generation of knowledge-based CAD systems: a study in biologically inspired design. Comput Aided Des 44(10):879–900

Helms M, Vattam S, Goel A (2009) Biologically inspired design: process and products. Des Stud 30:606–622

Helms M, Vattam S, Goel A, Yen J (2011) Enhanced human learning using structure-behavior-function models. Interactive learning environments. In: Proceedings of 11th IEEE international conference on advanced learning technologies, Athens, Georgia, USA, July 2011, pp 239–243

Hofstadter D (1996) Fluid concepts and creative analogies: computer models of the fundamental mechanisms of thought. Basic Books, NY

Holyoak KJ, Thagard P (1995) Mental leaps: analogy in creative thought. MIT Press, Cambridge

Keane M (1988) Analogical problem solving. Ellis Horwood, Chichester

Kolodner J (1993) Case-based reasoning. Morgan Kaufmann, San Francisco

Lave J, Wenger E (1991) Situated learning: legitimate peripheral participation. Cambridge University Press, Cambridge

Lee H, Lee BP, Messersmith PB (2007) A reversible wet/dry adhesive inspired by mussels and geckos. Nature 448:338–341

Long J (2012) Darwin's devices. What evolving robots can teach us about the history of life and the future of technology. Basic Books, New York

Markman AB, Gentner D (1993) Structural alignment during similarity comparisons. Cog Psych 25:431–467

Miklosovic DS, Murray MM, Howle LE, Fish FE (2004) Leading-edge tubercles delay stall on humpback whale (*Megaptera novaeangliae*) flippers. Phys Fluids 16:L39–L42

Nagel R, Midha P, Tinsley A, Stone R, McAdams D, Shu L (2008) Exploring the use of functional models in biomimetic concept design. J Mech Des 130:1–13

Nagel J, Nagel R, Stone R, McAdams D (2010) Function-based, biologically inspired concept generation. AIEDAM 24:521–535

Nakrani S, Tovey C (2007) From honeybees to internet servers: management of internet hosting centers. Bioinsp and Biomim 2:S182–S197

National Research Council (2009) A new biology for the 21st century. http://www.nap.edu/catalog.php?record_id=12764. National Academies Press, Washington

National Research Council (2011) A Framework for twenty-first century science education: practices, cross-cutting concepts and core ideas. http://www.nap.edu/catalog.php?record_id=13165. National Academies Press, Washington

Nersessian NJ (2008) Creating scientific concepts. MIT Press, Cambridge

Pahl G, Beitz W, Feldhusen J, Grote K (2007) Engineering design: a systematic approach. English edition: Wallace K, Blessing L (trans: 3rd edn.) Springer

Potyrailo RA, Ghiradella H, Vertiatchikh A, Dovidenko K, Cournoyer JR, Olson E (2007) *Morpho* butterfly wing scales demonstrate highly selective vapour response. Nat Photonics 1:123–128

Rasmussen J (1985) The role of hierarchical knowledge representation in decision making and system management. IEEE Trans Syst Man Cybern 15:234–243

Sarkar P, Chakrabarti A (2008) The effect of representation of triggers on design outcomes. AIEDAM 22:101–116

Sarkar P, Phaneendra S, Chakrabarti A (2008) Developing engineering products using inspiration from nature. J Comp Info Sci Eng 8(3):1–9

Sartori J, Pal U, Chakrabarti A (2010) A methodology for supporting "transfer" in biomimetic design. AIEDAM 24:483–505

Shu L (2010) A Natural-language approach to biomimetic design. AIEDAM 24:507–519

Silk E, Schunn C (2008) Core concepts in engineering as a basis for understanding and improving K-12 engineering education in the United States. Paper presented to the National academy of engineering/National research council workshop on K-12 engineering education, Washington, DC

Simon H (1996) Sciences of the Artificial. 3rd Ed, MIT Press

Simonton D (2004) Creativity in science: chance, logic, genius, and zeitgeist. Cambridge University Press, Cambridge

Speth E, Dirnbeck M, Murrilo L, Momsen J, Long T (2011) Introductory (but not simple) biology: student's representation of complex causal relationships in biological systems. Paper presented to the first meeting of the society for the advancement of biology education research, Minneapolis

Tero A, Takagi S, Saigusa T, Ito K, Bebber DP, Fricker MD, Yumiki K, Kobayashi R, Nakagaki T (2010) Rules for biologically inspired adaptive network design. Science 327:439

Vattam S, Goel A (2011) Foraging for inspiration: understanding and supporting the information seeking practices of biologically inspired designers. In: Proceedings ASME DETC conference on design theory and methods, Washington, DC

Vattam S, Helms M, Goel A (2007) Biologically inspired innovation in engineering design: a cognitive study. Technical Report GIT-GVU-07-07, graphics, visualization and usability center, Georgia Institute of Technology

Vattam S, Helms M, Goel A (2010a) A content account of creative analogies in biologically inspired design. AIEDAM 24:467–481

Vattam S, Wiltgen B, Helms M, Goel A, Yen J (2010b) DANE: fostering creativity in and through biologically inspired design. In: Proceedings of first international conference on design creativity, Kobe, Japan

Vattam S, Goel A, Rugaber S, Hmelo-Silver C, Jordan R, Gray S, Sinha S (2011) Understanding complex natural systems by articulating structure-behavior-function models. J Educ Tech Soc 14:66–81

Vincent J, Bogatyreva O, Bogatyrev N, Bowyer A, Pahl A (2006) Biomimetics: its practice and theory. J R Soc Interface 3:471–482

Vogel S (2000) Cat's Paws and Catapults: mechanical worlds of nature and people. W.W. Norton and Co, New York

Vygotsky L (1978) Mind in society: the development of the higher psychological processes. The Harvard University Press, Cambridge

Weissburg M, Tovey C, Yen J (2010) Enhancing innovation through biologically inspired design. Adv Nat Sci 3:145–167

Wilson EO (1984) Biophilia. The Harvard University Press, Cambridge

Wilson J, Rosen D, Nelson B, Yen J (2010) The effects of biological examples in idea generation. Des Stud 31:169–186

Yen J, Helms M, Goel A, Vattam S (2010) Evaluating biological systems for their potential in engineering design. Adv Nat Sci 3:168–187

Yen J, Weissburg M, Helms M, Goel A (2011) Biologically inspired design: a tool for interdisciplinary education. In: Bar-Cohen Y (ed) Biomimetics: nature-based innovation. Taylor & Francis, Boca Raton

Chapter 8
Supporting Analogical Transfer in Biologically Inspired Design

Amaresh Chakrabarti

Abstract Biologically inspired design (BID) is an emergent area of research for understanding design with biological mechanisms as inspiration, and for supporting systematic BID for developing creative designs. Understanding and supporting the processes of *analogical transfer*, whereby potential biological material is identified and adapted to solve engineering problems, is the focus of this chapter. Two questions are asked: At what level does analogical transfer take place? How to support analogical transfer? Our empirical studies show that transfer generally takes place at four levels of abstraction: state change, organ, attribute, and part. When unaided, BID is dominated by transfer at part, attribute, and organ levels, which reduces potential for creativity. This led to development of new guidelines for supporting systematic analogical transfer, an Integrated Framework for designing to encourage transfer at each level of abstraction, and a computational tool called 'Idea-Inspire' to provide analogically relevant biological stimuli for inspiration at any of these levels. Comparative studies using these interventions show significant increase in the number of transferred designs when aided by these interventions and a shift in the majority of the transfer to state change and organ levels, thereby increasing the potential for greater creativity.

Keywords Analogical transfer · Biomimetics · Biological stimuli · Engineering design · Guidelines · Idea-inspire · GEMS of SAPPhIRE · Novelty · Usefulness · Creativity · Technical product development

A. Chakrabarti (✉)
Innovation, Design Study and Sustainability Laboratory (IdeasLab), Centre for Product Design and Manufacturing, Indian Institute of Science, Bangalore 560012, India
e-mail: ac123@cpdm.iisc.ernet.in

A. K. Goel et al. (eds.), *Biologically Inspired Design*,
DOI: 10.1007/978-1-4471-5248-4_8,

8.1 Introduction

Biologically inspired design (BID) is an emergent area of research that focuses on understanding, and supporting designing systematically, with biological mechanisms as inspiration. BID is interesting because nature has 'invented' a plethora of mechanisms that use a variety of phenomena to achieve, in a resource-effective manner, numerous functions that have their equivalent in engineering problems. Gordon (1961) identified biology as a major source of analogies for engineering design.

BID has the potential to enhance creativity due to various reasons; using biological systems may inspire generation of:

- *a large number of designs;* previous research shows that generating a large number of designs enhances creativity (Chakrabarti et al. 1992; Chakrabarti and Tang 1996; Fricke 1996; Sarkar and Chakrabarti 2007a);
- *designs that use a wide variety of actions and phenomena*; previous research shows that designs with greater variety at higher levels of abstraction such as actions and phenomena lead to designs of greater novelty (Srinivasan and Chakrabarti 2010a). Also, cross-domain inspiration is identified as a valuable source of analogical transfer (Benami and Jin 2002; Hon and Zeiner 2004; Tseng et al. 2008).
- *designs that utilise resources efficiently*; it is noted that resource effectiveness is an important criterion for designs of value (Ulrich and Seering 1990; Chakrabarti 2000, 2001, 2004a; Chakrabarti and Singh 2007).

Engineers have often been inspired by biological mechanisms. However, these inspirations traditionally resulted from serendipity and individual interest, and rarely from systematic study (Chakrabarti et al. 2005). BID research promises to expand the understanding of how BID is currently carried out, how this can be enhanced with systematic processes as framework and biological mechanisms as stimuli, for developing novel and valuable—that is, creative designs (Sarkar and Chakrabarti 2007b, 2011).

8.2 BID Research: A Brief Overview

The DRM framework proposed by Blessing and Chakrabarti (2009) describes research into design in terms of the following four stages:

- *Research Clarification.* This is the stage at which the overall goals of the research, criteria for assessing research success (e.g. which goals and aspects of designing could be understood or improved with the results of this research), and the main research questions and hypotheses are developed.
- *Descriptive Study I.* During this stage, an understanding of (the current state of) designing, and the factors that influence designing and its success, is developed.

This understanding of designing 'as is' provides the basis for identifying possible aspects of designing that could be improved.

- *Prescriptive Study*. During this stage, specific aspects of (current) designing are focused on for improvement, and some form of support—which can be guidelines, methods, tools, or standards—is developed in order to improve these aspects of designing. The support is envisaged to influence and change current designing into a more desired 'future' designing, so as to influence success criteria in a more positive manner.
- *Descriptive Study II*. During this stage, the support developed is evaluated for its effectiveness in changing designing as desired and assessing its influence on success. Depending on the results, iterative steps are undertaken to improve the support (going back to Prescriptive Study), or even improve the understanding of current designing (going back to Descriptive Study I).

According to the DRM framework, a specific programme or piece of research into design can be described as a combination of a subset of these four stages, with varying emphasis on the stages. DRM categorises design research broadly into two categories: Descriptive Studies—where the main focus is on developing research criteria and understanding of designing 'as is', and Prescriptive Studies—where the main focus is on developing and evaluating support so as to change designing 'as should be'. Using the DRM framework, BID can be categorised into descriptive (i.e. as is) studies and prescriptive (i.e. as should be) studies (Chakrabarti and Shu 2010).

Descriptive studies into BID are relatively few and recent (Chakrabarti and Shu 2010). Helms et al. (2009) analyse the processes followed in BID projects, where student designers carry out given design tasks in order to develop BID. A major finding is that in BID, both problem and solution decompositions are transferred; this is consistent with the earlier findings that problems and solutions co-evolve in designing—and therefore both are design outcomes (Nidamarthi et al. 1997; Chakrabarti et al. 2004b). Vattam et al. (2008) report that generation of compound solutions are achieved in these processes through two related approaches: analogy and problem decomposition. Vattam et al. (2010a) argue that while BID is inherently analogical in nature, understanding of its analogical basis is currently limited. They present an observational study of a series of BID sessions towards developing a content theory of creative analogies in the context of BID. Sartori et al. (2010) analyse a collection of published biomimetic design cases, and identify the generic levels of abstraction at which biomimetic transfer occurs in design (further discussed in Sect. 8.3).

There are many prescriptive studies in this area. Some focus on biomimetic processes, some on databases and guidelines, and others on support tools.

Hill (1997, 2005) proposes an orientation model for supporting biomimetics projects; it has two steps: goal setting and solution identification. Based on contradicting demands identified in goal setting, solution identification consists of: (1) determining the basic function(s) underlying the contradicting demands; (2) identifying relevant biological structures with similar functional characteristics;

(3) compiling the identified biological structures and analysing each to extract underlying principles and make preliminary solution associations; (4) transferring these preliminary solutions into technical solutions according to the requirements and conditions of the goal; (5) varying and combining relevant characteristics of these solutions; (6) enlisting alternatives of each characteristic into a morphological table and identifying possible combinations of these characteristics; (7) using common evaluation methods, evaluating the solution elements or completing variants to select the best; and (8) elaborating the chosen solution. Schild et al. (2004) propose a systematic method for identifying analogue solutions to a given problem, with the following steps: (1) problem formulation at an appropriate level of abstraction; (2) evaluation; (3) search for analogies; and (4) verification and evaluation. Gramann (2004) recommends the following biomimetic process to support technical problem solving: (1) formulate a search objective; (2) search for and assign a set of relevant biological systems; (3) analyse these biological systems; and (4) evaluate.

Based on an analysis of the above processes, Sartori et al. (2010) construct a generic model of the biomimetic process with these steps: (1) formulate search objectives; (2) search for biological analogues; (3) analyse biological analogues; and (4) transfer these analogues to the technical domain.

Several databases of biological systems have been developed to aid the biomimetic process. Catalogue sheets by Hill (1997, 2005) aim to capture knowledge about biological structures and functions. With the goal of developing bio-TRIZ, the database of biological effects by Vincent et al. (2002, 2006) uses TRIZ methods (Terninko et al. 1998; Mann 2001). Further, Vincent et al. developed a framework for capturing biological data in a technology-compatible manner, to support designers in BID. One issue with these approaches, as Chakrabarti et al. (2011) point out, is the distribution of biological functionality over several levels of scale and complexity; as Sartori et al. (2010) argue, developing an appropriate functional representation of biological systems suitable for engineering design seems to be a major, unresolved issue. Neither in Hill's catalogue sheets nor in Vincent et al.'s database is there an explicit relationship between function and structure of biological systems with an objective basis. A possible resolution to this is the basis provided by the SAPPhIRE model of causality (Chakrabarti et al. 2005). SAPPhIRE model uses seven constructs to explain how system-functions are achieved, see further details in Sect. 8.4.2. A database of over 1,000 entries of biological and technical systems structured using SAPPhIRE model has been developed for a software tool Idea-Inspire (see further in this section) to retrieve relevant entries as inspiration for a given design problem (further details in Sect. 8.5). There are other approaches of cataloguing biological systems, such as function-based approaches (e.g. Vakili et al. 2007; Nagel et al. 2010), approaches that use reverse engineering and ontologies (Wilson and Rosen 2007; Wilson et al. 2009), or SBF model-based approaches (Vattam et al. 2010b).

The database approach to biomimetics is not without criticism; Gramann (2004), for instance, criticises these due to the vast amount of and variety in biological knowledge to be captured—a massive task. The natural language-based approach (e.g. by Hacco and Shu 2002; Chiu and Shu 2007; Cheong et al. 2010) is a possible resolution to this issue. Their approach uses natural language processing methods to analyse biological information already available in existing resources, in order to extract relevant biological phenomena that can be transferred by designers to the target domain by applying analogical reasoning. A comprehensive review of this approach can be found in (Shu 2010).

According to Chakrabarti and Shu (2010), each approach has pros and cons. While the natural language approach avoids the effort involved in structuring and populating databases, it needs effort in developing appropriate search processes for identifying meaningful information from the natural language resources, and transfer may be more difficult. In contrast, the database approach needs substantial effort into prestructuring information and populating databases, so that search for relevant information in the use phase becomes easier, and transfer less difficult.

A set of guidelines for biomimetic transfer has been suggested by Sartori et al. (2010); various guidelines for composition of biological systems have been proposed by Vattam et al. (2010a). A number of software tools have also been developed. For example, Idea-Inspire is an interactive, biomimetic-inspiration tool that uses the database approach, structures the entries using the SAPPhIRE model of causality (Chakrabarti et al. 2005; Sarkar et al. 2008; Srinivasan et al. 2011), and provides biological systems as stimuli for ideation. DANE is a tool, also based on the database approach, which uses SBF model for structuring information (Vattam et al. 2010b).

In summary, current research in biomimetic support is focused on providing three types of outputs:

- *stimuli* for biomimetic inspiration (e.g. Chakrabarti et al. 2005).
- *framework* for systematic BID (e.g. Srinivasan and Chakrabarti 2010b).
- *guidelines* for biomimetic transfer (e.g. Sartori et al. 2010).

In this chapter, we provide an overview of our research into each of these.

8.3 BID Process and 'Transfer'

According to Sartori et al. (2010), the BID process has four general steps:

1. formulate search objectives;
2. search for biological analogues;
3. analyse biological analogues;
4. Transfer relevant knowledge to the target domain.

Our research is focused on the following:

- provide biological inspiration—Step 2 in the BID process (Sect. 8.5);
- provide a systematic design process for supporting BID (Sect. 8.6);
- provide systematic guidelines for transfer—Step 4 in the BID process (Sect. 8.7).

8.4 SAPPhIRE Model, Creativity, and an Integrated Model of Designing

For understanding the work in the rest of the chapter, knowledge of the following is essential and is therefore briefly described below.

8.4.1 Creativity

Since the focus of the work reported in this chapter is on understanding designing, especially BID, in terms of its influence on the various aspects of creativity, an understanding of creativity and its various aspects is crucial. Based on the 'common definition' of creativity (Sarkar and Chakrabarti 2007b, 2011), creativity in design 'occurs through a process by which an agent uses its ability to generate ideas, products or solutions that are novel and useful'. This 'common' definition is developed after analysing an extensive set of definitions of creativity from literature, taking into account all encompassing features that constitute these definitions. In this definition, the two major aspects of creativity are 'novelty' and 'usefulness'; novelty means 'being recent and original to the society', while usefulness refers to the 'utilitarian value to the society'.

Novelty is assessed using a measure that estimates how different the idea, product, or solution under assessment is from existing ideas, products, or solutions. The assessment is based on the following assumption: if the product is different from existing products in the functionality itself, that is, invents a new function, it is of the highest novelty. If that is not the case, and the product is structurally no different from existing products, then there is no novelty. If, however, it is in between, SAPPhIRE model of causality (see Sect. 8.4.2) can be used to assess how different the product is from existing products (Sarkar and Chakrabarti 2011). More recently, a variant of this measure has been developed based entirely on the SAPPhIRE model (Srinivasan and Chakrabarti 2010a).

Usefulness is assessed based on the assumption that, the value of a technical product lies in its usefulness to the society. Usefulness can be assessed using the extent to which it is used, or likely to be used. In this sense, usefulness of a product is reflected in the extent of use of the product. This is assessed using these sub-measures: how beneficial the use of the product is (for instance, a life-saving drug is more beneficial than a lifestyle drug), how long the product is used or its effect lasts (for instance, a drug whose intended effects last a lifetime is more useful than one that has to be taken once every day to remain effective), how many people use, or benefit from the product (for instance, a car which runs the same distance with all seats full is more useful than one that runs similar distance with fewer seats occupied) (Sarkar and Chakrabarti 2011).

Finally, technical creativity is taken as a product of the degree of novelty and degree of usefulness; a product with a high degree of both is considered the most creative, a product lacking in substantial novelty or usefulness is less creative, and a product lacking in both is the least creative (Sarkar and Chakrabarti 2011).

All the above measures have been empirically validated by comparing the responses of experienced designers on the relative novelty, usefulness, and creativity of multiple series of products against those estimated using the measures. The assumption, similar to that made by Amabile (1983, 1996) has been that, experts are the best judges for creativity, and therefore, any measure of design creativity should match the collective, intuitive opinions of experienced designers.

8.4.2 SAPPhIRE Model of Causality

As mentioned in Sect. 8.2, SAPPhIRE (acronymed from **S**tate change–**A**ction–**P**art–**Ph**enomenon–**I**nput–o**R**gan–**E**ffect) model of causality (Chakrabarti et al. 2005) was developed for explaining functioning of an entity that uses physical phenomena for attaining its functions. It explains how an entity works, through causal descriptions of how part-level characteristics of the entity and its surroundings trigger physical phenomena that change the state of the entity and its surroundings. The constructs of SAPPhIRE model are (Ranjan et al. 2012):

Parts	Physical components and interfaces that constitute the entity of interest and its surroundings.
Physical phenomenon	An interaction between the entity and its surroundings.
State	A property of the entity (or its surroundings) that is involved in an interaction.
Physical effect	A principle of nature that underlies and governs an interaction.
Organ	A set of properties and conditions of the entity and its surroundings required for an interaction between them.
Input	A physical variable that crosses the boundary of the entity and is essential for an interaction between the entity and its surroundings.
Action	An abstract description or high-level interpretation of an interaction between the entity and its surroundings.

The relationships among these constructs are as follows: Parts (P) of an entity and its surroundings create organs (R), which are the structural requirements for a physical effect (E). A physical effect is activated by various inputs (I) on the organs and creates a physical phenomenon (Ph), and changes the state (S) of the entity. The changes of state are interpreted as actions (A), as new inputs, or as changes that create or activate parts (see Fig. 8.1).

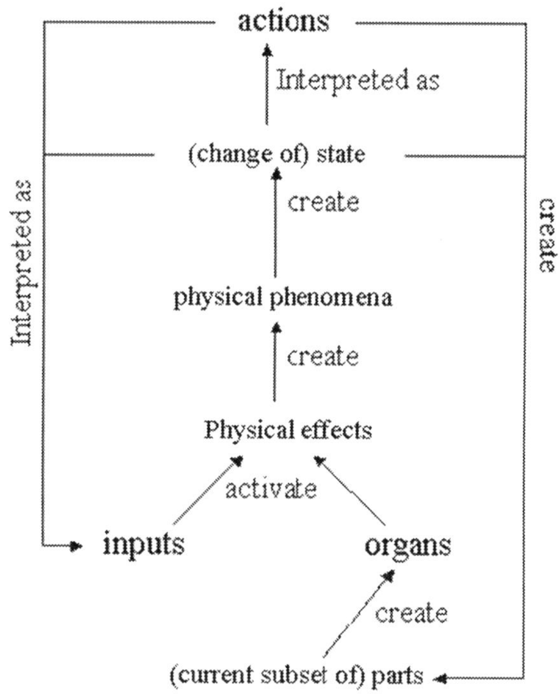

Fig. 8.1 SAPPhIRE model of causality (*Source* Chakrabarti et al. 2005)

The use of these constructs together and their links with functionality provides a richer description of behaviour. Action, state change, and input constitute the higher levels of abstraction. Physical phenomenon and effect comprise the intermediate levels of abstraction. Organs and parts constitute the lower levels of abstraction (Srinivasan and Chakrabarti 2010a). This model of causality explains the functioning of a system as follows (Srinivasan and Chakrabarti 2009): a set of components and interfaces that constitute a system and its environment (parts) creates a set of properties and conditions of the system and its environment (organs). When the system and its environment are not in equilibrium, there is a transfer of a physical quantity in the form of material, energy, or signal (input) across the system boundary. This physical quantity, in combination with a particular set of properties and conditions (organs), activates a principle (physical effect). Activation of this principle creates an interaction between the system and its environment (physical phenomenon). The interaction between the system and its environment creates a change in property of the system (state change). The change in property can be interpreted at a higher level of abstraction (action).

The example below is used for clarifying the model. Let us assume that some water is kept in an electric kettle (parts). The relevant properties and conditions that are available from the parts include the fluidic property of the water, the electrical resistance properties of the coil, the heat transfer properties of the water and the coil, etc. (organs). When the kettle is switched on, electric current flows through the coil (input), and due to the resistance properties of the coil, activates the resistive-heating effect (effect). This leads to exchange of heat within the coil (phenomenon) which leads to an increase in its temperature (state change) that may be interpreted as heating of the coil (action). The temperature difference between the coil and the water (input), along with the heat transfer properties of the water and the coil, activates heat conduction effect (effect), which leads to heat exchange between the coil and the water (phenomenon) that leads to an increase in the temperature of the water (state change), which is interpreted as heating of water (action).

8.4.3 Integrated Model of Designing

While a model of causality explains how an existing entity satisfies its goals, that is, its intended effects on the surroundings, designing is a process in which, starting from its goals, descriptions for an appropriate entity have to be worked out. Designing, therefore, should involve working out the details of the intended entity at the various levels of abstraction that a model of causality requires, but in reverse order, so as to end up with sufficient detail about its lowest—part level of abstraction. Based on this assumption, a model of designing has been developed that uses the levels of outcome abstraction, provided by the SAPPhIRE model, for an entity to be designed. It is called the *Integrated Model of designing* (also called the GEMS of SAPPhIRE as Req-Sol model), where GEMS(**G**enerate–**E**valuate–**M**odify–**S**elect/reject) activities are applied on SAPPhIRE levels of outcome which evolve as Req-Sol (**R**equirements or **S**olutions), see (Srinivasan and Chakrabarti 2010b). The Integrated Model of designing has been empirically validated using video–protocol studies of designing sessions that occurred well before the model was proposed. This was to ensure that the way designers worked could not have been biased by the knowledge of this model. A two-way comparison was used for validation: whether all constructs of the model were present in the events that constituted the designing sessions, and whether all events could be described using the constructs of the model. The model was found to represent the natural processes used by designers for conceptual design of technical products.

8.4.4 SAPPhIRE Abstraction Levels, Number of Ideas, and Novelty

A series of designing sessions were analysed using the Integrated Model of designing and the proposed measures of creativity (Srinivasan and Chakrabarti 2010a, b). The intention was to see what aspects of designing influenced which aspects of creativity. In particular, two aspects of design outcomes were investigated: the *number of ideas* explored and the *levels of abstraction* (in this case,

SAPPhIRE levels of outcome abstraction) at which these ideas belonged. Similarly, two aspects of creativity were investigated: *variety of solutions* generated and *novelty of solutions* generated. Variety is used to mean how different, on average, the solutions generated by a design team or individual in a given designing session have been from one another; novelty of a solution space is used to mean how different on average the solutions generated by a designer or design team have been from the solutions that already existed before the designing session started.

The number of ideas generated had a significant correlation with both variety and novelty of the solution space; this means that generating more ideas is likely to lead to greater variety and novelty. It was also found that the number of ideas at a higher level of outcome abstraction correlated more strongly to the variety and novelty of the solution space; this means that being able to generate more ideas at higher levels of outcome abstraction should have a more significant impact on variety and novelty of solutions generated than with a large number of ideas at a lower level of outcome abstraction (Srinivasan and Chakrabarti 2010a).

Analyses of the designing sessions led to another interesting finding. For the six designing sessions that were analysed using the Integrated Model, the number of ideas generated at the various levels of outcome abstraction, starting from action and input (extreme left in the plot in Fig. 8.2), through the intermediate levels to the part level (extreme right) was plotted; the six cases are marked as 1–6. For every single designing session analysed, each of which involved a different combination of design problem and designers, the distribution of the number of ideas generated at the various levels of abstraction has been remarkably similar. While the number of ideas at the action and part levels has been relatively high for every design session, the number of ideas generated at the intermediate levels has been substantially lower. Since for carrying out every action, multiple state changes could be utilised; for every state change to be carried out, multiple phenomena could be used; and so on for each level of abstraction, we expected that

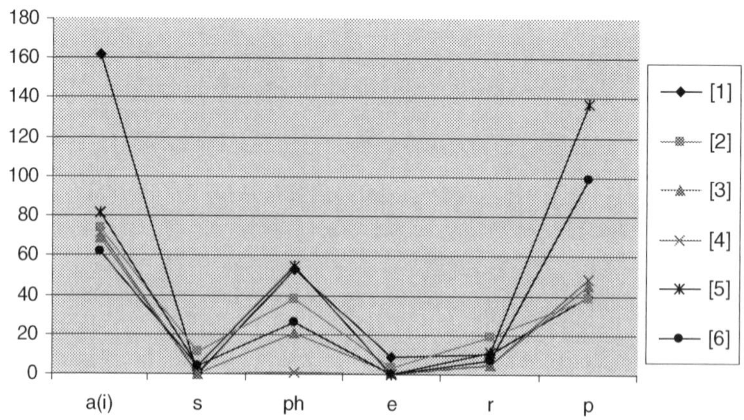

Fig. 8.2 'Bathtub' curve (*Source* Srinivasan and Chakrabarti 2010a)

the number of ideas should have increased monotonically as the designers progressed from the higher levels to the lower levels of abstraction. However, this did not happen. Why did this not happen, and how did that affect creativity?

We argue that there are two possible reasons as to why, instead of monotonically increasing as one progressed from higher to lower levels of abstraction, the number of ideas generated followed a 'bathtub' curve as in the Fig. 8.2:

- Conjecture 1: The high number of action-level ideas comes from the design brief. The high number of part-level ideas comes from designer competence in mapping action-level ideas to part-level ideas. The relatively small number of ideas at the remaining levels comes from a *lack of designer competence* at mapping ideas at action or part levels to those at these intermediate levels.
- Conjecture 2: The high number of action-level ideas comes from the design brief. The high number of part-level ideas comes from designer competence in mapping action-level ideas to part-level ideas, and the *belief that this would produce less risky and more feasible solutions fast*, which going into the fundamentals (via intermediate levels) might sacrifice.

If Conjecture 1 is true, improving designer competence would impact novelty directly as there would be more number and variety of ideas generated. This would impact usefulness indirectly, since more ideas with greater novelty would stand a greater chance for stumbling onto solutions with substantially higher usefulness. If Conjecture 2 is true, this designer attitude would have an indirect, incremental impact on usefulness while sacrificing variety and novelty. Even if Conjecture 2 is true, this designer attitude would be harder to change if there is lack in designer competence in exploring ideas across the levels. Hence we take Conjecture 1 to be more reasonable and develop support to improve designer competence to explore ideas across all levels of outcome abstraction.

Supporting this requires two kinds of knowledge: product (or, domain) knowledge—knowledge about how various entities work, and process knowledge—knowledge with which domain knowledge can be modified (Srinivasan et al. 2011); both are important for designing. In order to provide BID-related domain knowledge, a tool called Idea-Inspire has been developed, which can be used to provide biological and technical stimuli, see Sect. 8.5. In order to provide process knowledge, a systematic approach is necessary that provides a process framework, within which this domain knowledge can be provided in a situated manner. A framework for this purpose has been developed and discussed in Sect. 8.6.

8.5 Supporting BID Using Stimuli: Idea-Inspire

Idea-Inspire (Chakrabarti et al. 2005) is a software tool that is used for supporting ideation using knowledge about biological and technical systems as stimuli. Idea-Inspire has three components.

The first component is a *database of entries*, each of which explains how an entity achieves its goals; the information is provided using multiple modes: video/animation, photographs or drawings, natural language description, and a SAPPhIRE model–based description, which is primarily used for computer-based search for relevant entries (Chakrabarti et al. 2005).

The second component is an *analogical search algorithm*, which takes a user description of the problem, described at one or a combination of SAPPhIRE levels of outcome abstraction, and retrieves a series of entries that are relevant to the problem description.

The third component is a *graphical user interface*, through which users interact with the software, in terms of specifying the problem, receiving the list of relevant entries, or looking at the details of these entries. Figures 8.3 and 8.4 show, respectively, the GUI and one entry from Idea-Inspire.

Ideation effectiveness of Idea-Inspire has been evaluated in two stages. While details have been given in Sarkar et al. (2008), an excerpt is provided below. In the first stage of evaluation, three designers were asked to generate as many ideas as possible to solve given design problems without any support; when they were 'exhausted' (according to them), they were asked to generate further ideas using entries retrieved by Idea-Inspire as stimuli. On average, each designer produced 165 % additional ideas using Idea-Inspire entries as stimuli. This indicated enhancement of 'fluency' of the designers, which is often used as a measure of

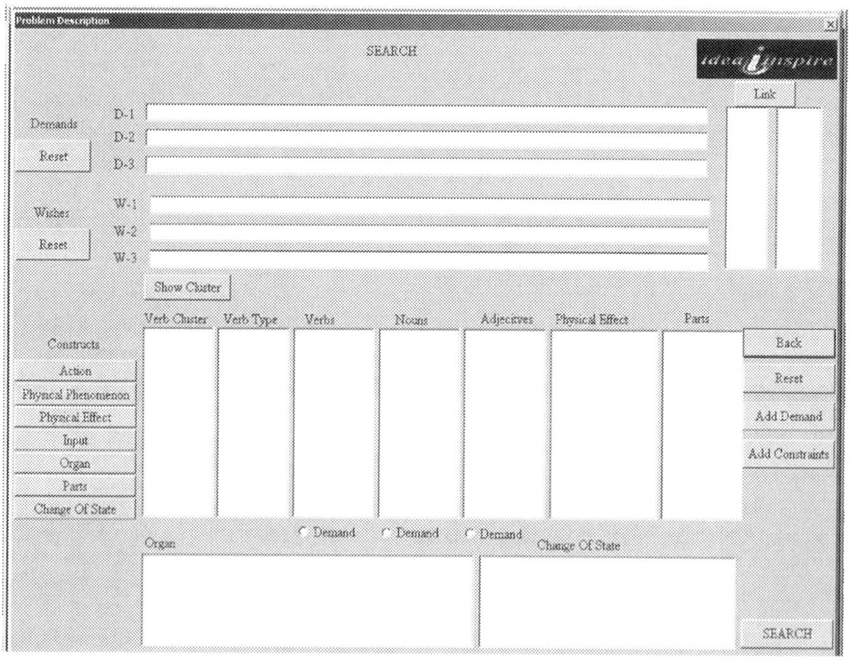

Fig. 8.3 Graphical user interface for idea-inspire

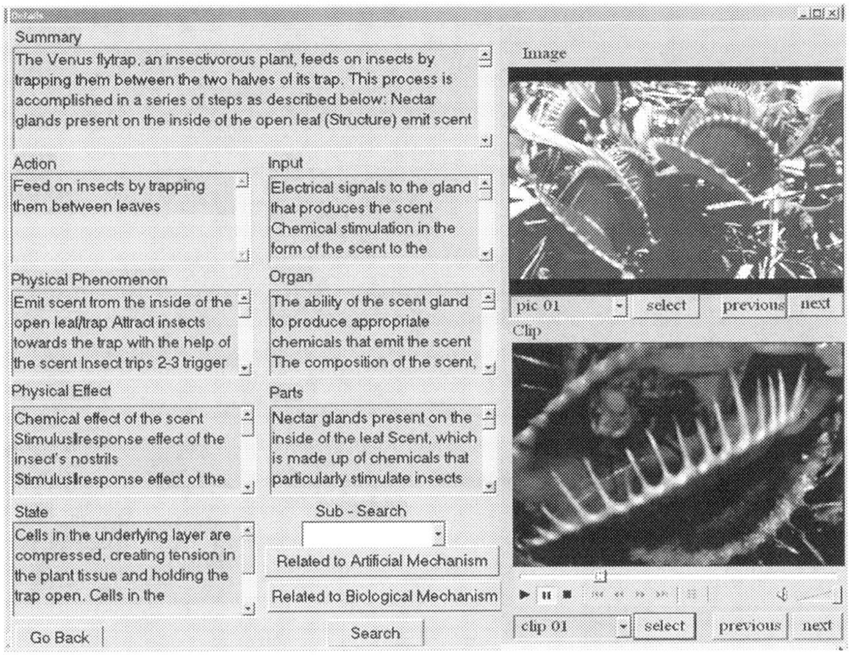

Fig. 8.4 An entry from idea-inspire (*Source* Chakrabarti et al. 2005)

novelty (e.g. Csikszentmihalyi 1997; Sternberg and Lubart 1999), which in turn is a measure of creativity according to the 'common' definition of creativity.

More recently, influence of fluency on novelty in the context of conceptual design has been empirically verified by Srinivasan and Chakrabarti (2010a); they found that the number of ideas explored strongly correlates with the variety of the solutions generated, which correlates with the novelty of the solutions generated. It was also found, in the Idea-Inspire study, that about 42 % of the ideas generated by the designers with stimuli from Idea-Inspire were selected by themselves as worth developing further; this points to the likely usefulness of these ideas—the other indicator for creativity according to the common definition.

In the second stage of evaluation, twelve masters-level design students with an undergraduate degree in engineering and up to a year of industrial experience were asked to use Idea-Inspire in their respective, mandatory design projects. Feedback from these students, on the usefulness and usability of the software for their purpose, indicated both an enhancement in the number of new ideas generated and ease of use of the software.

8.6 Systematic Design Process: An Integrated Framework of Designing

As discussed in Sect. 8.4.4., empirical studies of the conceptual design process (Srinivasan and Chakrabarti 2010b) revealed that, while designers explored the action and part levels of outcome abstraction extensively, they did not adequately explore the intermediate levels of abstraction.

From earlier studies (e.g. Srinivasan and Chakrabarti 2010a), it was known that exploration at higher levels of abstraction has a greater impact on novelty of solutions generated. It therefore became clear that a systematic approach providing process knowledge was necessary to encourage designers to explore ideas at all levels of abstraction. The Integrated Framework for designing (also called the GEMS of SAPPhIRE framework, see Srinivasan and Chakrabarti 2010b; Srinivasan et al. 2011) has been developed for this purpose. The framework prescribes designers to apply GEMS to requirements at each SAPPhIRE level of outcome abstraction (this stage is called requirement development stage or RDS) and apply GEMS to solutions at each SAPPhIRE level of outcome abstraction (this stage is called solution development stage or SDS), with the aim of exploring as many ideas as possible for requirements and solutions at each level of outcome abstraction (for details, see Srinivasan et al. 2011). Figure 8.5 depicts these two stages: RDS followed by SDS, where G (generate), E (evaluate), M (modify), and S (select/reject) activities are prescribed to be applied to requirements (shown as 'req') and solutions (shown as 'sol'), respectively, at all levels of outcome

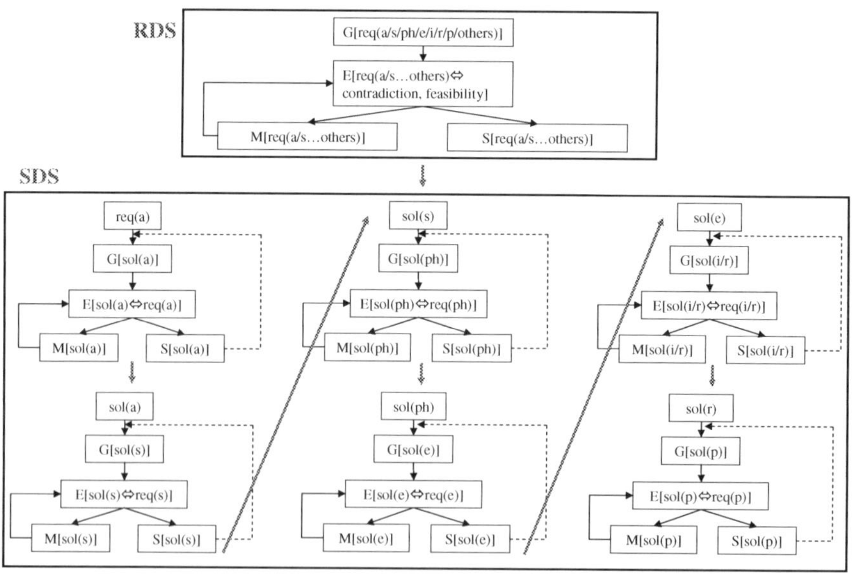

Fig. 8.5 Integrated Framework for designing (*Source* Srinivasan et al. 2011)

Table 8.1 Number of ideas generated when various types of support are used: *d* stands for designer; *f* for Integrated Framework; *i* for Idea-Inspire; *B* for biologically inspired ideas; *NB* for non-biologically inspired ideas

	Mobility			Steering			Handling gradient			Stability			Overall		
	B	NB	total	B	NB	total	B	NB	total	B	NB	total	B	NB	total
d	0	0	0	0	0	0	1	4	5	0	2	2	1	6	7
f + d	0	7	7	0	5	5	0	0	0	0	3	3	0	15	15
i + d	0	0	0	0	0	0	2	0	2	0	0	0	2	0	2
f + i + d	11	3	14	3	5	8	2	1	3	1	0	1	17	9	26
	11	10	21	3	10	13	5	5	10	1	5	6	20	30	50

(*Source* Srinivasan et al. 2011)

abstraction, while going from action level (shown as (a)), through state change (shown as (s)), phenomena (ph), effect (e), organ (r) levels, to part (p) level, in that order. The framework proposes that designers generate, evaluate, modify as necessary, and finally select appropriate requirements at all levels of SAPPhIRE abstraction and then do the same for solutions, working gradually at more concrete levels of the solution as they progress from action to part level (Fig. 8.5).

Two types of evaluations were used to test the effectiveness of the Integrated Framework: comparative study in laboratory setting where various design teams used various interventions (no support, framework only, and both framework and Idea-Inspire), see Srinivasan (2011). On average, number, variety, and novelty increased in the cases in which either the Integrated Framework was used alone, or used along with Idea-Inspire, compared to when no support was used.

In the second study, a group of designers used the Integrated Framework and Idea-Inspire during an industrial project (Srinivasan et al. 2011). Designers initially developed some ideas on their own and then started using the Integrated Framework; they had the freedom to decide when they wanted to use Idea-Inspire and when not, within the Integrated Framework. The results of evaluation are detailed in Srinivasan et al. (2011) and summarised here (see Table 8.1) as follows: Designers generated very few ideas on their own—most of them non-biological ideas. Using the framework alone, this number doubled although still using only non-biological ideas as inspiration. When they used Idea-Inspire only, few ideas were generated—but all of these were inspired by biological stimuli. When the Integrated Framework and Idea-Inspire were used together, the number quadrupled, with a dominance of biological ideas. This indicates that using the Integrated Framework and Idea-Inspire together substantially improves ideation and balances the generation of ideas that are inspired by biological as well as non-biological stimuli.

8.7 Guidelines: SAPPhIRE-Based Guidelines for 'Transfer'

While various guidelines are provided in literature as to how various steps in BID are (to be) carried out, relatively little is specified as to how transfer is carried out. The definition of 'transfer', taken from Sartori et al. (2010) and based on Schmidt (2005), is as follows: it is the reproduction of information from a model of a biological system in a model or prototype for a technical system.

According to Schild et al. (2004), there are four possible levels of transfer: transfer of technology to a new context, transfer of structure, partial transfer of functional principles, and use of an analogy as idea stimulus. According to Vattam et al. (2010a), transfer is guided by the sub-functions identified. However, literature provides little further detail as to what kind of knowledge is transferred within each single sub-function, or what the steps of transfer are.

In order to understand the process of transfer is greater detail, we asked the following questions (Sartori et al. 2010):

- What kind of knowledge is transferred in biomimetics?
- How can the step 'transfer' in the biomimetic process be supported?

SAPPhIRE model of causality (Chakrabarti et al. 2005) is used to analyse twenty existing cases of biomimetic transfer from literature, to understand the types of knowledge used, and the generic process of transfer. The levels of abstraction at which transfer took place are identified, and a guideline is developed to enhance fluency of transfer. Based on this analysis, five types of transfers have been identified to have taken place (for details, see Sartori et al. 2010):

- Transfer parts: This is direct mimicking of a biological system in a technical system. The same materials are used and arranged in the technical system in the same way as in biology. This is transferred at the lowest level of abstraction.
- Transfer organs: This involves developing a technical system with similar organs as in its biological analogue.
- Transfer attributes: This involves developing a technical system with the same or similar attributes as its biological analogue. Attributes are properties of the biological analogue that are not clearly connected to any physical effects.
- Transfer state changes: This involves using a state change of a biological analogue by the technical system in order to achieve an analogue action.
- Resulting (or incidental) transfer: This is the case where the technical system resulting from an associated transfer provides adequate means for a new action to be also supported.

Note that the above types mark the highest level of abstraction at which transfer was found explicitly to occur; in many of these cases what was transferred, implicitly or explicitly, was not just the knowledge for that level but a combination of knowledge from multiple levels of abstraction including that level; in a broader sense, these are similar to 'design patterns' of Alexander (1977), which were proposed by Goel and Bhatta (2004) to be useful in analogy-based designing.

Transfers that were carried out in above the twenty cases are distributed as follows. Most transfers took place at the part, attribute, and organ levels, with very few at state change levels. Since variety at a higher level of outcome abstraction has a greater impact on novelty, this leaves scope for improvement. Hence, a new, SAPPhIRE-based set of guidelines for analogical transfer has been developed.

Two sets of guidelines have been developed (Sartori et al. 2010). The 'Generic Guideline' has been developed to encapsulate the essential steps for carrying out BID, and the recommendations specific to each of these steps as found from existing literature. The Generic Guideline therefore encapsulates the current wisdom as to how to carry out BID, as weaned out of existing literature.

The second, new 'Guideline with SAPPhIRE' has been proposed to follow the same generic steps as in the Generic Guideline, but with specific recommendations for using SAPPhIRE constructs within these steps. This guideline recommends that the four classes of transfer identified be systematically used in the analysis and transfer steps. This, it is argued, should increase the number of BID alternatives generated, see details in Sartori et al. (2010).

A series of design studies have then been carried out to evaluate these guidelines, by comparing the performance of designers when using SAPPhIRE-based guideline in carrying out BID, with that when using the Generic Guideline. The study has been carried out in both India and Germany, using two different groups of designers. In both the countries, the number of concepts produced when SAPPhIRE guidelines were used was higher, by between 50 and 100 % from when Generic Guideline was used. Not all of these solutions were biologically inspired or feasible; for the concepts that were biologically inspired and feasible, the average increase in the number of concepts produced was 60 % when SAPPhIRE-based guidelines were used. More importantly, the percentage of transfer at the higher levels of abstraction (from part and attribute to organ and state change) improved steadily from the cases in literature to when the Generic Guideline was used, and even further when the SAPPhIRE guidelines were used. This indicates that there is potential for greater novelty in using these guidelines.

8.8 Summary and Conclusions

This chapter provides a brief overview of descriptive and prescriptive BID research and then elaborates on descriptive and prescriptive BID research in the research group of the author. The work produced measures for creativity and models of causality and designing, which were used to understand how the outcome levels of abstraction at which ideas are explored in current ways of designing influence creativity, and at what levels of outcome abstraction do analogical transfer currently take place. Using these as the bases, the following have been developed: an Integrated Framework for designing, a set of guidelines for supporting analogical transfer, and a database and associated software tool for providing biological stimuli for ideation. Comparative, empirical studies using these as interventions indicate substantial potential for their use as support in enhancing creativity in developing technical products.

References

Alexander, C. (1977) A pattern language: towns, buildings, construction. Oxford University Press, Oxford

Amabile TM (1996) Creativity in context. Westview Press, Colorado

Amabile TM (1983) The social psychology of creativity. Springer, New York

Benami O, Jin Y (2002) Creative stimulation in conceptual design. In: Proceedings of ASME design engineering technical conferences and computers and information in engineering conference (DETC/CIE), Montreal, DETC2002/DTM-34023

Blessing LTM, Chakrabarti A (2009) DRM, a design research methodology. Springer, London

Chakrabarti A, Bligh TP, Holden T (1992) Towards a decision-support framework for the embodiment phase of mechanical design. AI in Eng 7(1):21–36

Chakrabarti A (2000) Increasing efficiency of compositional synthesis by improving the database of its building blocks, AI EDAM, vol 14. Cambridge University Press, Cambridge, pp 403–414

Chakrabarti A (2001) Improving efficiency of procedures for compositional synthesis using bi-directional search. AI EDAM 15(1):67–80

Chakrabarti A (2004) A new approach to structure sharing. ASME JCISE 1(1):1–78

Chakrabarti A, Shu LH (2010) Guest editorial: biologically inspired design. AI EDAM 24:453–454

Chakrabarti A, Tang, MX (1996) Generating conceptual solutions on FuncSION: evolution of a functional synthesiser. In: Gero S, Sudweeks F (eds) Artificial intelligencedesign'96. Kluwer Academic Publishers, Dordrecht, pp 603–622

Chakrabarti A, Singh V (2007) A method for structure sharing to enhance resource-effectiveness. J Eng Des 18(1):73–91

Chakrabarti A, Cagan J, Shea K, Campbell M, Stone R, Vergas-Hernandez N, Wood K (2011) Computer based design synthesis research: an overview. ASME JCISE 11(2):021003-1–021003-10

Chakrabarti A, Morgenstern S, Knaab H (2004) Identification and application of requirements and their impact on the design process: a protocol study. Res Eng Design 15:22–39

Chakrabarti A, Sarkar P, Leelavathamma B, Nataraju B (2005) A functional representation for aiding biomimetic and artificial inspiration of new ideas. AI EDAM 19(2):113–132

CheongH, ChiuI, Shu LH (2010) Extraction and transfer of biological analogies for creative concept generation. In: Proceedings of ASME international design engineering technical conferences and computers and information in engineering conference, Montreal, QC, Canada, 15–18 Aug 2010, DETC2010-29006

Chiu I, Shu LH (2007) Biomimetic design through natural language analysis to facilitate cross-domain information retrieval. AI EDAM 21:45–59

Csikszentmihalyi M (1997) Finding flow: the psychology of engagement with everyday life. BasicBooks, New York

Fricke G (1996) Successful individual approaches in engineering design. Res Eng Design 8(3):151–165

Goel AK, Bhatta SR (2004) Use of design patterns in analogy-based design. Adv Eng Inform 18(2):85–94

Gordon WJJ (1961) Synectics. Harper & Row, NY

Gramann J (2004) Problemmodelle und Bionik als Methode. Dissertation, TU, Munich
Hacco E, Shu L (2002) Biomimetic concept generation applied to design for remanufacture. In: Proceedings of ASME design engineering technical conferences and computers and information in engineering conference, Montreal, 29 Sept to 2 Oct 2002, DETC2002/DFM-34177
Helms ME, Vattam SS, Goel AK (2009) Biologically inspired design: process and products. Des Stud 30(5):606–622
Hill B (1997) Innovationsquelle natur: naturorientierte innovationsstrategie für entwickler. Konstrukteure und Designer, Shaker, Aachen, Germany
Hill, B. (2005) Goal setting through contradiction analysis in the bionics-oriented construction process. CIM, 14, 19, 59–65. Blackwell Publishing Ltd, Oxford/USA, p 33
Hon KKB, Zeiner J (2004) Knowledge brokering for assisting the generation of automotive product design. Annals CIRP 53(1):159–162
Mann D (2001) System operator tutorial—9—windows on the world. The TRIZ Journal. http://www.triz-journal.com/archives/2001/09/c/index.htm as at 20 April 2010
Nagel JKS, Nagel RL, Stone RB, McAdams DA (2010) Function-based, biologically-inspired concept generation. AI EDAM 24(4):521
Nidamarthi S, Chakrabarti A, Bligh TP (1997) The significance of co-evolving requirements and solutions in the design process. In: International conference of engineering design pp 227–230
Ranjan BSC, Srinivasan V, Chakrabarti A (2012) An extended, Integrated Model of designing. In: Horváth AA, Behrendt M, Rusák Z (eds.) TMCE 2012, Karlsruhe, Germany
Sarkar P, Chakrabarti A (2011) Assessing design creativity. Design stud 32(4): 348–383
Sarkar P, Phaneendra S, Chakrabarti A (2008) Developing engineering products using inspiration from nature. ASME J Inf Sci Eng 8:3
Sarkar P, Chakrabarti A (2007a) Understanding search in design. In: International conference on engineering design (ICED07), Paris, France
Sarkar P, Chakrabarti A (2007b) Development of a method for assessing design creativity. In: International conference on engineering design (ICED07), Paris, France
Sartori J, Pal U, Chakrabarti A (2010) A methodology for supporting "transfer" in biomimetic design. AI EDAM 24:483–505
Schild K, Herstatt C, Lüthje C (2004) How to use analogies for breakthrough innovations. Institute of Technology and Innovation Management, Technical University of Hamburg
Schmidt JC (2005) Bionik und Interdisziplinarität. Wege zu einer bionischen Zirkulationstheorie der Interdisziplinarität. In: Rossmann T, Tropea C. (Hrsg.): Bionik. Aktuelle Forschungsergebnisse aus Natur-, Ingenieur- und Geisteswissenschaften. Springer, Berlin, pp 219–246
Shu LH (2010) A natural-language approach to biomimetic design. Artif Intell Eng Des Anal Manuf 24:507–519
Srinivasan V, Chakrabarti A (2010a) Investigating novelty-outcome relationships in engineering design. AI EDAM 24(2):161–178
Srinivasan V (2011) Supporting novelty in conceptual phase of engineering design, Ph D Thesis, Indian Institute of Science, Bangalore, India
Srinivasan V, Chakrabarti A (2010b) An Integrated Model of designing. JCISE 10(3):031013
Srinivasan V, Chakrabarti A (2009) SAPPhIRE—an approach to analysis and synthesis. In: Proceedings of international conference on engineering design [ICED09], Stanford, USA
Srinivasan V, Chakrabarti A, Pal U, Ranjan BSC, Ojha SP, Ranganath R (2011) Supporting process and product knowledge in biomimetic design. Int J Design Eng 4(2):132–158
Sternberg RJIT, Lubart TI (1999) The concept of creativity: prospects and paradigms. In Sternberg RJ (ed) Handbook of creativity. CUP, NY, pp 3–16
Terninko J, Zusman A, Zlotin B (1998) Systematic innovation. An introduction to TRIZ (Theory of Inventive Problem Solving). St. Lucie Press, Boca Raton
Tseng I, Moss J, Cagan J, Kotovsky K (2008) The role of timing and analogical similarity in the stimulation of idea generation in design. Des Stud 29:203–221
Ulrich KT, Seering WP (1990) Function sharing in mechanical design. Des Stud 11(4):223–234

Vakili V, Chiu I, Shu LH, McAdams D, Stone R (2007) Including functional models of biological phenomena as design stimuli. In: Proceedings of ASME international design eng. technical conferences and computers and information in engineering conference, Las Vegas, NV, USA, Sept 4–7 2007, DETC2007-35776 (DTM)

Vattam SS, Helms ME, Goel AK (2008): Compound analogical design: interaction between problem decomposition and analogical transfer in biologically inspired design. In: Proceedings of 3rd international conference on design computing and cognition, Atlanta, Springer, Berlin, pp 377–396

Vattam S, Wiltgen B, Helms M, Goel AK, Yen J (2010a) DANE: fostering creativity in and through biologically inspired design. In: Procedings of 1st International conference on design creativity (ICDC2010), Kobe, Japan

Vattam SS, Helms ME, Goel AK (2010b) A content account of creative analogies in biologically inspired design. AI EDAM 24:467–481

Vincent J, Mann D (2002) Systematic technology transfer from biology to engineering. Philos Trans R Soc Phys Sci 360:159–173

Vincent JFV, Bogatyreva OA, Bogatyrev NR, Bowyer A, Pahl AK (2006) Biomimetics: its practice and theory. J R Soc Interface 3:471–482

Wilson J, Chang P, Yim S., Rosen D (2009) Developing a bio-inspired design repository using ontologies. In: Proceedings of ASME design eng. technical conferences and computers and information in engineering conference, IDETC/CIE, San Diego, DETC2009-87272

Wilson JO, Rosen D (2007) Systematic reverse engineering of biological systems. In: Proceedings of ASME design engineering technical conferences and computers and information in engineering conference (IDETC/CIE), Las Vegas, DETC2007/DTM-35395

Chapter 9
Overcoming Cognitive Challenges in Bioinspired Design and Analogy

Julie S. Linsey and Vimal K. Viswanathan

Abstract Bioinspired design and analogy are powerful tools for innovation. Engineers face many cognitive challenges when seeking to employ design by analogy and bioinspired design. This chapter presents known difficulties engineers must overcome for bioinspired design and summarizes the cognitive psychology, multi-media learning and design evidence for the cognitive challenges. A number of cognitive challenges block a designer from being effective when using design by analogy. The challenges range from retrieving appropriate analogues based on deep similarities to the challenge of seeing multiple solutions based on a single analogue, to becoming fixated on initial solutions. Like any other idea generation process, design fixation limits the solution space explored during design by analogy and bioinspired design. There are empirically proven strategies for mitigating design fixation ranging from presenting uncommon examples to abstractions and categories of solutions. From research on multimedia learning and design, additional heuristics applicable to the design of new bioinspired tools have also been identified. These include annotations directly next to ambiguous or unfamiliar representations to enhance communication and make learning easier. Design heuristics and principles are presented after each section of the relevant research. The chapter ends with the summary of the cognitive design heuristics for bioinspired design methods and tools. This set of heuristics can be used as guidelines for researchers developing new methods and support tools for bioinspired design.

Keywords Design by analogy · Cognitive biases · Fixation

J. S. Linsey (✉) · V. K. Viswanathan
Georgia Institute of Technology, Atlanta, USA
e-mail: julie.linsey@me.gatech.edu

V. K. Viswanathan
e-mail: vimkucv@gmail.com

A. K. Goel et al. (eds.), *Biologically Inspired Design*,
DOI: 10.1007/978-1-4471-5248-4_9,

9.1 Introduction

Bioinspired design and innovation through analogy are powerful tools for design, as demonstrated by the anecdotal and empirical data (Casakin and Goldschmidt 1999; Leclercq and Heylighen 2002; Christensen and Schunn 2007; Basalla 1988). Bioinspired design is a specific type of design by analogy which is based on analogies from nature. Currently, significant work aims to create computer tools and design methods to enhance and support this critical process (Hacco and Shu 2002; Chiu and Shu 2007; Chakrabarti et al. 2005a, b; McAdams and Wood 2002; Cheong et al. 2011; Goel and Bhatta 2004; Linsey et al. 2012; Oriakhi et al. 2011; Linsey et al. 2008a; Helms et al. 2009; Nagel et al. 2010; Vattam et al. 2010). Designers who are attempting to implement bioinspired design and design by analogy face a number of cognitive biases and challenges. Effective tools for bioinspired design and analogy must be effectively designed to overcome these cognitive biases and challenges. This chapter will describe the cognitive considerations that design researchers must be aware of and then list a set of heuristics for analogy tool design.

This chapter begins with a cognitive description of the design by analogy process and the associated challenges including retrieving appropriate analogues, creating multiple solutions based on a given analogue, and selecting design features to copy. The next section then discusses the road blocks created by design fixation and ways to mitigate its effects. The final section presents other considerations and heuristics from research in multimedia learning and design. The chapter ends with a summary of the cognitive challenges and the associated heuristics and principles.

9.2 Analogy

The anecdotal and empirical evidence consistently demonstrates that professional designers often use analogies and it is a powerful tool for innovation (Casakin and Goldschmidt 1999; Leclercq and Heylighen 2002; Christensen and Schunn 2007; Basalla 1988). Further, controlled experimental studies indicate that the naturalistic studies are likely underestimating the actual frequency of analogy use, since naturalistic protocol studies rely on either the designers reporting the use of analogy or some direct indication of it (e.g., stating a plane is like a bird) (Linsey 2007; Linsey et al. 2008b).

9.2.1 Cognitive Process Model for Analogical Reasoning

Psychology has sought to understand the cognitive processes people use to create and understand analogies (Falkenhainer et al. 1989; Gentner and Markman 1997; Hummel and Holyoak 1997). Figure 9.1 shows the basic steps involved in

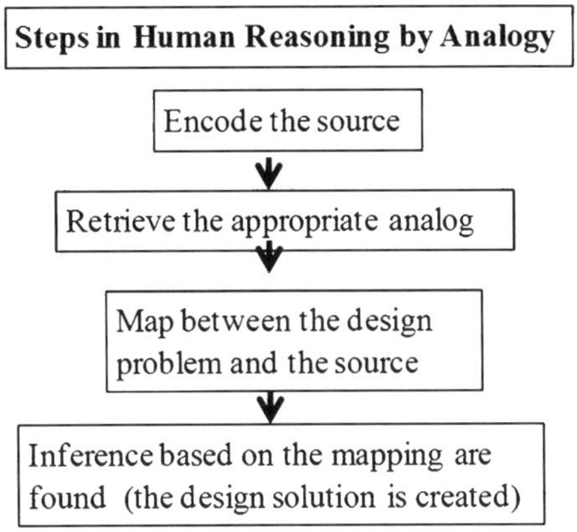

Fig. 9.1 Steps in human reasoning by analogy processes

reasoning by analogy. Analogy is generally viewed as a comparison between two products in which their relational or causal structure match (Holyoak and Thagard 1989; Gentner 1983; Falkenhainer et al. 1989). The problem domain is the *target* of the analogy and the domain of prior knowledge that provides a potential solution to the problem is called the *source*. Research on analogy suggests that people first find a mapping between the relations in the source and the target. On the basis of this mapping, aspects of the target may be re-represented to make them more similar to the source. Furthermore, inferences about the target (i.e., potential solutions) may be made based on the similarity of the target to the source. The potential for creative problem solving is clearest when the two domains being compared are very different on the surface, though the same process of comparison can also be used for domains that share significant surface similarity (Gentner and Markman 1997).

Designers face many challenges during the analogical reasoning process. Retrieving an appropriate analogue from memory has often been shown to be the most challenging step (Holyoak and Thagard 1989; Gentner 1983; Gentner and Markman 1997; Falkenhainer et al. 1989; Holyoak and Koh 1987; Gick and Holyoak 1980; Markman and Gentner 1993b). Search engines and other bioinspired design tools currently in development can assist with this (Hacco and Shu 2002; Chiu and Shu 2007; Chakrabarti et al. 2005a; Chakrabarti et al. 2005b; McAdams and Wood 2002; Cheong et al. 2011; Goel and Bhatta 2004). People also tend to focus too much on surface features instead of deep similarities (Gentner and Landers 1985; Gick and Holyoak 1980; Helms et al. 2009). For example, when using a bird as an analogue for human flight, the color of the bird is a surface feature. Even once an effective analogue has been retrieved from either memory or with computer tools, studies have shown that there is a difficulty in selecting appropriate design features to map from the analogue to the problem (Linsey et al. 2006; Linsey et al. 2007; Mak and Shu 2008; Cheong and Shu 2009; Helms et al. 2009). Inaccurate models, both in scientific knowledge and the designers' knowledge can cause the wrong features to be selected and mapped.

Linsey et al. (Linsey et al. 2007) found that senior undergraduate mechanical engineers who were required to use an airplane as an analogy and to map design features that cause lift would occasionally select incorrect features such as the smoothness of the plane or the propellers and try to map these features to the solution rather than the shape of the wing for lift. This was in spite of the fact that these students had taken and passed fluid dynamics where this material would have been covered.

Heuristics

- *Deep Connections*—Encourage deep and functional connections between the analogue and the problem.
- *Matching Features*—Highlight appropriate features that should be mapped from the analogue to the problem.

9.2.2 Inaccurate Mental Models and Naïve Physics

Since at least da Vinci's time, birds were recognized as good analogues for human flight, but no one found a practical solution. The Wright brothers recognized that birds' twisting of their wings for control was an effective means to successful control of their plane, rather than feathers or flapping being the key to success. Often due to designers' inaccurate mental models, an appropriate analogy will be identified, but an effective solution will not result. People have mental models of the world that are highly efficient for the tasks they frequently complete; but generally inaccurate unless significant experience or training has occurred (Markman 1999; McAfee and Proffitt 1991; Chi et al. 1981). The Wright brothers had a rough theory of aerodynamics providing them with a better mental model of flight and also guiding them to select appropriate features from birds.

Significant research in psychology has sought to understand how people reason about the physical world and this directly impacts the concepts engineers develop along with their evaluation. The areas of mental models and Naïve physics seek to understand how people reason about the world around them (e.g., Gentner and Stevens 1983; Forbus 1984; Kuipers 1994). Mental models of physical systems are internal mental representations of external systems (Markman 1999). The area of Naïve physics has produced some rather surprising findings about people's mental models which have direct implications for engineering design research. An important finding is people's mental models of physical phenomena, which they observe frequently, can be surprisingly inaccurate. For example, when asked to draw the water level of a glass tipped on its side, over 40 % will not be able to indicate that the water line will be parallel to the horizontal plane (Markman 1999; McAfee and Proffitt 1991). Similarly, inaccurate mental models of home-heating systems have also been observed. Many people have the mental model of a home thermostat like that of a car's accelerator: the higher the thermostat is set, the faster the house is heated (Kempton 1986). In reality, most home-heating and home-cooling systems are either on or off. This mental model has a direct real-world impact on people's behavior and the environment. People with this mental model tend to change the thermostat frequently throughout the day which is inefficient and wastes energy. Limits in the capability of people's mental models have also been shown with highly trained scientists (Hutchins 1995).

Naïve physics' implications for engineering design are that engineers' mental models are not likely to be highly accurate unless they have been repeatedly tested through either extensive experience in situations where more accurate mental models were required or through education. This is true for phenomena that engineers may have observed many times. The challenge is even greater for aspects of biology that engineers are not at all familiar with. This means that for very innovative concepts, engineers are not likely to have accurate mental models of behavior and will judge a concept inaccurately. Very viable and effective concepts may be overlooked due to inaccurate mental models. Helms et al. (2009) document one of the common errors with student teams is that they often miss the significance of the underlying principles of the biological analogue and then oversimplify the complex functions of the analogue.

Heuristic

- Supplement designers' mental models—assist designers by supplementing their erroneous mental models. Provide physical prototypes, virtual prototypes and other external models that test and provide feedback to the designers on their mental models. Provide other means to test and evaluate designers' mental models.

9.2.3 Distant Domain Analogies are Initially Bypassed

Another challenge engineers face is that distant domain analogies are ignored unless the design problem is open (unsolved). Engineers tend to ignore distant domain analogies unless they have spent time attempting to solve a problem and are having difficulty, an "open" problem. If the analogous information is more distantly related to the problem, it is more likely to be used to solve the problem if it is presented when there is an open design problem to be completed rather than before work has started on the problem (Tseng et al. 2008).

Heuristic

- *Present distant domains analogues later*—Present more distant domain analogues after the designer has spent time working on the problem and it is still unsolved.
- *Encourage distant domain analogy use*—Encourage designers to use distant domain analogues by suggesting they create new solutions for every example

presented or by highlighting how distant domain analogues are related to the problem.

- *Provide several far-domain examples*—Present more than one far-domain example.

9.2.4 Single Inference Bias and Principles for Creating Multiple Solutions

Another bias designers face is that it is often difficult to create many solutions based on a single analogue (Krawczyk et al. 2005; Holyoak and Thagard 1989; Gadwal and Linsey 2010a). For most tasks in everyday life, a single, most likely to work solution is desired, for example, if you walk up to a door, you probably want to try opening it the same way you did last time. The fundamental purpose of analogy is to generate plausible and useful inferences. In order to obtain useful inferences from analogical reasoning, analogical mappings have to be constrained, otherwise too many inferences are possible (Krawczyk et al. 2005; Holyoak and Thagard 1989). Very often multiple inferences based on a single analogue do exist. For example, the analogy between a bird and human flight is a case where the wrong inference was made for a long time. There have also been a number of wall climbing devices and adhesives based on the gecko lizard which is able to quickly scale walls (Gadwal and Linsey 2010a). In general, it is difficult to develop multiple inferences (solutions) from a single analogue (Krawczyk et al. 2005; Holyoak and Thagard 1989). Engineers can develop multiple inferences when instructed to do so, but it is not an easy task (Gadwal and Linsey 2010a). On average, participants were only able to create three solutions in 30 min, indicating it is a difficult task and engineers need methods to assist them.

To assists engineers, Gadwal and Linsey created a set of principles from existing examples of multiple solutions based on a single analogue (Fig. 9.2, Gadwal and Linsey 2010b). They asked designers to generate ideas for a device dispensing flour efficiently. The provided a child's toy device that can hold and release substances, as shown in Fig. 9.2, as an analogue for the problem. The set of principles were derived based on the ideas generated by the participants. The principles are as follows:

- *Change the scale of the feature mappings*
  - *Map at the same scale.*
  - *Map at different scales*: Substitute physical principles (e.g., magnetic instead for van der Waals).
- *Abstract the problem*
  - *Change the abstraction of the design problem* (e.g., "wall climbing" to the more general case of "adhesion"). The WordTree Method can facilitate this process (see Linsey et al. 2008a, 2012 for more details). Figure 9.3 displays

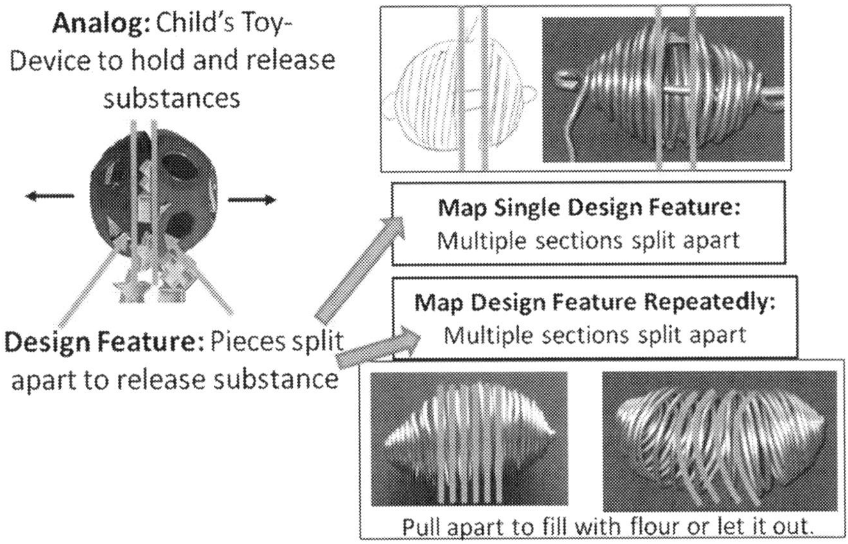

Fig. 9.2 Above are two solutions for a device to sprinkle flour based on an analogue to a child's toy. These solutions illustrate the design principle for multiple solutions of repeating a design feature (Gadwal and Linsey 2010b)

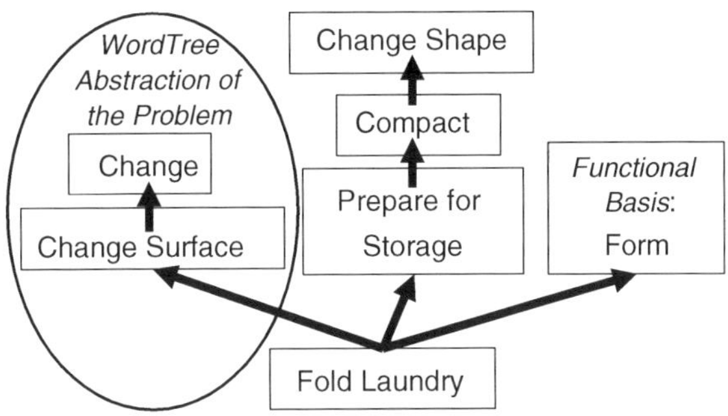

Fig. 9.3 Examples of abstraction of the design problem of folding laundry

three different abstraction of the problem of designing a device to fold laundry. Using the WordTree Method, this can be abstracted to "change surface" and even more generally to just change. Fold could also be thought of more abstractly as a problem to "prepare laundry for storage," "compact" or most abstractly as "changing shape." The functional basis (Otto and Wood 2001) results in an abstract to the function "form."

- Choose different features
  - List properties and design features. Map various features into solutions. For example, Fig. 9.4 shows the design problem of creating a lightweight travel

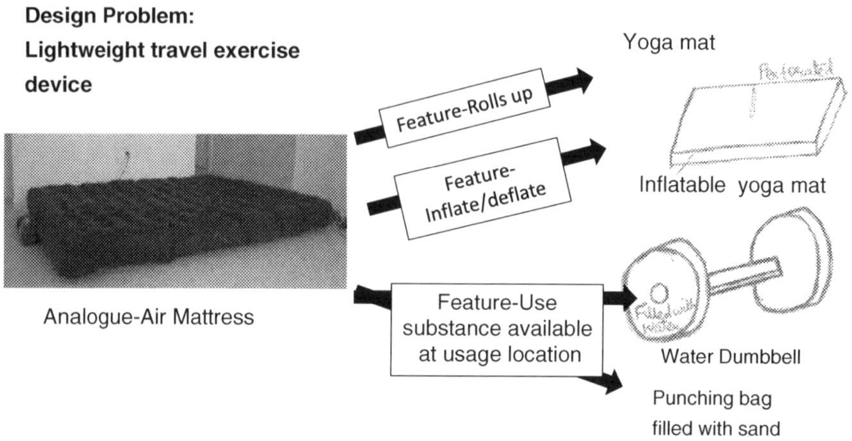

Fig. 9.4 Examples of multiple solutions created based on different features of the analogue and by abstracting the analogue features

exercise device and the analogue of an inflatable mattress (lightweight and easy to store). In Fig. 9.4, the design features of "rolls up" and "inflate/deflate" are mapped to solutions for portable exercise equipment.)

- Abstract the features. The WordTree Method can facilitate this. For example, "inflate/deflate" can be abstracted to "increase in size", "use available substance at use location to make the device functional", "expand", etc.
- Repeat a design feature or take a repeated design feature and use it only once. Figure 9.2 demonstrates this design principle through the design of device to sprinkle flour based on an analogy to the child's toy. The design in the top right of the figure uses the design feature once to release the flour. The design in the bottom right repeats the same design feature.

Heuristic

- *Multiple Inferences*—Assist users in creating multiple inferences. Use multiple analogies design principles.

9.2.5 Alignable Differences

Data from cognitive psychology indicates that individuals systematically and unintentionally ignore information that they believe to be important when it is non-alignable with what they know about other options (Lindemann and Markman 1996). For example, a bicycle and a motorcycle have many alignable differences (one uses a motor and the other a human power, motorcycles travel longer distances and are more expensive) whereas a car and a phone have very few alignable differences.

It is actually hard to say what is different about a car and a phone. When making a selection and when finding similarities during the analogical reasoning process, individuals will focus more on alignable differences than non-alignable ones even though in most design situations both are likely equally important.

Heuristic

- *Alignable and Non-alignable*—Illustrate both alignable and important non-alignable differences between the analogue and the problem

9.2.6 Biases and Challenges Still Requiring Heuristics to Overcome Them

There are additional challenges that currently do not have tools to assist in overcoming them, but are still important. Confirmation bias is the tendency to accept information that is consistent with current beliefs while ignoring information discredits or is contrary to current beliefs. Within design by analogy and bioinspired design, (Hallihan et al. 2012) demonstrate that designers suffer confirmation bias during concept generation. This likely leads designers not to re-evaluate the initial analogies they select, nor evaluate their initial analogical mappings. Helms et al. (2009) observe design teams sticking to their first analogue which is likely due to a confirmation bias.

Helms et al. (2009) have documented other common errors that student teams encounter. Professionally likely face many of the same challenges. Many of these errors have been referenced in the previous sections with heuristics for overcoming them, but a few of the identified common errors have not been discussed. One error is using the "off-the-shelf" biological solution, for example when designing a device to shell peanuts, designers suggest training squirrels to shell the peanut rather than using the principles a squirrel implements. Designers also often map the wrong features from the analogue to the inspired solution frequently due to superficial matches. Another issue is designers also transfer additional features which are not required for the problem domain.

9.2.7 Approaches to Increasing Success in Design by Analogy

While much work needs to be done in order to identify approaches for overcoming the cognitive challenges during analogical reasoning, a few proven approaches have been identified. Wherever possible, the high-level principle between the analogue and the problem should be presented. Presenting both the abstract principle and the analogue significantly increases success rates (Thompson et al. 2000; Namy and Gentner 2002b; Mak and Shu 2004). The presence of the analogue helps designers in understanding the underlying principle and this facilitates the analogical transfer of that principle instead of the surface features in the analogue.

Unfortunately, the abstract principle is not often known prior to the solution being created. Presenting at least two analogues also increases success rate. To maximize success, these analogues should be diverse (Clark and Mayer 2008). The examples should be different from each other to facilitate identification of the principle and the unimportant surface features should be different. Prior work has consistently shown that presenting two analogues rather than only one increases the rate at which the problem will be solved (Namy and Gentner 2002a; Markman and Gentner 1993a; Keane 1988; Gick and Holyoak 1980; Lopez 2011; Gadwal and Linsey 2010a). What is unknown is how many examples combined with the abstract principle is most effective, what the optimal number is in the presence of extraneous information, or how different the examples should be.

Most of the prior work with the exception of Lopez et al. (Lopez 2011; Gadwal and Linsey 2010a; Lopez et al. 2012) evaluated the effect in multiple analogues with very little extraneous information that participants could focus on. In realistic situations, designers must filter out the extraneous information to identify the critical features which should be aligned and mapped between the analogues and the solution. Said study did measure the effect of presenting participants with multiple analogues both with and without significant extraneous products (Fig. 9.5). They implemented an experiment with two factors: the number of analogues (four levels: 1, 2, 3, or 5 products) and the level of extraneous products (two levels: none or 3 extraneous products per analogue). Participants were presented with a set of products and then asked to solve a design problem. After solving the design problem, participants were also asked to identify the principle from the analogues that lead to the solution. Consistent with prior studies, without extraneous products being present, two analogues were optimal to obtain success in solving the design problem (Fig. 9.5). When a combination of both useful and extraneous analogues was presented, three analogues were optimal. Consistent with prior studies, multiple analogues that share the same principle are effective for increasing analogical transfer.

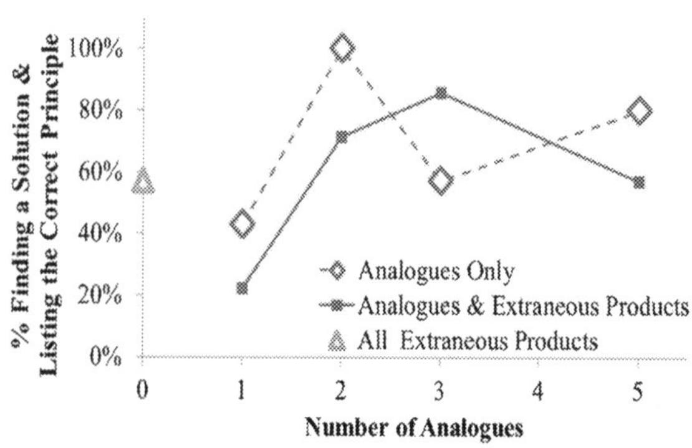

Fig. 9.5 Percentage of participants finding a solution based on the analogues and correctly identifying the principle

Heuristics

- *Abstract and Specific*—Present both the abstract principle and specific analogue examples to maximize analogical transfer.
- *Diverse examples*—Provide several diverse examples for far transfer.
- *Multiple Analogues*—Present two to three analogues that share the same principle.

9.3 Design Fixation

Any tool or method to support and enhance the bioinspired design process must minimize design fixation. The existing literature demonstrates the susceptibility of both experts and novices to design fixation (Jansson and Smith 1991; Wiley 1998). Design fixation causes the blind adherence of designers to presented examples or their initial ideas (Jansson and Smith 1991). Fixation narrows the solution space where designers conceive their ideas and decreases creativity. Design fixation is a common problem for practicing engineers, design faculty, and engineering students (Purcell and Gero 1996; Jansson and Smith 1991; Linsey et al. 2010; Christensen and Schunn 2007).

There are a number of factors that likely influence design fixation including the sunk cost effect, unusual solutions, abstraction and categories of solutions (Table 9.1). Current and prior work indicate that sunk cost has great potential to cause design fixation (Viswanathan and Linsey 2010; Viswanathan and Linsey in review). Sunk cost basically indicates that once a decision is made or significant effort is applied, people tend not to change the course of action (Kahneman and Tversky 1979a; Arkes and Blumer 1985). This indicates that designers should be more fixated when significant effort has gone into a project. In design, it has been observed that once prototypes are built, designers make very few design changes (Christensen and Schunn 2007). Sunk cost's impact is further described in the next section.

Two experimental studies attempting to identify the benefits of brainstorming groups presented participants with uncommon examples (ideas participants rarely

Table 9.1 Factors that likely influence design fixation

	Factors	Examples
Increases fixation	Sunk cost	Time-consuming and expensive prototypes
Decreases fixation	Unusual solutions	Analogies to nature
	Abstraction	Functions, design principles such as TRIZ or design for X
	Categories of solutions	Energy sources, checklists
	Focusing on ineffective features	Providing the causal reason why a design is ineffective

thought of spontaneously) and found that more ideas were generated compared to those who received common examples, indicating that fixation is probably reduced (Dugosh and Paulus 2005; Perttula and Sipila 2007). Providing uncommon examples like biological analogues can reduce design fixation. Commonly identified biological analogues will likely not have this effect, but most biological analogues would be uncommon examples.

Limited direct experimental work exists on identifying approaches for overcoming design fixation, but two studies have directly identified approaches: Chrysikou and Weisberg (2005), Linsey et al. (2010). First, Chrysikou and Weisberg (2005) replicated a study by Jansson and Smith (1991) and also presented participants with warnings on features they should not replicate. Chrysikou and Weisberg (2005) did not observe participants fixing to the example like the Jansson and Smith study did. The Jansson and Smith study also told the participants which features to not duplicate. A likely reason the Chrysikou and Weisberg study did not see fixation is because participants in this study were forced to focus more on what they should not do.

Linsey et al. (2010) reduced design fixation through providing participants with a combination of unusual solutions (distant domain analogies), abstractions of the design problem (functions of a device), and categories of solutions (categories of natural energy sources). In the Linsey et al. study, an experimental group is provided with a fixating example (Fixation Group), whereas another group is provided with the same example and the defixation materials (Defixation Group). A third group did not receive an example (Control). Linsey et al. found that all the defixation materials in combination reduced experts' fixation; however, did not identify which individual factors reduced design fixation. A follow-up study found that fixation was only reduced for experts (design faculty) but not for more novice engineers (senior mechanical undergraduates) (Viswanathan and Linsey 2012, 2013, Fig. 9.6). When a fixating example was present, the quantity of non-redundant ideas generated by both faculty and novice designers was reduced, indicating fixation. When the defixation materials were present, the quantity increases for faculty designers, but not for novices. This indicates that defixation

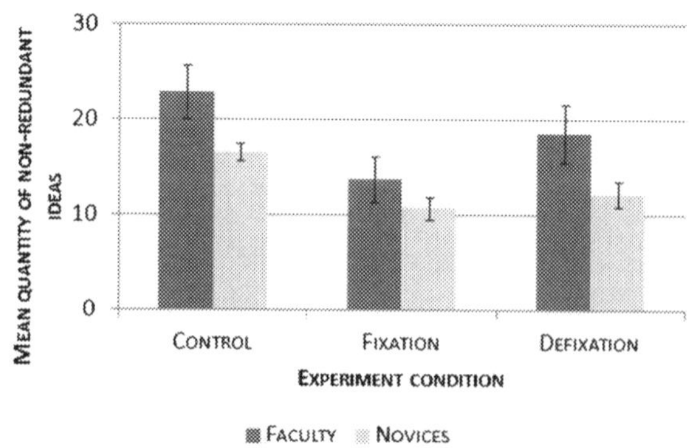

Fig. 9.6 The defixation materials only reduced fixation for the experts (*Error bars* ±1 S.E)

materials mitigated fixation for faculty designers, but those materials were not equally effective on novices. Uncommon examples, abstractions, and categories likely reduce design fixation.

Numerous design books also advocate the use of abstraction and categories in design (Otto and Wood 2001; Antonsson and Cagan 2001; Pahl and Beitz 1996). Suggested abstractions include, but are not limited to, functional, graph grammars, black boxes, and bond graphs. Useful categories include the functional basis, idea generators and other checklist methods, physical principles, and classifying schemes (e.g., working geometry, and working motions). Idea generation methods such as Mind Maps and Morph Matrices also highlight categories.

Much further, work is required to study these influences on design fixation and to identify more ways to overcome design fixation. The effect sizes of each of these factors also need to be measured, so that factors with greatest impact can be focused on. The types of abstractions that reduce design fixation also need to be determined. Ideally, bioinspired design tools would be able to help designers in recognizing their design fixation and then provide stimuli to reduce that fixation. For identifying design fixation in a design session, Gero (2011) has developed a metric based on linkography from coding design protocol. Gero's metric does not measure fixation relative to a control group; so his approach, unlike the measures often implemented in fixation studies, could be incorporated into a computational tool to identify when a designer is fixated and then provide examples to reduce fixation.

Heuristics

- *Uncommon Examples*—Common examples cause design fixation and uncommon ones do not. Unusual example (analogues) should be provided whenever possible.
- *Causal Reason for Ineffectiveness*—Explain why certain features are undesirable in the design.
- *Categories of Solutions*—Categories of solutions can help designers identify associated principles.
- *Encourage Abstraction*—Encourage thinking in terms of high-level principles instead of surface features.

9.3.1 Sunk Cost Effect

Identified by behavioral economics, the sunk cost effect manifests a greater tendency to continue in a selected path, after significant money, time, or effort is invested in that path, even when an alternate path is more beneficial for the future endeavors (Arkes and Blumer 1985). Good decisions should be based on the expected costs of the choices in the future not past sunk costs (Keeney and Raiffa 1993; Holcomb and Evans 1987). However, in actual practice, sunk costs do affect decisions, due to the sunk cost effect (Kahneman and Tversky 1979b). Some good examples of this effect are portrayed by Thaler (1980). The resale prices of cars are generally guided by the current market price, whereas the sellers always decide based on the original buying price.

The sunk cost effect can fixate designers to their initial ideas, especially when they spend more time or effect (costs) on those ideas. In engineering design, the cost can be money, time, or effort that designers spend to solve a problem. Once significant investment of these resources is made into a particular solution path, designers tend to fixate on that path. In engineering design, the generation of highly novel ideas is important and this requires "out-of-the box" thinking. The adherence to one selected solution path can hinder this target. This can be especially true when designers build physical models of their ideas during idea generation. If this building process takes longer, the chances of fixation is also greater.

A controlled study conducted by the Viswanathan and Linsey (Viswanathan and Linsey 2011; Viswanathan and Linsey in review) has shown evidence supporting the presence of the sunk cost effect in engineering idea generation. In that study, novice designers were instructed to generate as many ideas as possible for a small object that could securely bind ten sheets of paper together without damaging them. The participants were randomly assigned to five different conditions. The first condition was sketching only condition, in which the participants sketched their ideas. In the second condition, metal building, the participants sketched their ideas and built those with steel wire. The third condition was a plastic building condition, in which the participants sketched their ideas and molded those out of plastic. Building ideas with plastic consumes more time as compared to building with metal, making the associated sunk cost higher for plastic building. The fourth and fifth conditions were metal constrained sketching and plastic constrained sketching, respectively. In these conditions, the participants were told that they would build their ideas in the second half of the experiment and were instructed to sketch their ideas. These constrained sketching conditions isolated

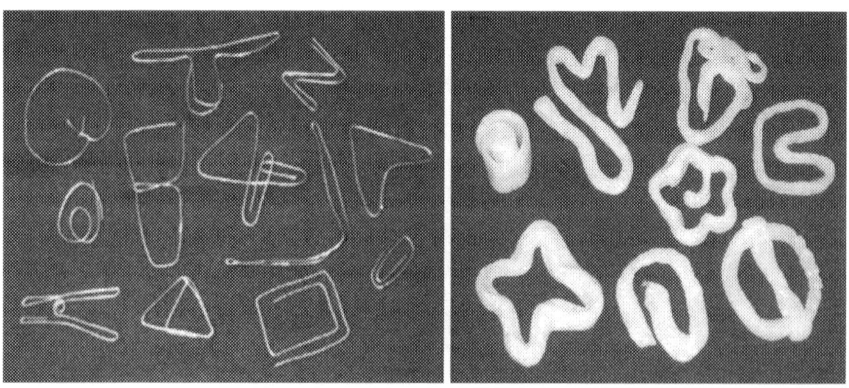

Fig. 9.7 Example physical models built by the participants in the study

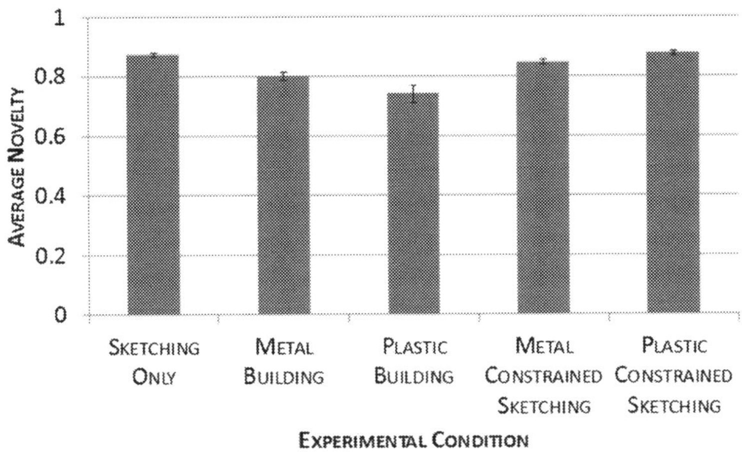

Fig. 9.8 Variation of average novelty across the conditions (*Error bars* ±1 S.E.)

any possible effects of implicit constraints imposed by the building materials and processes. Figure 9.7 shows some example solutions generated by the participants.

The presence of design fixation in the ideas generated by participants was measured using novelty and variety metrics (Shah et al. 2000; Linsey et al. 2005; Linsey et al. 2011). Design fixation is believed to cause ideas to be less novel and for participants to search a smaller portion of the design space. Novelty measured how frequently a particular idea occurred. The ideas were sorted into bins of very similar ideas and novelty was calculated as one minus the frequency of ideas in each bin. Variety measures the span of the entire solution space that a participant's

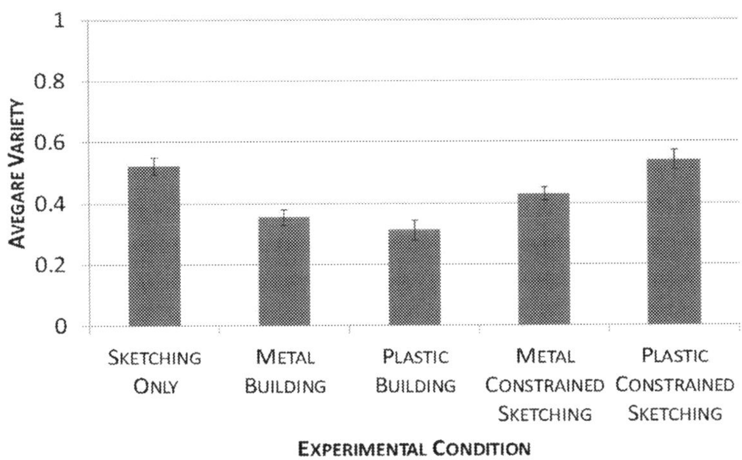

Fig. 9.9 Variation of average variety across the conditions (*Error bars* ±1 S.E.)

ideas cover. This is measured as the ratio of the number of bins that the participant's ideas occupy to the total number of bins. Figures 9.8, 9.9 show the variation of these measures across the five conditions. Participants who built their ideas with metal generated ideas with a lower average novelty and variety as those in the Sketching Only Condition, indicating design fixation. The data clearly showed a further reduction in novelty of ideas when they built those ideas with plastic. As building with plastic consumed more time as compared to building with metal, the sunk cost effect could be causing this reduction. Comparing the Sketching Only Condition with the constrained sketching conditions, there was no effect of implicit constraints observed on novelty and variety.

Heuristic

- *Minimize Sunk Cost*: Minimize sunk cost (time, effort, money, etc.) to encourage changes in the course of action and minimize fixation. Tools and methods should require little time, money, and effort from the designer in order to be effective.

9.4 Other Considerations

Much work has been done in the area of human computer interaction on creating effective computer tools. Much more applicable work exists beyond the scope of this chapter. Within this area, research on learning in multi-media environments and from design research provides guidelines applicable for the development of new methods and computer-based tools for bioinspired design. Biological phenomena are generally outside a designer's knowledge domain; therefore, before an effective solution can be created based on the biological phenomena, the designer must first learn about the biology to some degree. Engineering design research has demonstrated that ambiguous or unfamiliar representations should be annotated with short phrases to enhance communication (Hisarciklilar and Boujut 2009; Linsey et al. 2011). Information, such as a graphic and annotation that must be integrated, should not be separated. Images and labels should be placed close to each other (Clark and Mayer 2008). Figure 9.10 demonstrates the principles of annotation and contiguity. The images on the left show two examples that do not follow these principles and the image on the right shows a clear description of the flat spring outlined in the image (an ambiguous representation) placed directly next to the image.

Researchers also must be aware that much depends on the knowledge and experiences of the user (Clark and Mayer 2008). Certain representations will be effective for different groups of people based on their prior experiences. A number of studies in engineering design demonstrate cognitive differences between novices and experts in design (Ball et al. 2004; Casakin and Goldschmidt 1999;

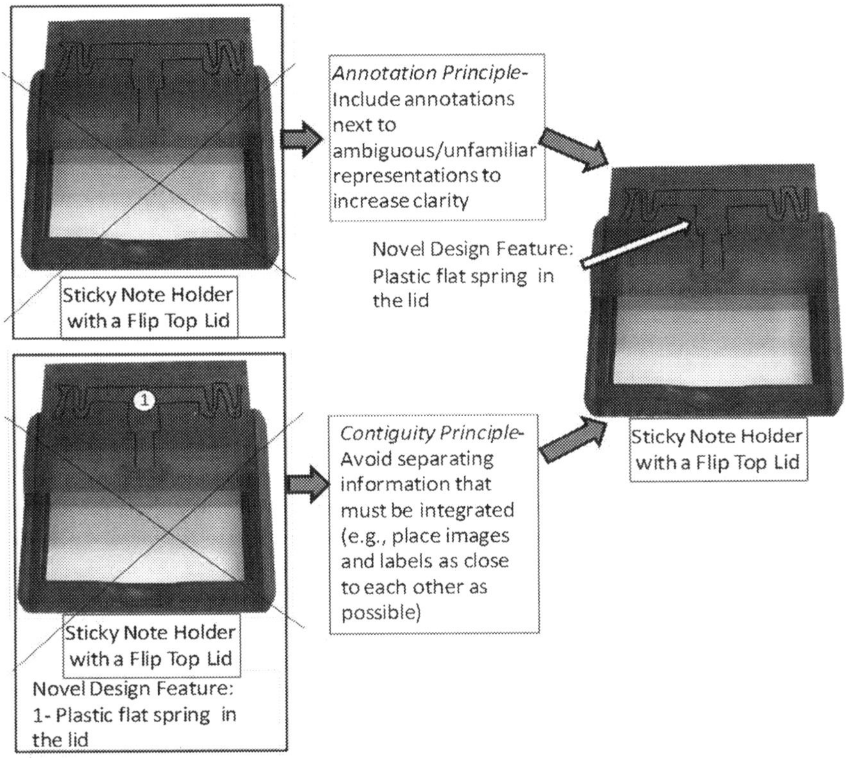

Fig. 9.10 Illustration of the annotation and contiguity principles

Chi et al. 1981; Moss et al. 2006). Experimental findings often differ based on expertise level and discipline expertise. Purcell and Gero (1996) found that experts in mechanical engineering (senior undergraduates) would fixate on presented examples while experts in industrial design (senior undergraduates) did not fixate. The impact of intervention to assist designers can also be impacted by expertise. Viswanathan and Linsey (2012, 2013) found that defixation materials such as analogies, categories of solutions, and abstractions effectively reduced experts' fixation but did not assist novices.

Further challenging the situation, users' perceptions about the effectiveness of an idea generation method are often inconsistent with the quantitative outcomes. Users believe that brainstorming, where ideas are shared verbally with a group then recorded, is more effective than brainwriting where individuals simultaneously write down their ideas and then exchange them (Paulus and Yang 2000). In Paulus and Yang's study, people generated more ideas during brainwriting than brainstorming but believe that brainstorming is more effective for idea generation. The inaccuracy of productivity perceptions has also been found with experts in design. Linsey et al. (2010) in a study on design fixation presented experts with poor design examples. The experts inaccurately believed that the example did not hurt their idea generation process. The experts were somewhat accurate at perceiving that the defixation materials alleviated their fixation.

Heuristics

- *Annotation*—Include annotations next to ambiguous/unfamiliar representations to increase clarity.
- *Contiguity*—Avoid separating information that must be integrated (do not separate images and labels and place them as close to each other as possible).
- *User Dependence*—A representation's effect depends on the prior knowledge of the user.
- *User Perceptions*—Measure both user perceptions and quantitative outcomes when assessing a method's effectiveness.

9.5 Summary

This chapter presents the known cognitive challenges that designers face when attempting to implement design by analogy and bioinspired design. Cognitive research in design, psychology, and multi-media learning has identified these cognitive challenges. Table 9.2a, b summarizes the known cognitive challenges that designers face along with providing a summary of the principles and heuristics that methods and tool developers can implement to assist designers in overcoming the cognitive challenges. Design by analogy is a powerful but not easy cognitive process. Designers need assistance in retrieving appropriate analogues based on deep and functional connections. Due often to inaccurate mental models of how the analogue works or the non-alignable differences, designers may select the wrong features from the analogue to implement to find a solution. In addition, designers tend to initially unintentionally ignore distant domain analogues and then are only able to find a few inferences (solutions) based on each analogue even though numerous solutions may actually exist. Design fixation, initial examples or solutions limiting creativity, further inhibits bioinspired design.

While significant challenges exist, there is much that developers of new design methods and tools can do to assist designers. Computer-based designer tools can retrieve analogues based on deep connections between the problem and analogue such as functional or other relational similarity. Designers need to be encouraged to identify multiple solutions (inferences) based on each analogue presented. Presenting diverse examples, multiple analogues that share the same principle, a diverse set of analogues and the abstract principles will also facilitate the design by analogy process. To minimize design fixation, designers need uncommon solutions, categories of solutions and abstractions. New tools also need to not require significant sunk costs of time, money, or effort from designers in order to minimize design fixation.

Table 9.2 Summary of cognitive challenges that designers face during design by analogy and heuristics to guide the design of tools to assist designers

	Cognitive biases and challenges	Principles and heuristics
a		
Challenges in analogy	Retrieving based on unimportant (surface) features	Deep connections—encourage deep and functional connections between the analogue and the problem
	Selecting the wrong features to map	Matching features—highlight appropriate features that should be mapped from the analogue to the problem
	Inaccurate mental models—individuals' mental models of the world are cognitively efficient but often inaccurate	Supplement designers' inaccurate mental models (e.g., external models, feedback, and testing)
	Distant domains are initially bypassed	Present distant domains later. Encourage distant domain use. Provide several far-domain examples
	Single inference—designers only infer a single analogue for a problem	Multiple inferences—assist users in creating multiple inferences. Use multiple solution design principles
	Alignable differences—individuals focus more on alignable differences than non-alignable ones. Individuals systematically ignore important information when it is non-alignable	Alignable and non-alignable—illustrate both alignable and important non-alignable differences between the analogue and the problem
	Analogical inference is cognitively difficult	Diverse examples—provide several diverse examples for far transfer
		Multiple analogues—present two to three analogues that share the same principle whenever possible
		Abstract and specific—present both the abstract principle and specific analogue examples to maximize analogical transfer

(continued)

Table 9.2 (continued)

	Cognitive biases and challenges	Principles and heuristics
	Confirmation bias	
	"Off-the-shelf" solutions	
	Transferring extraneous features	
b		
Design fixation	Designers fixate	Uncommon examples—common examples cause design fixation and uncommon ones do not
		Causal reason for ineffectiveness—explain why certain features are undesirable in the design
		Categories of solutions—categories of solutions can help designers identify associated principles
		Encourage abstraction—encourage thinking in terms of high-level principles instead of surface features
	Sunk cost effect—designers fixate when associated sunk cost is high	Minimize sunk cost—minimize sunk cost (time, effort, money, etc.) to encourage changes in the course of action and minimize fixation. Tools and methods should require little time, money, and effort from the designer in order to be effective
Other considerations	Other considerations for the design of tools and methods	Annotation—include annotations next to ambiguous/unfamiliar representations to increase clarity
		Contiguity—avoid separating information that must be integrated (do not separate images and labels and place as close to each other as possible)
		User dependence—a representation effects depend on the prior knowledge of the user
	Inaccurate user perceptions—user's perceptions of the impact of methods are often inaccurate when compared with the quantitative outcomes	User perceptions—measure both user perceptions and quantitative outcomes when assessing a method's effectiveness

The chapter summarizes only the known cognitive biases and many more likely exist. Most of the challenges and heuristics in the chapter are based on a limited number of studies and much more research needs to be done. Almost nothing is known about the relative effect sizes or the impact of multiple principles in combination. There is little data on the influence of expertise in regards to developing new tools or with the presented cognitive challenges and heuristics.

References

Antonsson EK, Cagan J (eds) (2001) Formal engineering design synthesis. Cambridge University Press, Cambridge

Arkes HR, Blumer C (1985) The psychology of sunk cost. Organ Behav Hum Decis Process 35(1):124–140

Ball LJ, Ormerod TC, Morley NJ (2004) Spontaneous analogising in engineering design: a comparative analysis of experts and novices. Des Stud 25(5):495–508

Basalla G (1988) The evolution of technology. Cambridge University Press, Cambridge

Casakin H, Goldschmidt G (1999) Expertise and the use of visual analogy: implications for design education. Des Stud 20(2):153–175

Chakrabarti A, Sarkar P, Leelavathamma B, Nataraju BS (2005a) A behavioural model for representing biological and artificial systems for inspiring novel designs. Paper presented at the international conference on engineering design, Melbourne

Chakrabarti A, Sarkar P, Leelavathamma B, Nataraju BS (2005b) A functional representation for aiding biomimetic and artificial inspiration of new ideas. AIEDAM 19(2):113–132

Cheong H, Shu LH (2009) Effective analogical transfer using biological descriptions retrieved with functional and biologically meaningful keywords. Paper presented at the ASME international design engineering technical conferences and computers and information in engineering conference, San Diego

Cheong H, Chiu I, Shu LH, Stone RB, McAdams DA (2011) Biologically meaningful keywords for functional terms of the functional basis. J Mech Des 133:021007. doi:http://dx.doi.org.lib-ezproxy.tamu.edu:2048/10.1115/1.4003249

Chi MTH, Feltovich PJ, Glaser R (1981) Categorization and representation of physics problems by experts and novices. Cogn Sci 5:121–152

Chiu I, Shu LH (2007) Biomimetic design through natural language analysis to facilitate cross-domain information retrieval. AIEDAM 21(1):45–59

Christensen BT, Schunn C (2007) The relationship of analogical distance to analogical function and pre-inventive structures: the case of engineering design. Mem Cogn 35(1):29–38

Chrysikou EG, Weisberg RW (2005) Following the wrong footsteps: fixation effects of pictorial examples in a design problem-solving task. J Exp Psychol Learn Mem Cogn 31(5):1134–1148

Clark R, Mayer RE (2008) E-learning and the science of instruction: proven guidelines for consumers and designers of multimedia learning, 2nd edn. Wiley, San Francisco

Dugosh KL, Paulus PB (2005) Cognitive and social comparison processes in brainstorming. J Exp Soc Psychol 41:313–320

Falkenhainer BF, Forbus KD, Gentner D (1989) The structure mapping engine: algorithm and examples. Artif Intell 41(1):1–63

Forbus KD (1984) Qualitative process theory. Artif Intell 24:85–168

Gadwal A, Linsey J (2010a) Exploring multiple solutions and multiple analogies to support innovative design. Paper presented at the international conference on design, computing and cognition, Stuttgart

Gadwal A, Linsey J (2010b) Inspiring multiple solutions from a single analogy. Paper presented at the 13th AIAA/ISSMO multidisciplinary analysis optimization (MAO) conference, Fort Worth

Gentner D (1983) Structure mapping—a theoretical framework. Cogn Sci 7:155–177

Gentner D, Landers R (1985) Analogical remindings: a good match is hard to find. Paper presented at the international conference on systems, man, and cybernetics, Tucson

Gentner D, Markman AB (1997) Structure mapping in analogy and similarity. Am Psychol 52:45–56

Gentner D, Stevens AL (1983) Mental models Lawrence Erlbaum associates. Hillsdale, New Jersey

Gero JS (2011) Fixation and commitment while designing and its measurement. J Creative Behav 45(2):108–115

Gick ML, Holyoak KJ (1980) Analogical problem solving. Cogn Psychol 12:306–355

Goel AK, Bhatta SR (2004) Use of design patterns in analogy-based design. Adv Eng Inform 18(2):85–94

Hacco E, Shu LH (2002) Biomimetic concept generation applied to design for remanufacture. Paper presented at the ASME design engineering technical conferences and computer and information in engineering conference, Montreal

Hallihan GM, Cheong H, Shu LH (2012) Confirmation and cognitive bias in design cognition. A paper presented at ASME design engineering technical conferences, Chicago

Helms M, Vattam SS, Goel AK (2009) Biologically inspired design: process and products. Des Stud 30(5):606–622

Hisarciklilar O, Boujut JF (2009) An annotation model to reduce ambiguity in design communication. Res Eng Des 20:171–184

Holcomb JH, Evans DA (1987) The effect of sunk costs on uncertain decisions in experimental markets. J Behav Econ 16(3):59–66

Holyoak KJ, Koh K (1987) Surface and structural similarity in analogical transfer. Mem Cogn 15(4):332–340

Holyoak KJ, Thagard P (1989) Analogical mapping by constraint satisfaction. Cogn Sci 13(3):295–355

Hummel JE, Holyoak KJ (1997) Distributed representations of structure: a theory of analogical access and mapping. Psychol Rev 104(3):427–466

Hutchins E (1995) Cognition in the wild. MIT Press, Cambridge

Jansson D, Smith S (1991) Design fixation. Des Stud 12(1):3–11

Kahneman D, Tversky A (1979a) Prospect theory: an analysis of decision under risk. Econometrica 47(2):263–291

Kahneman D, Tversky A (1979b) Prospect theory: an analysis of decision under risk. Econometrica J Econometric Soc 47:263–291

Keane MT (1988) Analogical problem solving. Wiley, New York

Keeney RL, Raiffa H (1993) Decisions with multiple objectives: preferences and value tradeoffs. Cambridge University Press, Cambridge

Kempton W (1986) Two theories of home heat control. Cognitive Sci 10(1):75–90

Krawczyk D, Holyoak K, Hummel J (2005) The one-to-one constraint in analogical mapping and inference. Cogn Sci 29(5):797–806

Kuipers B (1994) Qualitative reasoning: modeling and simulation with incomplete knowledge. The MIT Press, Cambridge

Leclercq P, Heylighen A (2002) 5, 8 Analogies per hour. In: Gero JS (ed) Artificial intelligence in design '02, pp 285–303

Lindemann PG, Markman AB (1996) Alignability and attribute importance in choice. Paper presented at the 18th annual meeting of the cognitive science society, Mahwah

Linsey J (2007) Design-by-analogy and representation in innovative engineering concept generation. The University of Texas, Austin

Linsey J, Green MG, Murphy JT, Wood KL, Markman AB (2005) Collaborating to success: an experimental study of group idea generation techniques. Paper presented at the ASME design theory and methodology conference, Long Beach

Linsey J, Murphy JT, Wood KL, Markman AB, Kurtoglu T (2006) Representing analogies: increasing the probability of success. Paper presented at the ASME design theory and methodology conference, Philadelphia

Linsey J, Laux J, Clauss EF, Wood K, Markman A (2007) Increasing innovation: a trilogy of experiments towards a design-by-analogy method paper presented at the asme design theory and methodology conference, Las Vegas

Linsey J, Wood K, Markman A (2008a) Increasing innovation: presentation and evaluation of the wordtree design-by-analogy method. Paper presented at the ASME IDETC design theory and methodology conference, New York

Linsey J, Wood K, Markman A (2008b) Modality and representation in analogy. AIEDAM 22(2):85–100

Linsey J, Tseng I, Fu K, Cagan J, Wood K, Schunn C (2010) A study of design fixation, its mitigation and perception in engineering design faculty. ASME J Mech Des 132 (4):041003-041001-041012

Linsey J, Clauss EF, Kurtoglu T, Murphy JT, Wood KL, Markman AB (2011) An experimental study of group idea generation techniques: understanding the roles of idea representation and viewing methods. ASME J Mech Des 133 (3):031008-031001-031008-031015

Linsey J, Markman A, Wood K (2012) Design by analogy: a study of the wordtree method for problem re-representation. ASME J Mech Des (in press

Lopez R (2011) Characterizing the effects of noise and domain distance in analogous design. MS Thesis, Texas A&M University, College Station, TX

Lopez R, Smith S, Linsey J (2012) The effects of multiple analogies in the presence of extraneous information. Working paper

Mak TW, Shu LH (2004) Abstraction of biological analogies for design. CIRP Ann 53(1):1170120

Mak T, Shu L (2008) Using descriptions of biological phenomena for idea generation. Res Eng Des 19(1):21–28. doi:10.1007/s00163-007-0041-y

Markman A (1999) Knowledge representation. Lawrence Erlbaum Associates, Mahwah

Markman AB, Gentner D (1993a) Structural alignment during similarity comparisions. Cogn Psychol 25:431–467

Markman AB, Gentner D (1993b) Structural alignment during similarity comparisons. Cogn Psychol 25(4):431–467

McAdams D, Wood K (2002) A quantitative similarity metric for design by analogy. ASME J Mech Des 124(2):173–182

McAfee EA, Proffitt DR (1991) Understanding the surface orientation of liquids. Cogn Psychol 23:483–514

Moss J, Kotovsky K, Cagan J (2006) The role of functionality in the mental representations of engineering students: some differences in the early stages of expertise. Cogn Sci 30(1):65–93

Nagel JKS, Stone RB, McAdams DA (2010) An engineering-to-biology thesaurus for engineering design. Proceedings of 2010 ASME IDETC/CIE, Montreal

Namy L, Gentner D (2002a) Making a silk purse out of two sow's ears: young children's use of comparison in category learning. Int J Exp Psychol Gen 131(1):5–15

Namy LL, Gentner D (2002b) Making a silk purse out of two sow's ears: young children's use of comparison in category learning. J Exp Psychol Gen 131(1):5–15

Oriakhi E, Linsey J, Peng X (2011) Design-by-analogy using the wordtree method and an automated wordtree generating tool. Paper presented at the international conference on engineering design, Copenhagen

Otto K, Wood K (2001) Product design: techniques in reverse engineering and new product development. Prentice Hall, Upper Saddle River

Pahl G, Beitz W (1996) Engineering design—a systematic approach, 2nd edn. Springer, New York

Paulus PB, Yang HC (2000) Idea generation in groups: a basis for creativity in organizations. Organ Behav Hum Decis Process 82(1):76–87

Perttula M, Sipila P (2007) The idea exposure paradigm in design idea generation. J Eng Des 18(1):93–102

Purcell AT, Gero JS (1996) Design and other types of fixation. Des Stud 17(4):363–383

Shah JJ, Kulkarni SV, Vargas-Hernandez N (2000) Evaluation of idea generation methods for conceptual design: effectiveness metrics and design of experiments. ASME Trans J Mech Des 122(4):377–384

Thaler R (1980) Toward a positive theory of consumer choice. J Econ Behav Organ 1(1):39–60

Thompson L, Gentner D, Loewenstein J (2000) Avoiding missed opportunities in managerial life: analogical training more powerful than individual case training. Organ Behav Hum Decis Process 82(1):60–75

Tseng I, Moss J, Cagan J, Kotovsky K (2008) The role of timing and analogical similarity in the stimulation of idea generation in design. Des Stud 29(3):203–221

Vattam S, Wiltgen B, Helms M, Goel A, Yen J (2010) Dane: fostering creativity in and through biologically inspired design. In: proceedings first international conference on design creativity, pp 115–122

Viswanathan VK, Linsey JS (2010) Physical models in the idea generation process: hindrance or help? Paper presented at the ASME international design engineering technical conferences, Montreal

Viswanathan V, Linsey J (2011) Design fixation in physical modeling: an investigation on the role of sunk cost. Paper presented at the ASME international design engineering technical conferences, Washington D.C

Viswanathan VK, Linsey JS (2012) A study on the role of expertise in design fixation and its mitigation. Paper presented at the ASME international design engineering technical conferences, Chicago

Viswanathan V, Linsey J (2013) Design fixation and its mitigation: a study on the role of expertise. ASME J Mech Des 135(5):051008-1–051008-15

Viswanathan V, Linsey JS The role of sunk cost in engineering idea generation: an experimental investigation. ASME J Mech Des (in review)

Wiley J (1998) Expertise as mental set: the effects of domain knowledge in creative problem solving. Mem Cogn 26(4):716–730

Chapter 10
An Engineering Approach to Utilizing Bio-Inspiration in Robotics Applications

Saytandra K. Gupta, Wojciech Bejgerowski, John Gerdes, James Hopkins, Lengfeng Lee, Madusudanan Sathia Narayanan, Frank Mendel and Venkat Krovi

Abstract As an interdisciplinary field lying at the intersection of biology and robotics, bio-inspired robotics has seen considerable interest from a research, education, and ultimately application perspective. The scope of this interest has ranged from creating and operating walking, crawling, and flying robots based on biological counterparts; to evaluating biological algorithms for potential engineering applications; to deconstructing the functioning of living organisms from the macro- to the microlevels; and ultimately to constructing the next generation of bio-inspired robots from the bottom up. Significant synergies are forthcoming from such an approach, but numerous limitations still exist. We begin with the

S. K. Gupta (✉) · W. Bejgerowski · J. Gerdes · J. Hopkins
Department of Mechanical Engineering and Institute for Systems Research, University of Maryland, College Park, USA
e-mail: skgupta@umd.edu

W. Bejgerowski
e-mail: wojbej@umd.edu

J. Gerdes
e-mail: jgerdes@umd.edu

J. Hopkins
e-mail: james.k.hopkins@navy.mil

L. Lee · M. S. Narayanan · V. Krovi
Department of Mechanical and Aerospace Engineering, University at Buffalo (SUNY), Buffalo, NY 14260, USA
e-mail: llee3@buffalo.edu

M. S. Narayanan
e-mail: ms329@buffalo.edu

V. Krovi
e-mail: vkrovi@buffalo.edu

F. Mendel
Pathology and Anatomical Sciences, University at Buffalo (SUNY), Buffalo, NY 14260, USA
e-mail: fcmendel@buffalo.edu

A. K. Goel et al. (eds.), *Biologically Inspired Design*,
DOI: 10.1007/978-1-4471-5248-4_10,

discussion of general principles behind taking inspiration from a biological source and converting it into implementable engineering concepts using case studies of bird-inspired robots, snake-inspired robots, and mastication robots. These case studies describe how useful features of the biological creatures were selected and simplified so that they can be implemented using the existing technologies.

Keywords Robotics · Bio-mimetic systems · Kinematics · Robotic design · Parallel manipulators

10.1 Introduction

> The problem of biology is not to stand aghast at the complexity but to conquer it—Sydney Brenner.

Robot designers are often presented with challenging scenarios, where meeting design requirements through traditional engineering approaches becomes difficult. In such cases, robot designers may be interested in taking inspiration from biology; the biological world is full of extraordinary systems that routinely solve extremely complex manipulation and locomotion problems very effectively and efficiently. For example, the musculoskeletal structures of animals have evolved over millennia to enable energy-efficient and effective mobility in a variety of terrains. Locomotion research has long taken advantage of inspiration from these biological analogs: using natural shapes for improved aero- and hydrodynamics, passive dynamic systems for energy efficiency, neuromuscular synergies for simplifying coordination- and control-challenges.

However, biological systems have evolved to exploit structural and organizational principles, spanning multiple physical scales and operational modes, in order to realize functional performance gains in efficiency, passivity, decentralization, and adaptability. Biological structures are well known for using flexible/deformable structures, capable of adapting, bending, and twisting in response to changing environmental conditions, instead of the typical rigid/non-compliant engineering counterparts. Their underlying multi-scale and nonlinear nature makes it a challenge to analyze their ultimate functional behavior using either a bottom-up substructuring framework or a top-down subsystem-based approach.

Therefore, faithfully, mimicking biology alone does not appear to be a realistic option for systematic engineering design of robots because of many existing limitations including that (1) engineered materials are unable to match properties of multi-functional living materials; (2) engineered subsystems cannot realize highly multi-modal characteristics, for example, compliant joints present in animals; (3) analyzing and simulating such multi-physics-based designs is a very difficult task; (4) finally, manufacturing complex structures found in nature remains a great challenge even today. Overcoming these challenges requires an interdisciplinary research approach within a bio-inspired paradigm. The research approach must apply novel engineering methodologies for controlled in vivo experimentation with biological systems while simultaneously evaluating biological hypotheses and observations to inform design and control of artificial engineered systems (Cardelli 2005).

In recent years, every scientific arena has benefited from the ubiquitous availability of computational power and advanced computational tools. Computer simulation can now be used to calculate the kinematic, dynamic, and finite element analysis (FEA)-based responses of a prototype and visualize the results in a 3D interactive virtual environment. By permitting designers to realistically, accurately, and quantitatively prototype and test multiple intermediate models within a virtual environment, virtual prototyping (VP), also known as simulation-based design (SBD), has rapidly gained popularity and become a crucial part of most engineering design processes (Bhatt et al. 2003). The usefulness of such a VP exercise is limited only by the fidelity of the model and the accuracy of the results. Oftentimes, there are many effects such as friction, contact that are very simplistically modeled (for computational efficiency or for the lack of more accurate models) and can only be accurately determined by physical testing. For such situations, hardware-in-the-loop (HIL) frameworks permit a quick replacement of the virtual model by the actual physical prototype to permit experimental testing. The area of bio-inspired robotics has benefited significantly from the use of both VP and HIL technologies.

While engineering-related fields have witnessed the greatest benefits, these advances have percolated down far slower into other arenas. In particular, the lack of significant and useful computational tools in traditional biological sciences such as anatomy hinders the ability of scientists to effectively and rapidly test various hypotheses in a rigorous and quantitative manner (Narayanan et al. 2008). Hence, in this chapter, we adopt a pragmatic approach to using bio-inspiration for design and control of robotic systems in a staged manner as

- Analysis of functional requirements and evaluation of feasibility of seeking biological inspiration.
- Identification of the biological feature that offers an opportunity for significant performance improvement.
- Engineering prototypical bio-inspired design incorporating parametric features and exploiting rapid manufacturing.
- Analysis, refinement, and optimization of the parametric design which may lead to significant natural feature exaggeration due to engineering constraints.

The above-described approach aims to identify a promising feature in nature and then exploit the feature based on the engineering constraints and available technology. The resulting solution may not at all resemble the original source of inspiration, but this approach leads to vastly improved design of robots as noted in the three illustrative case studies.

The first case study describes how compliant flapping mechanisms observed in birds can be used to realize a highly maneuverable and quiet miniature air vehicle. The second case study describes how rectilinear gaits used by snakes can be used to realize robots with small cross-section, achieve significant forward velocity, and operate on a wide variety of terrains. The final case study investigates vertebrate mastication performance by using parallel manipulator architectures. In each of the case studies, we will examine different aspects of (1) obtaining functional requirements; (2) taking inspiration from nature; (3) manufacturing of the bio-inspired design; (4) evaluating the performance for optimization and refinement; and (5) lessons learned.

10.2 Bird-Inspired Miniature Air Vehicles

10.2.1 Functional Requirements

Miniature air vehicles (MAVs) are beginning to gain popularity in a variety of applications. The combination of small size, portability, and ease of use provides users with new opportunities to utilize MAVs in a variety of tasks. Traditionally, MAVs use fixed wings or rotary wings to achieve flight. Each traditional style offers unique advantages and disadvantages. For example, rotary-wing MAVs are ideal for hovering and complex maneuvers, while fixed-wing MAVs provide higher efficiency and speeds. Flapping-wing MAVs offer a compromise between rotary- and fixed-wing designs while reducing the noise associated with a high-speed propeller or rotor.

Our goal was to create a new MAV platform that was sufficiently maneuverable to enable flight inside a building. We also wanted this platform to be quiet during operation compared to rotary-wing platforms. The target flight weight for the platform was under 100 g. The primary objective was to maximize payload capacity. The secondary objective was to minimize the manufacturing cost.

The functional requirement of the drive mechanism was to generate flapping motion. The functional requirement for the wings was to generate lift and thrust forces using a flapping motion.

10.2.2 Taking Inspiration from Nature

Flapping motion used by birds tends to be quiet in operation and highly maneuverable, so we derived our biological inspiration from avian flight (Gerdes et al. 2012). For this case study, a variety of flying animals were observed that were in a similar size range to establish common traits which could improve performance. A well-known property of aerodynamics is the increasing importance of unsteady aerodynamics as Reynolds number is reduced, which generally corresponds to smaller flying animals (Dickinson and Gotz 1993). Due to the high frequency of wing beats in smaller flying animals, kinetic energy recovery is required to maintain efficient operation. To harness these effects, these animals employ resonant bodily structures that store and release inertial energy with elastic flight muscles, connecting tissues, and thorax structures. As the wings are displaced from the equilibrium position, a body structure such as a thorax or a system of joints and flight muscles is elastically deformed, storing potential energy. This technique reduces the energy consumed by the flight muscles by partially eliminating inertial loads (Madangopal et al. 2005; Madangopal et al. 2006).

Due to limitations of the existing actuator and battery technologies, we chose to taking our inspiration from avian flight instead of inspect flight. Birds that provided inspiration included pigeons and doves. The benefits of elastic flapping structures in nature clearly provide useful benefits for efficient operation. Flying animals have also been shown to exhibit elastic storage of wing inertia to reduce power requirements for flight (Ellington 1985). Hence, we decided to utilize elastically deforming flapping wings in our design and use compliant drive mechanism. Due to space restrictions, this case study will only focus on compliant drive mechanism.

10.2.3 Description of Bio-Inspired Design

Compliant mechanisms, defined as flexible structures that elastically deform to produce a desired force or displacement, have recently been integrated into many mechanical systems to provide improved functionality, simpler manufacturing, and greater reliability (Bejgerowski et al. 2009; Gouker et al. 2006; Howell 2001; Mueller et al. 2009). Compliant mechanisms generally consist of rigid links with compliant joints added in strategic locations in order to control the force or displacement. Advantages over traditional designs with rigid-body-articulated joints include (1) reduction in wear between joint members, (2) reduction in backlash, and (3) potential energy storage in deflected members (Howell 2001). Using compliant mechanisms also reduces the noise generated by mating components in operation. Additionally, localized compliance in the mechanism can be used to eliminate joints from the assembly, resulting in reduced part count and improved system precision.

Our MAV design is powered by an electric motor; therefore, it uses a crank–rocker mechanism to achieve flapping action. The mechanism uses localized compliance, so it consists of rigid links connected by compliant joints. The schematic diagram of the design is shown in Fig. 10.1.

An outer frame provides crash protection and houses the gearbox which reduces the flapping speed and torque requirement on the motor to acceptable levels. The rigid-body revolute joints used in the crank are steel pins due to the very large oscillatory loads experienced during flapping. A prismatic joint maintains left–right symmetry in the mechanism; otherwise, large torque from the motor is able to cause a bulk deformation mode that results in asymmetric flapping and reduced force output. The layers of the gearbox structure are held together with lightweight

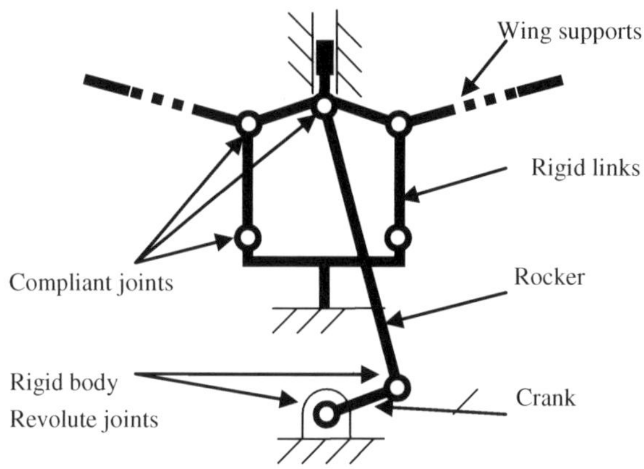

Fig. 10.1 Schematic diagram of compliant mechanism used for flapping-wing action

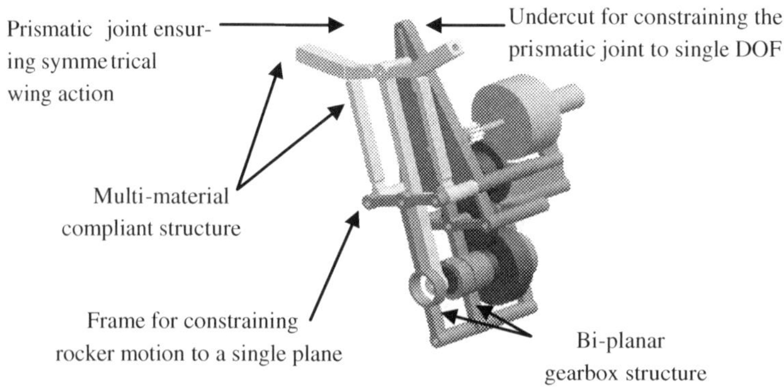

Fig. 10.2 MAV drive mechanism design

carbon fiber rods, press fit into the holes distributed around the mechanism. The design of the mechanism is shown in Fig. 10.2.

After identifying the basic shape of the mechanism, the next step was to determine detailed dimensions of the mechanism. Considering the functional requirements and constraints on the overall size of the MAV, it was important to first identify the constrained and free dimensions of the mechanism design, illustrated in Fig. 10.3. The constrained dimensions were identified from the functional requirements of the MAV. The design of the mechanism required the rocker operational envelope to be placed between the wing arm supports. For the required flapping range of 65°, the minimum separation between the supporting members was constrained to 19 mm. Inherent stability was desired to simplify flight control, so the relative angle on the wing arms was designed to be 15°, resulting in a stroke plane extending from 52.5 to $-12.5°$ with an average dihedral of 20°. The length of the crank and the rocker was determined to be 4.1 and

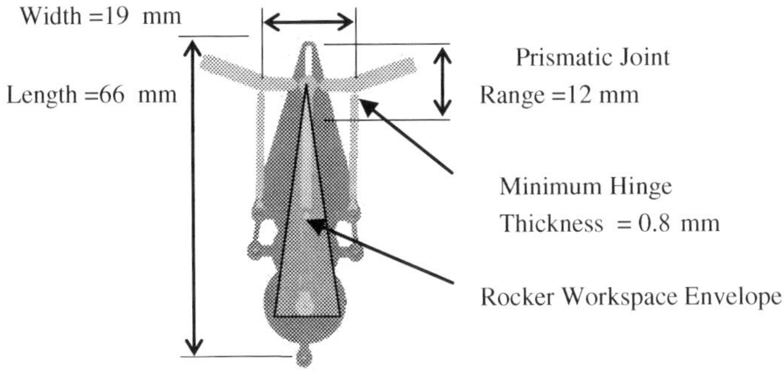

Fig. 10.3 Constrained dimensions of the design

45.7 mm, respectively. The range of flapping motion also determined the range for the prismatic joint to be at least 12 mm to account for elastic deformations of the structure in operation due to loading. The gear axis separation and the range of motion for the prismatic joint determined the minimum length of the mechanism to be 66 mm.

The next step was to optimize the design of the compliant hinges in the mechanism. In miniature multi-material compliant mechanisms, compliant joints are required to transfer relatively large loads while allowing for the required degrees of freedom (DOF). The load transferred by the miniature compliant hinge not only determines the sizing of the hinge cross-section, but also influences the required level of bonding between the link and joint materials selected to create rigid and flexible portions of the structure. Since many polymers do not chemically bond during the molding stage (Gouker et al. 2006; Rotheiser 2004), restricting the material choices to just chemically compatible pairs significantly reduces the design space. Therefore, to expand the design space, it was necessary to consider physical (i.e., mechanical) bonding through interlocking features to ensure a robust interconnection between the materials.

A parameterized hinge model was used for a simulation-based design optimization using design variables and constraints that describe the geometry completely. To evaluate the hinge design, the parameterized geometry was represented as a 3D solid model. The model was assigned material properties corresponding to the polymer used to mold the hinge prototype, high-impact polypropylene (HIPP). The simulation of wing forces' propagation into the mechanism parts was performed using a dynamics model, modeled using MSC Adams View R3 software. Each iteration required incremental design changes and remeshing. The final dimensions of the compliant hinges were chosen to minimize weight.

10.2.4 Manufacturing of Bio-Inspired Design

As described in the description of the bio-inspired design, multi-material injection molding (MMM) was chosen for manufacturing and assembly. The choice of a specific MMM method to realize compliant structures with miniature hinges is a trade-off between the mold complexity and the cycle time. Several morphing cavity methods have been developed (Priyadarshi et al. 2007; Bejgerowski et al. 2011), offering a high level of automation. The morphing cavity method needs to satisfy the assembly and disassembly constraints imposed on the mold pieces; the complexity of the part shape drastically affects the complexity of the molds and hence significantly increases the tooling cost and lead times. To reduce the mold complexity, while maintaining the generality of the manufacturing approach, the cavity transfer method (Gouker et al. 2006) was employed in this design.

The manufacturing process sequence developed consists of two successive injections of material. The compliant HIPP hinges were first molded and inserted into the cavity for the second-stage mold. For the second stage, polyamide 6,6 (85 % volume) with short glass fibers (15 % volume) was used to provide the rigid links in the mechanism. As the second-stage mold encapsulated the first stage, assembly was automatically completed, leaving demolding and final assembly with the rest of the MAV as the only remaining steps.

Figure 10.4 shows a photograph of the multi-material compliant MAV drive frame molded using the cavity transfer method described. Careful visual inspection of the demolded part revealed no molding defects such as deep weld lines, excessive warping, or displaced compliant joints. The differential shrinkage of the part could not be completely eliminated. Nevertheless, since it did not affect the functionality of the mechanism, we concluded that the shrinkage in the structure was functionally acceptable.

After successfully accomplishing the molding of the multi-material compliant drive frame, the MAV was assembled using the molded and machined parts along with off-the-shelf components, including gears, rods, and motor. The body of the MAV consisted of the drive mechanism with wings attached on the front and the tail of servomotor at the back, both connected with two carbon fiber rods and a foam body. The foam body housed the motor speed controller and remote control

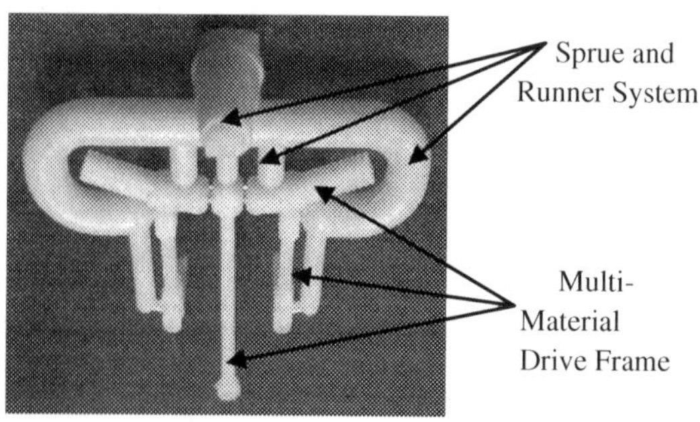

Fig. 10.4 Multi-material molding of MAV flapping mechanism

Fig. 10.5 MAV final assembly

receiver, as well as provided a rear anchoring point for the wings. The assembly of the MAV is shown in Fig. 10.5.

10.2.5 Performance Results

Upon completion of final assembly, the MAV was evaluated in terms of flight performance and functioning of mechanism. During the FEA and dynamic simulations, it was determined that the most likely failure points were the left and right interfaces between the rocker and the wing supports, as these joints supported the majority of the aerodynamic and inertial loads. High-speed photographic analysis showed minimal out-of-plane motion of these joints, and extended testing revealed good fatigue life. Additionally, flight testing in gusty conditions, as well as some crash landings, confirmed the robustness of these key mechanism points. Test flights indoors and outdoors revealed good controllability, passive stability, and excellent endurance. Previous versions of the MAV that used all rigid mechanism

Table 10.1 MAV test flight results

Parameter	Unit	Value
Overall max. weight	g	72.5
Payload capacity	g	31.0
Flapping frequency	Hz	5.2
Wing area	cm^2	920.2
Wing span	cm	63.5
Flight duration[a]	Min	15
Flight velocity	m/s	3.0

[a] With 21.1 g 3 × 250 m Ah-cell li-po battery pack

links were only capable of about half the endurance exhibited by this model, thus confirming the value of compliant mechanisms in boosting the efficiency of operation (Bejgerowski et al. 2011). Results are summarized in Table 10.1.

10.2.6 Lessons Learned

This case study illustrated that the use of compliance in mechanisms and wings can be very beneficial from the performance point of view. The idea of using compliance came by observing birds. But ultimately, the use of compliance in MAVS was incorporated very differently from the original source of biological inspiration. In the drive mechanism, it was implemented using a combination of hard and soft polymer materials that led to compliant joints. In case of wings, compliance was exaggerated significantly to produce large deformations during the flapping cycle. This ultimately led to significantly improved performance in terms of payload-carrying capacity.

The design of the drive mechanism required use of the in-mold assembly process, an emerging manufacturing process. This process enabled reduction in the number of assembly operations and use of lightweight polymer composite materials. This process also enabled realization of parts with small features. This combination ultimately led to a lightweight drive mechanism design with improved transmission efficiency. This process is also scalable in nature and can be used during high-volume production runs to realize the design at a low cost.

10.3 Snake-Inspired Robot

10.3.1 Functional Requirements

Search and rescue applications often require use of robots that need to go through very small spaces. Traditional robot platforms are not useful in such applications. Our goal was to create a new robotic platform that has a cross-section of less than 5,000 mm^2. We also wanted this robot to be able to reach a peak speed of at least 2 km/h. This robot had to be able to work on rugged terrains with significant variations in coefficient of friction.

The main functional requirement for the robot was to be able to locomote over a wide variety of terrains. It required the robot to be able to move forward, backward, and turn. The robot also needed to be able to traverse through small passages. These requirements eliminated wheeled and legged locomotion.

10.3.2 Taking Inspiration from Nature

Because of their long, slender, limbless bodies, snakes possess the ability to traverse small enclosed terrains, such as small holes, tunnels, and gaps, which would prohibit most legged animals. Another natural advantage for snakes is an ability known as terrainability, which is the ability of an animal to traverse rough or difficult terrain. This ability allows a snake to crawl over rugged, non-level terrain as effortlessly as it can traverse open, smooth terrains, and environments, allowing much more maneuverability compared to conventional concepts. Because of these two considerations, we derived our biological inspiration from snakes in designing our new robot platform.

In snakes, rectilinear locomotion is described as the whole snake moving forward along a straight line, sliding against the terrain as illustrated in Fig. 10.6 (Lissmann 1950). Among the various snake-inspired robot gaits, rectilinear-gait-based motion has demonstrated favorable results through many useful features. However, the majority of snake-inspired robot platforms which utilize rectilinear motion are relatively slow in speed (Hopkins et al. 2009). Higher speeds in snake-inspired robot platforms are possible, if rectilinear-gait-based motion was applied in a straight line, as observed in natural snakes, and executed at high rates. Hence, we decided to choose rectilinear gait for inspiration.

10.3.3 Description of Bio-Inspired Design

In order to achieve the functional requirements described above and take advantage of the benefits of rectilinear motion observed in nature, the new robot platform must be able to perform high-speed linear expansion/contraction and pivoting

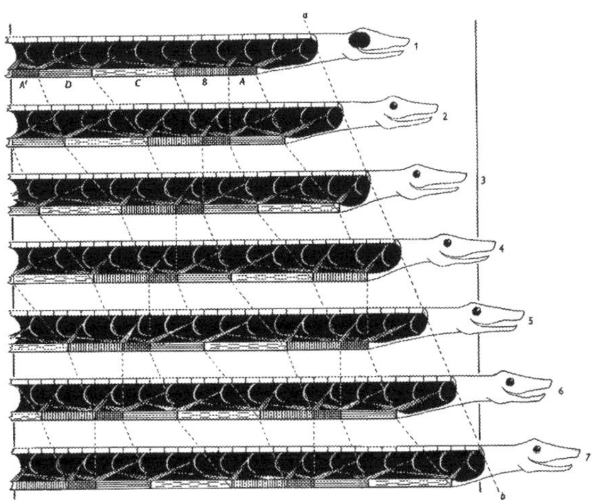

Fig. 10.6 Rectilinear motion (Lissmann 1950)

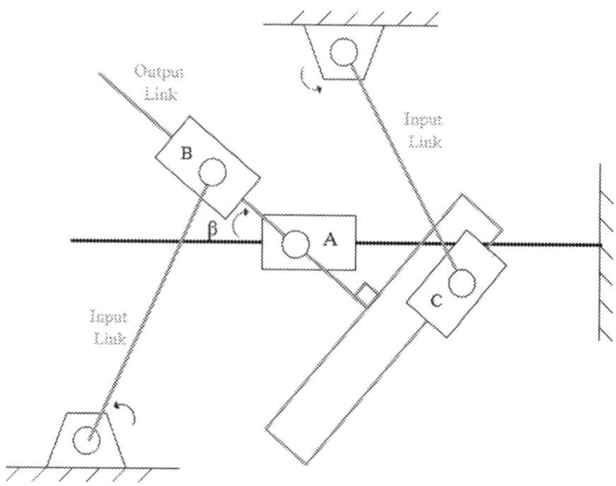

Fig. 10.7 Parallel mechanism concept

motions between segments. This is achieved by developing a mechanism that mimics the joint motion observed in Fig. 10.6 while providing the bending articulation observed in traditional snake-inspired robot designs. A solution was found in the form of a new parallel mechanism. Advantages over simply adding independent prismatic joints to traditional snake-inspired robot architecture include (1) high-speed linear motion, (2) compact design, and (3) scalable design. Due to the wide variety of torque–speed ranges for servomotors, the speed and torque capabilities for the 2-DOF mechanism are modifiable without impacting the physical size of the mechanism.

The conceptual design of the new parallel mechanism, introduced by (Hopkins and Gupta 2012) and illustrated in Fig. 10.7, couples the output from two planar, independently powered scotch yoke–like mechanisms. This mechanism differs from a standard parallel mechanism in that the input paths do not mirror one another. The two input links are independently actuated by a powered revolute joint mounted to the base link and are connected to the output link through a passive prismatic joint and passive revolute joint in series. A third path between the base link and the output link is established through an additional passive prismatic and revolute joint set acting along the x-axis of the base link, permitting pivoting motion for the output link while resisting motion along the base link's y-axis. A key feature in this mechanism is that the sliding axes of the passive prismatic joints of the output link remain perpendicular to one another throughout the full range of motion, as seen in Fig. 10.7. Due to this simple but unique arrangement, the constraints imposed by the orientation of the prismatic joints prevent any motion of the output link, while the powered revolute joints are held stationary.

The detailed design of the parallel mechanism concept utilizes slotted holes and sliding pin joints to replicate the functions of passive prismatic and revolute joints. These features allow for fewer parts, fewer assemblies, and a more compact design. Each parallel mechanism, pictured in Fig. 10.8, is composed of two

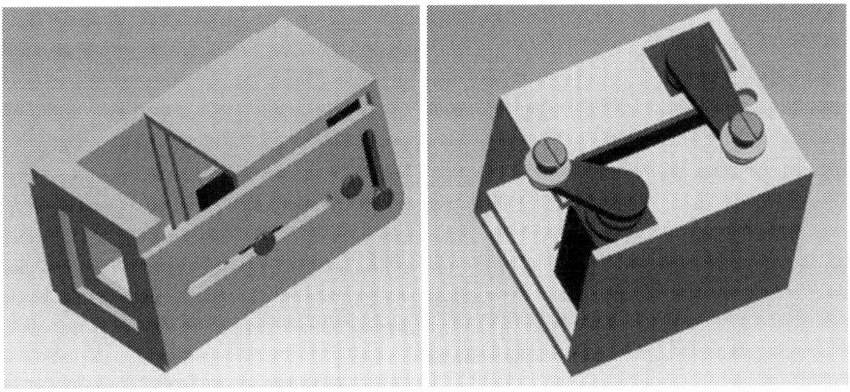

Fig. 10.8 CAD model of parallel mechanism

servomotors with servo arms attached to the output shafts acting as the input links to the mechanism. Each servo arm is attached to the output link of the mechanism (a U bracket) through a slotted hole and pin joint (see left image in Fig. 10.8).

Essentially, the complete robot is a serial collection of modules. Thus, a modular structure was devised in which two identical parallel mechanisms were assembled in a single module. The two mechanisms are assembled serially in a single housing, with the mechanisms' orientation offset 90° apart about the x-axis (direction of the linear expansion) of the module. Both mechanisms contribute to the total linear displacement of the adjacent module, while one mechanism is capable of providing yawing motion and the other provides pitching motion. This assembly provides the potential for full spatial motion for the robot through the fact that the modules are able to lift as well as pivot horizontally. In addition, this configuration allows all modules to contribute to the expansion–contraction capability of the robot, significantly increasing its speed.

In addition to the modular design, the other important design aspect of this snake-inspired robot architecture is a variable friction force concept used to provide anchoring points on the terminal ends of the robot to enable locomotion. The variable friction force concept is a simple yet effective method of anchoring one end of the robot to the terrain to provide a counter to the reaction forces of the powered joints of the modules during forward or turning gaits. In nature, the anchoring is accomplished by redistributing more of the animal's body weight across the surface of the foot to increase the friction force between the foot and the terrain. This concept adopts a similar approach. The surface of the friction anchor is covered in a material with a much higher coefficient of friction than the rest of the robot's housing material. The friction anchor is placed in contact with the terrain by the action of a powered revolute joint as illustrated in Fig. 10.9. The friction force, a function of the normal force between the anchor and the terrain, is increased or decreased by varying the angle, θ_{FA}, of the revolute joint which changes the amount of the module's weight being supported by the friction anchor.

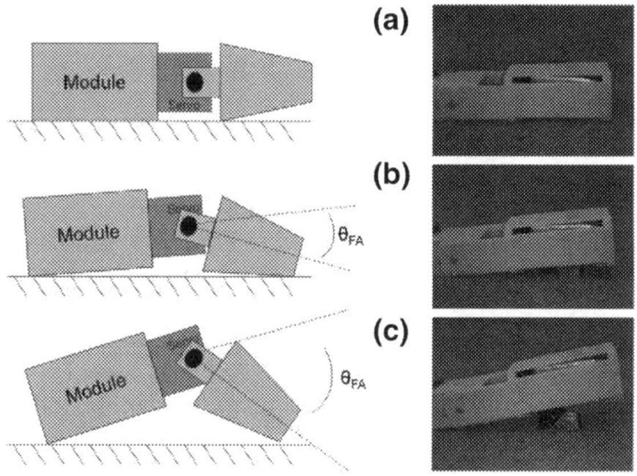

Fig. 10.9 Kinematic representation and prototype anchor

Figure 10.9a depicts the friction anchor in its nominal position, with the anchor's high coefficient of friction surface not in contact with the terrain, allowing the terminal end of the robot to freely slide. Figure 10.9b depicts the friction anchor surface in contact with the terrain with only a slight change in θ_{FA}, useful in low reaction force gaits. Figure 10.9c depicts a large change in θ_{FA}, useful in high reaction force gaits.

The force normal between the friction anchor pad and the terrain may be modeled as a function of θ_{FA} and the extension of robot using the solid model illustrated in Fig. 10.10. In addition to θ_{FA} and the prismatic extension, the model also incorporates a vertical joint at the location of the friction anchor pad. This joint is not actuated throughout the analysis and only serves as analysis point to compute the force normal at the pad location, using analytical dynamic techniques. The Lagrangian formulation is then used to find the inverse dynamic equations of motion for the closed loop model in Fig. 10.10 (Tsai 1999).

Fig. 10.10 Friction anchor model

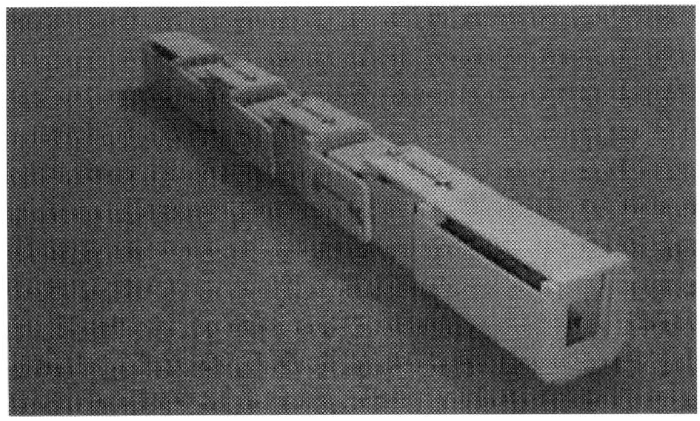

Fig. 10.11 FDM fabricated snake-inspired robot prototype

10.3.4 Manufacturing of Bio-Inspired Design

Additive manufacturing techniques are often used to automatically construct physical objects from computer-aided design models. Prominent additive manufacturing techniques for thermoplastics include selective laser sintering (SLS), fused deposition modeling (FDM), and stereolithography (SLA). FDM offered the best compromise between cost, part size, accuracy, and part strength. Hence, it was used to realize this design (Hopkins and Gupta 2012). The use of FDM for the snake-inspired robot prototype reduces the number of parts needed for assembly and thereby reducing the weight. It also enables manufacturing of true 3D features. The robot prototype, pictured in Fig. 10.11, was fabricated using FDM manufacturing to demonstrate the new drive mechanism design and friction anchor mechanism.

10.3.5 Performance Results

The prototype, pictured in Fig. 10.11, was made primarily from ABS plastic. The robot has a 69.85 $\times$ 69.85 mm cross-section. The robot has a contracted length of 850.9 mm and a fully extended length of 1,143 mm. The total mass of the robot is approximately 1.36 kg. The robot prototype consists of three modules. The prototype also includes a friction anchor at both terminal ends. Each parallel mechanism is capable of a 90° range of motion and 48.68 mm of extension. Each mechanism consists of two standard-sized Hitec HS-985MG High Torque servomotors. They are capable of 12.40 kg-cm of maximum torque and a maximum speed of 0.13 s/60°. The prototype demonstrated a maximum forward velocity of 196.65 mm/s (0.71 km/h) and a maximum turning rate of 26.32°/s. By the nature of the snake-inspired robot's modular design, the forward velocity of the robot may be increased by the inclusion of additional modules. Therefore, while the demonstrated forward velocity does not yet meet the design goals, the velocity of the prototype can theoretically achieve the speed requirement with the inclusion of six additional modules.

10.3.6 Lessons Learned

This case study illustrated how rectilinear gaits and anisotropic friction exhibited by snakes can be utilized to develop a mobile robot. Both concepts were implemented in the robot in a very different form compared to their natural counterparts. The rectilinear motion was exaggerated dramatically to achieve the high-speed locomotion. This was necessary due to the inherent limitations of the existing actuator technologies and the desire to realize the design with a small number of joints due to cost considerations. The exaggerated rectilinear gait led to very good performance in terms of speed. Limitations of the existing engineering materials required implementing anisotropic friction concept by using the friction anchors that control friction force by applying pressure on the ground.

The final design had parts with complex shapes. So in order to minimize the number of assembly operations, the design was realized using a layered manufacturing process (e.g., fused deposition modeling). Currently, the housings are made of ABS plastic which meets the strength and durability necessary to withstand the impact forces generated in the modules during expansions and contractions; however, a second material must be added to the housing's outer surface to reduce the coefficient of friction between the robot and the terrain. In the future, the housing material should be chosen with the friction as a requirement to eliminate the need for additional materials.

10.4 Robotic Masticator Test Bed

10.4.1 Functional Requirements

The goal of this project was to develop an experimental test bed for analyzing the masticatory motions of animals (including humans) and establishing the quantitative relationship between relevant geometric parameters (tooth geometries, numbers, and types) as well as regimen parameters (joint forces, motions). Therefore, we seek to (1) design a general purpose mastication simulator based on parallel architectures, capable of producing the motions and forces encountered in mastications and (2) ensure its working with a range of test foods within controlled and carefully monitored scenarios (with uniform/non-uniform mastication cycles). Such a setup would enhance our understanding of observed mechano-bio-physiological synergies from varied perspectives. From a biological science perspective, it will aid in understanding the anatomical variability in mastication for intra- and inter-species comparative evaluations. It will enable understanding of the cumulative functioning of different muscle groups by studying how various animal populations preprocess their food by chewing and biting and how certain breeds of animals kill their prey with improved estimates of biting forces by performing hypothesis testing of behavioral analyses (Signore et al. 2005). From an

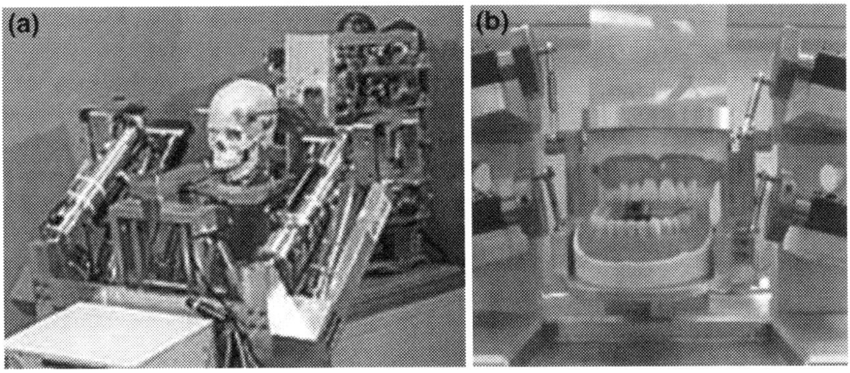

Fig. 10.12 Bio-mimetic human masticatory simulators: **a** Waseda University's Masticatory Robot Series (Waseda-University, 2003) and **b** BioMouth's Dental Simulator (University of Massey) (Xu et al. 2008)

engineering standpoint, such a study on masticatory performance would allow us to objectively assess "chewability or performance index" based on food properties, facilitate the parametric design of dentitions/prosthetic additions, and analyze the complex spatial kinematics of temporo-mandibular joint (TMJ)—the joint connecting the skull and jaw (Kapur and Soman 2006; Komagamine et al. 2011).

10.4.2 Taking Inspiration from Nature

Early simulator systems that were designed to mimic the jaw motions focused mainly on the "form" aspects (muscular architecture and geometry), with lesser emphasis on actual functional behaviors. As a result, such systems were used to reproduce masticatory motions of only a specific breed (humans or animals) and were too specialized to account for diverse species. Two of the most notable and relevant bio-mimetic systems developed for mastication are shown in Fig. 10.12. Therefore, building a generic mastication test bed is useful to enable better understanding of underlying muscular actuations across different species and to overcome the limitations in quantitative analysis of masticatory performance. The typical masticatory cycles, namely opening and closing, are found to be a non-uniform combination of both spatial translations and rotations through the entire chewing process (Koolstra 2002). The envelope of jaw motions corresponds to that of a 6-DOF joint (3 translational DOF + 3 rotational DOF), which implies that the robotic system must be a spatial manipulator (Narayanan 2008). Moreover, the biting force estimates cover a range of values (10–300 N) for small canines to humans to huge vertebrae animals (Signore et al. 2005). Therefore, the masticator should not only possess desirable workspace but also be capable of exerting high end-effector forces while simulating jaw masticatory cycles.

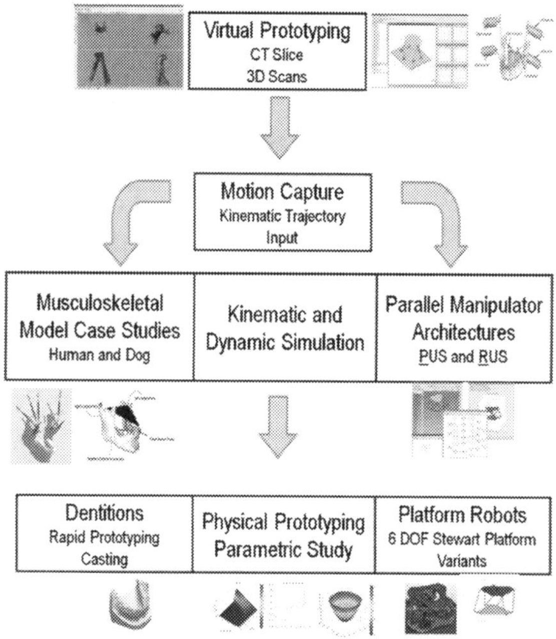

Fig. 10.13 A four-staged realization of our masticatory test bed project flowchart

10.4.3 Description of Bio-Inspired Design

Such diverse objectives pose significant challenges which we addressed in a four-staged realization approach summarized in Fig. 10.13. In our work, an optical passive marker-based MoCap system was used to capture the masticatory motions of animals. The typical setup used for such a study is shown in Fig. 10.14, and further details of various motion case studies (for canines and humans) are discussed elsewhere (Narayanan 2008; Kannan 2008; Signore 2006).

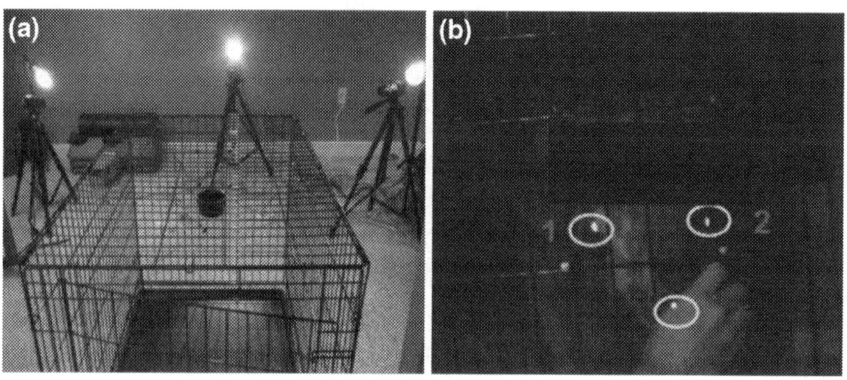

Fig. 10.14 **a** Animal motion capture setup; **b** Recorded frames with marker tracking

These motion studies provide an estimate of jaw motion envelope (Posselt envelope (Posselt 1957)) that needs to be contained completely within the workspace of our masticatory simulator. Parallel architectures are in general advantageous for most applications owing to their inherent stability, self-balanced structural characteristics and varying low-to-high stiffness, which can be closely attributed to biological systems. Kineto-static and quasi-static analyzes were then performed to estimate optimal geometric parameters as well as the actuator capacities required for candidate architectures. Thus, the sizing dimensions and geometry of various components of the parallel platform were obtained taking this into account which is detailed in (Narayanan et al. 2010). This enables us to realize our objective of functional performance matching (motion envelopes and bite forces that occur in a typical mastication cycle) rather than form-based (anatomical/structural/musculoskeletal) similarities with the corresponding biological systems (humans and animals).

10.4.4 Manufacturing of Bio-Inspired Design

Multiple spatial parallel manipulator (SPM) architectures considered for our masticatory simulator include conventional Stewart platform (Kannan 2008), 6-$\underline{R}$-U-S (revolute–universal–spherical with active revolute joints), and 6-$\underline{P}$-U-S (prismatic–universal–spherical with active prismatic joints) manipulators. However, due to the inherent advantages of locating actuators at the base (ground) in terms of actuator efforts and structural rigidity as well as benefits of using prismatic sliders compared to revolute actuators (Narayanan et al. 2010), the 6-$\underline{P}$-U-S system was considered superior than others.

A detailed 3D CAD model of the corresponding robotic platform is visualized in Fig. 10.15a. A comprehensive parametric study, based on maximizing workspace Fig. 10.15b, manipulability at various points in the workspace, and actuation limits, was undertaken to optimize design parameters for the 6-$\underline{P}$-U-S manipulator case (Shah et al. 2010). We then sought not only to accurately imitate the mastication cycle (motion and food resistance) but also to reproduce dentitions of different animals. For this purpose, CT scans and 3D laser scans of canine skulls

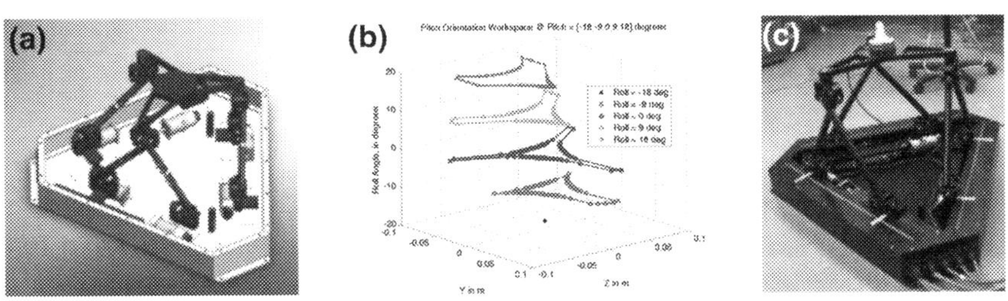

Fig. 10.15 6-DOF masticatory simulator system: **a** 3D CAD model; **b** Workspace; and **c** Physical prototype

and mandible were used to obtain highly detailed 3D models of canine jaws which were then fabricated into rapid prototypes and metal casts. Figure 10.16a shows the resulting dentitions for labrador and bulldog specimens. These physical prototypes were then mounted on the platform with the force transducer for conducting our mastication experiments.

10.4.5 Performance Results

The robotic simulator designed for this purpose, shown in Fig. 10.15c, is capable of measuring the bite forces during chewing cycle experiments using a 6-DOF force–torque transducer (ATI Delta) mounted under dentition prototypes (in Fig. 10.16). In the complete masticator setup (Fig. 10.17a), the platform was

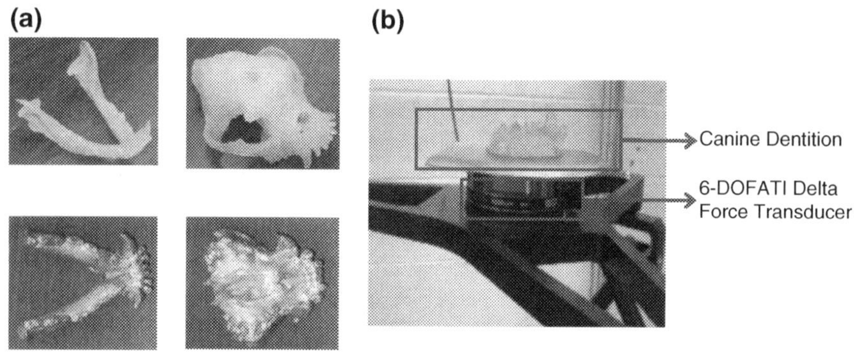

Fig. 10.16 **a** Rapid prototypes and metal casts of dentitions (*upper* and *lower* mandibles) and **b** Masticatory simulator mounted with bulldog jaw and force transducer

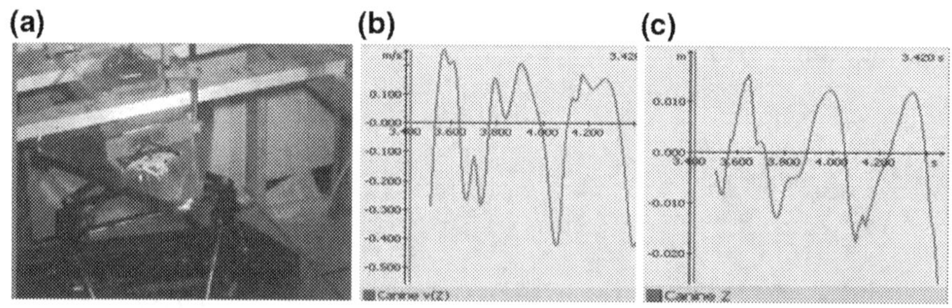

Fig. 10.17 Raw MoCap data **a** HIL testing; **b** Z-coordinate; and **c** Bite forces of canine chewing cycle

driven using canine MoCap trajectories for different types of food (hard to soft) with varied canine dentitions and resulting bite force estimates were measured in real time. Representative measured trajectory motions and bite forces, shown in Fig. 10.17b, c, now offer a quantitative data streams to evaluate masticatory performance/efficiency across a variety of test foods and dentitions.

10.4.6 Lessons Learned

A parallel architecture mastication simulator was designed and prototyped and ultimately tested by hardware-in-the-loop jaw motion-tracking experiments. The motion capture trajectories of various breeds of canines, from bulldogs to golden retrievers, chewing various forms of food were recorded. The mastication simulator proved to be capable of reproducing mastication cycles of varied breeds of canines, with their varied ranges of end-effector motions (Narayanan 2008; Kannan 2008). Further, by instrumenting the physical simulator with a force–torque transducer allowed direct estimation of bite forces while performing mastication experiments. While offering a good initial platform, this work has the potential to be extended in many directions. For example, the concept of developing a "chewability index" requires a more comprehensive motion analysis across different forms of animal foods as well as various breeds which will be pursued in the future. Further, our test bed now offers an automated means of performing both parametric sweep studies and careful design-of-experiment studies to help characterize these aspects. Finally, using a more robust motion-tracking method (electromagnetic) can overcome some of the challenges of marker-based motion capture (marker occlusion, non-adhesion, etc.).

10.5 Discussion

In this chapter, we summarized different approaches to successfully realize bio-inspired design and fabrication of robotic systems from an engineering perspective. In general, the approaches discussed here aimed to identify a promising feature in the nature and then analyze it using the engineering principles and scientific tools. Therefore, the resulting solution may not at all have resembled the original source of inspiration. However, we demonstrated by means of three specific bio-inspiration-based case studies that this approach can lead to novel engineered systems.

The first case study on MAV highlighted issues of design complexity—in achieving the challenging requirements with limited actuation and cost-effective manufacturing methods. The second case study involving snake-inspired robot explained critical issues in modular design, extended dynamic analysis and physical prototyping inherent in biological organisms. The final case study demonstrated the benefits of simulation-based design and its systematic application to analyze a bio-mechanical subsystem of the jaw masticator.

Significant progress has been made in taking inspiration from biology and incorporating these ideas into robotics. But numerous limitations still exist—on the one hand, the multi-scale irregularities, inhomogeneities, and nonlinearities inherent to biological systems still pose considerable technical challenges to complete characterization and understanding; on the other hand, the relationships between the distinct research vocabularies and approaches of biological and robotics researchers persist in creating roadblocks. Hence, there is a critical need not only to foster innovative research and technology but also to promote training of researchers and engagement of a community well-versed in biology, design, and robotics.

References

Bejgerowski W, Ananthanarayanan A, Mueller D, Gupta SK (2009) Integrated product and process design for a flapping wing drive mechanism. J Mech Des 131:6

Bejgerowski W, Gerdes JW, Gupta SK, Bruck HA (2011) Design and fabrication of miniature compliant hinges for multi-material compliant mechanisms. Int J Adv Manuf Technol 57(5):437–452

Bhatt RM, Tang CP, Lee L-F, Krovi V (2003) Web-based self-paced virtual prototyping tutorials. In: Proceedings of 2003 ASME design engineering technical conferences and computers and information in engineering conference (DETC/CIE), DETC2003/CIE-48201, Chicago, Illinois USA, Sep 2–6 2003

Cardelli L (2005) Biological systems as complex system. 7th framework progress information society technologies/future emerging technologies (position paper)

Dickinson M, Gotz K (1993) Unsteady aerodynamic performance of model wings at low reynolds numbers. J Exp Biol 174:45–64

Ellington CP (1985) Power and efficiency of insect flight muscle. J Exp Biol 115:293–304

Gerdes JW, Gupta SK, Wilkerson S (2012) A review of bird-inspired flapping wing miniature air vehicle designs. J Mech Robot 4:2

Gouker RM, Gupta SK, Bruck HA, Holzschuh T (2006) Manufacturing of multi-material compliant mechanisms using multi-material molding. Int J Adv Manuf Technol 30(11-12):1049–1075

Hopkins JK, Gupta SK (2012) Characterization of Forward Rectilinear-Gait Performance for a Snake-Inspired Robot. In: Proceedings of 2012 performance metrics for intelligent systems (PerMIS'12) workshop, College Park, MD

Hopkins JK, Spranklin BW, Gupta SK (2009) A survey of snake-inspired robot designs. Bioinspiration & Biomimetics 4:2

Howell LL (2001) Compliant mechanisms. Wiley-Interscience, Hoboken

Kannan S (2008) Quantitative analysis of masticatory performance in vertebrates. M.S. Thesis, University at Buffalo (SUNY)

Kapur KK, Soman SD (2006) Masticatory performance and efficiency in denture wearers. J Prosthet Dent 95(6):407–411

Komagamine Y, Kanazawa M, Minakuchi S, Uchida T, Sasaki Y (2011) Association between masticatory performance using a colour-changeable chewing gum and jaw movement. J Oral Rehabil 38(8):555–563

Koolstra JH (2002) Dynamics of the human masticatory motion. Crit Rev Oral Biol Med 13(4):11

Lissmann HW (1950) Rectilinear locomotion in a snake (Boa occidentalis). J Exp Biol 26(4):368–379

Madangopal R, Khan Z, Agrawal S (2005) Biologically inspired design of small flapping wing air vehicles using four-bar mechanisms and quasi-steady aerodynamics. J Mech Des 127(4):809–817

Madangopal R, Khan ZA, Agrawal SK (2006) Energetics-based design of small flapping-wing micro air vehicles. Ieee-Asme Trans Mechatronics 11(4):433–438

Mueller D, Gerdes J, Gupta SK (2009) Incorporation of passive wing folding in flapping wing miniature air vehicles. In: Proceedings of ASME mechanism and robotics conference, San Diego, CA, 2009

Narayanan MS (2008) Analysis of parallel manipulator architectures for use in masticatory studies, M. S. Thesis, University at Buffalo (SUNY)

Narayanan MS, Kannan S, Lee L-F, Mendel FC, Krovi V (2008) Virtual musculoskeletal scenario-testing case studies. In: Proceedings of 2008 IEEE virtual rehabilitation (VR), Vancouver, Canada: IEEE, Aug 25–27 2008

Narayanan MS, Chakravarty S, Shah H, Krovi V (2010) Kinematic-, workspace- and static-analysis of A 6-P-U-S parallel manipulator. In: Proceedings of ASME international design engineering technical conferences (DETC2010-28978), Montreal, Quebec, Canada, Aug 15–18 2010

Posselt U (1957) An analyzer for mandibular positions. J Prosthet Dent 7(3):368–374

Priyadarshi AK, Gupta SK, Gouker R, Krebs F, Shroeder M, Warth S (2007) Manufacturing multi-material articulated plastic products using in-mold assembly. Int J Adv Manuf Technol 32(3-4):350–365

Rotheiser J (2004) Joining of plastics, handbook for designers and engineers. Hanser Gardner, Cincinnati

Shah HL, Narayanan MS, Lee L, Krovi V (2010) CAD-enhanced workspace optimization for parallel manipulators: a case study. In: Proceedings of IEEE 2010 conference on automation science and engineering, Toronto, Ontario, Canada, Aug 21–24 2010

Signore MD (2006) A screw-theoretic framework for musculoskeletal system modeling and analysis. M.S. Thesis, University at Buffalo (SUNY)

Signore MJD, Krovi VN, Mendel FC (2005) Virtual prototyping and hardware-in-the-loop testing for musculoskeletal system analysis. In: Proceedings of IEEE international conference on mechatronics and automation, Niagara Falls, Canada, 394–9 2005

Tsai L-W (1999) Robot analysis: the mechanics of serial and parallel manipulators. A Wiley-Interscience Publication-Wiley, Hoboken

Waseda-University (2003) Dental robotics group [Online]. Available: http://www.takanishi.mech . waseda.ac.jp/research

Xu WL, Torrence JD, Chen BQ, Potgieter J, Bronlund JE, Pap JS (2008) Kinematics and experiments of a life sized masticatory robot for characterizing food texture. IEEE Trans Ind Electron 55(5):12

Chapter 11
An Ontology of Biomimetics

Julian F. V. Vincent

Abstract Statistical analysis of the mechanisms and processes in biological organisms (derived from published, peer-reviewed, research papers) reveals that there are 'design' rules which could be used to facilitate technical design, thus producing biologically inspired design without the necessity for the designer using such a system to invoke biology or biological expertise since this has already been done when the rules were extracted. Even so, this is not a necessary *and sufficient* condition for good design. Four principles derived from the Russian system TRIZ (widely used in technology as an objective system for solving problems inventively) are highlighted and summarised as Local Quality; Consolidation or Merging; Dynamics; Prior Cushioning. More design rules, derived in the same way, are needed to expand the importance of information (sensu lato) and materials, two aspects that the TRIZ system currently does not deal with adequately.

Keywords TRIZ · Inventive principle · Cluster analysis · Ontology · Local quality · Consolidation or merging · Dynamics · Prior cushioning · Information · Hierarchy

11.1 Introduction

There are several databases of 'biological design' of which the best known is 'AskNature', produced online by the Biomimicry Institute. Although beautifully presented, many of its sources are casual and not peer reviewed. Although it can be searched for specific words or ideas, it cannot derive relationships that might be implicit in the data. The European Space Agency had a database of biomimetics, but this seems not to be accessible. It was based on work done by the late Centre

J. F. V. Vincent (✉)
University of Bath, Bath, UK
e-mail: j.f.v.vincent@bath.ac.uk

A. K. Goel et al. (eds.), *Biologically Inspired Design*,
DOI: 10.1007/978-1-4471-5248-4_11,

for Biomimetic and Natural Technologies in Bath but suffered from being small, complex and inconsistent. Other databases are available but suffer from similar problems. Difficulties with a database are that it has to be designed before it can be used and it is difficult to expand its scope during its use; more importantly, a database can only record history and cannot deduce new relationships. In reality, these databases deal largely with the physical and chemical properties of organisms and biomimetically derived designs represent a small proportion. There is very little indication of how one can take a concept from its biological context and transfer it to an engineering or technical context, which can be very different and is the most difficult part of the design process.

There is a more basic problem. If the aim is to produce a bridge between engineering and biology, a comparison has to be made between a largely analytical topic and a purely descriptive topic (most modern engineering is reliant on mathematics; the differences between organisms can be transformed into numbers—leading to cladism—but this is a totally artificial transform). It is probably necessary to regard technology from a descriptive point of view, which slews the approach towards design but requires careful definition of what the technology is actually doing.

Used with care, lexicography enables new ideas and relationships to be derived from biological texts, but in its current form, it does not create a generally accessible database, being more of a search algorithm used in individual cases. Even so, with careful choice of words and their interrelations, using a thesaurus and a biological text, this approach produces good and unexpected results (Mak and Shu 2008). On its own, it lacks data storage—the methods and results need to be programmed into a search method. It has the advantage that it does not rely on specifically chosen information and so is much more general, choosing examples without any underlying assumptions.

11.2 A First Approximation

In the Centre for Natural and Biomimetic Technologies at the University of Bath, UK, we initially tried a very broad approach based on solving problems: given a certain problem, what change provides a solution? For instance, an insect or snake sheds its old skin easily by having a *structural* adaptation (a line of thinner material along the back which can split more easily). Or for a safety jacket to provide flotation, it can increase its volume (thus occupying more *space)* by inflation with gas. This is a very crude classification, and inevitably examples use more than one of these sources of change. For instance, the inflating jacket requires *energy* to compress the gas, but this can be said to occur at the size level of the gas molecules, whereas the inflating jacket is nearer a metre across. Many more examples (about 5, 000 from technology and 2, 500 from biology) were gleaned from the Internet and biological publications, and the solutions to the problems classified as changes in *structure, substance* (**things**), *energy, information* (**do things**), in *space* and *time* (**somewhere**). The examples were then plotted on to two graphs, one for technology and one for biology (Fig. 11.1a, b) in which the vertical axis represents all problems and the horizontal axis shows how these problems are distributed according to size (Vincent et al. 2006).

It is important to realise that these two graphs do NOT indicate the quantity of any of these six categories that is required to solve a problem—it is the CHANGE that is important. These are CONTROLLING variables, and their presence is in proportion to their frequency of use for producing change. There are many differences between these two graphs, but the most salient is at the μm level, which can be equated with materials processing. Technology uses *energy* as the main control variable and requires *substance* (i.e. material resources). Biology is very different, with energy the least important control parameter across the entire size scale—again note that this is importance in control. It is not to say that energy is not important for a living organism, just that it is not used very much when it

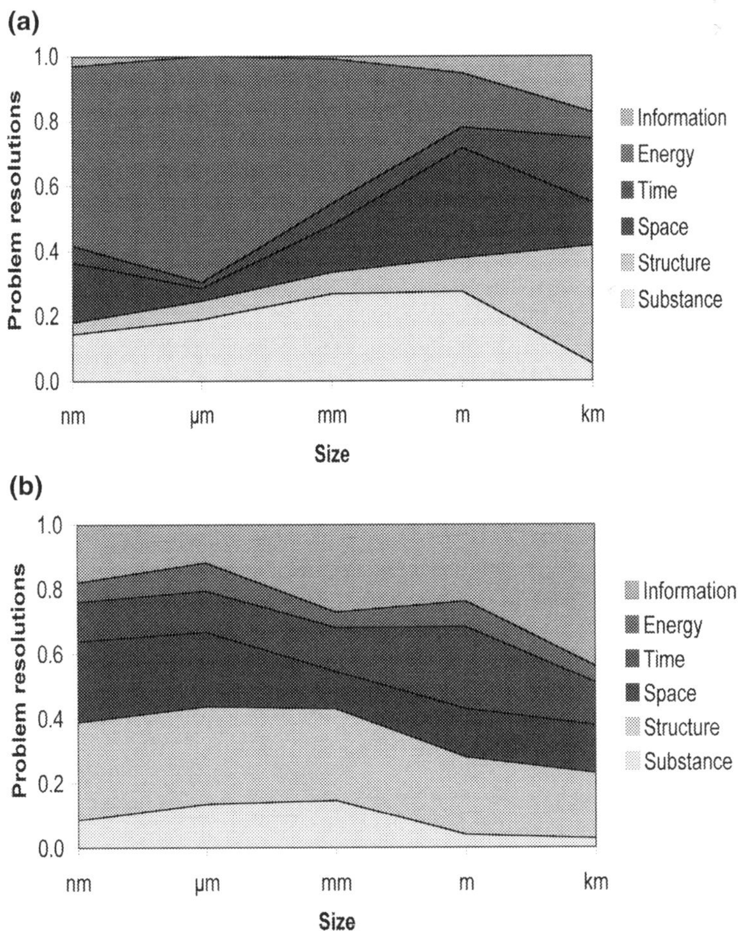

Fig. 11.1 **a**, **b** Different use of substance, structure, energy, space, time and information in solving problems in biology and technology

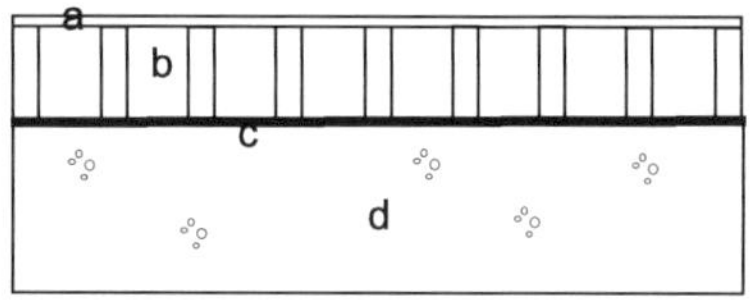

Fig. 11.2 Craig's roof insulation. *a* Short-wave reflector, long-wave transmitter. *b* Oriented honeycomb insulation. *c* Black layer for radiation. *d* Concrete heat sink

comes to managing change or adaptation (which may itself have energy constraints). There is also far less emphasis on input of raw material—in fact, nature uses many fewer raw materials than technology, probably in part because it recycles so much—and *information* (derived ultimately from DNA) is much more important. Although it is difficult to understand quite how information is expressed, the implication is that it informs the *structure* of the material leading to the wide range of properties which, when density is accounted for, provides much the same range of mechanical properties (or at least strength and stiffness) as technology. In other words, to a first approximation, the replacement of *energy* and *material* with *structure* and *information* implies a biomimetic transformation of materials processing, one of the most basic parts of engineering. This is important because it frees technology from believing that all such transformations necessarily require energy and material resource. Biology shows that careful assembly of molecular components (e.g. liquid crystalline structures) can produce acceptable results. This is crucial for the credibility of biomimetic materials processing, but still leaves open the means of effecting such a transformation, although for designers it is a definite direction in which to move.

An example of this transfer in thinking at a more general level is given by a novel form of roof insulation, developed with the precept that the clear night sky represents a heat dump at absolute zero (Craig et al. 2008). The normal solution for cooling a building in an area of extreme solar heating is to use an air conditioner—control by use of *energy*. Craig replaced *energy* with *information* and *structure* in his design and produced a novel form of insulation that keeps direct short-wave radiation from a concrete slab that forms the roof of the building, but allows the heat stored in the slab to be radiated away through an orientated honeycomb which forms the main part of the insulating layer (Fig. 11.2). Temperature differences in the building between day and night can be as large as 13C. The information is in the wavelengths and the orientation of the honeycomb and the structure is provided by the honeycomb and the organisation of the layers.

11.3 Ontologies

A more sophisticated and general, yet detailed, approach uses programming techniques developed for the Semantic Web. These are based on the resource description framework (RDF), a formal language for describing structured information. RDF allows exchange of information on the Web and so lends itself to a communal resource. Information on RDF is readily available (Hitzler et al. 2010; Robinson and Bauer 2011).

The most recent development of RDF uses OWL2. OWL stands for Web Ontology Language (don't ask…). The public domain editor developed by Stanford and Manchester Universities, ProtégéOWL, makes it relatively easy to organise and analyse information. A brief description of OWL2 serves to illustrate its advantage. Information is structured into a hierarchy of classes and subclasses that you define. A class can contain any number of subclasses at the same level, and the hierarchy of sub-, sub–sub-, etc., classes can go to great depth, although this is not always desirable. Classes are given suitable names, and their members are related to each other by a range of properties that you define. These properties relate objects to each other (Object Properties), or they relate specific data quantities to objects (Data Properties). Additions and changes can be made on the fly, so that the size, range and complexity of the ontology are increased or decreased as required. This, of course, is very different from a conventional database. Within broad limits an ontology can undergo great modification at all stages of its use and development. This is particularly convenient since the temptation is always to include too much information at the outset since one is never sure what information is necessary, sufficient or significant in any other way. With Protégé, the redundant information can easily be deleted or reformulated.

The heart of the ontology is the establishment of relationships between the classes and their members. Crucially, the reasoning is based on an 'open world' model as opposed to 'closed world' which is the case with simple databases. In a closed world assumption, if something is not stated to exist, then it is assumed not to exist. However, in an open world model, if something is not stated to exist, then it is impossible to say whether or not it does exist. A clear and explicit statement as to its existence or non-existence has therefore to be made. Large parts of ontology rely upon set theory. Classes can be defined as disjunct (i.e. non-overlapping) or intersecting (overlapping sectors). A Property can be functional, relating an item to only one other (e.g. Sally *hasBirthMother* Liz) or inverse functional (Liz *isBirthMotherOf* Sally). If we say that Liz *isBirthMotherOf* Sally and that Betty *is Birth MotherOf* Sally, we deduce that Liz and Betty are the same person. A Property can be transitive (any friend of yours is a friend of mine), symmetrical (I'm your friend and you are my friend), antisymmetrical (your friend is my friend but I'm not your friend) or reflexive (I know you and I know myself). All these properties can have their domain and range specified. This can ensure that relationships are clearly restricted and false inferences are difficult to make. In a hierarchical system, such as a living organism the ability to make Object Properties transitive is particularly useful, since it means that the properties of an organism can be retrieved from within any level of the hierarchy. This allows the establishment of a chain such as (amino acid)—(protein)—(collagen)—(bone)—(femur)—(endoskeleton). An example is shown in Figure 11.3, where collagen is explicitly stated as being part of a number of materials (in squares) but remaining reliance on collagen is implied within the rest of the relationships. Thus, using a

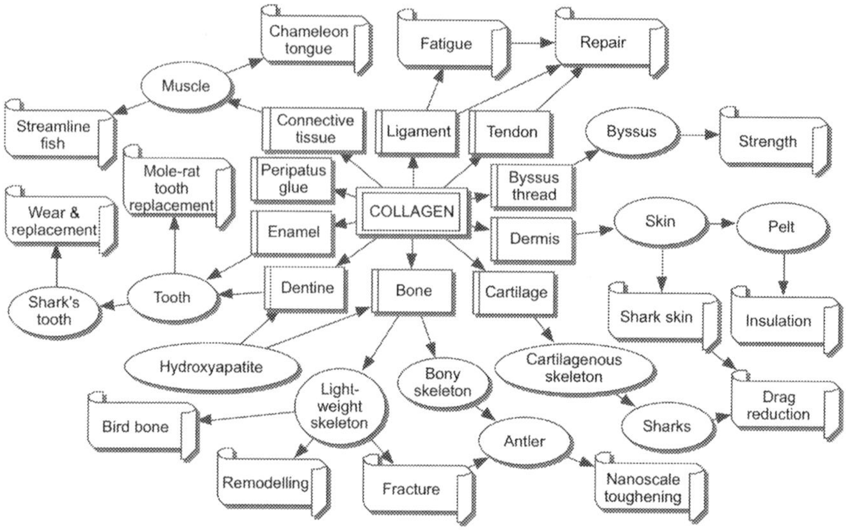

Fig. 11.3 Collagen can be inferred, by ontological reasoning, to be a component of many systems. The *rectangles* surrounding the central one represent the only materials stated to contain collagen. Topics at the ends of the chains are case studies of possible biomimetic significance. The example of lightweight skeleton is mentioned in the text

range of rules of relatedness, a Web of interrelations and properties is generated. These properties relate classes by means of restrictions 'some' (meaning 'at least one'—an existential restriction) and 'only' (a universal restriction).

There are more rules of relationship, for instance distinguishing between a *necessary* condition for an object to be a member of a class (which is termed primitive) and a *necessary and sufficient* condition, which makes the object a member of a defined class. Say we have a primitive class A for which membership is defined by the possession of characteristic A(c). This is not good enough for us to say that any object that has A(c) is a member although it could be. But if class A is made Definitive, then possession of A(c) can also be made a sufficient condition. This difference is useful if we have a member of class B that has the characteristic A(c), then we know that class B is subsumed by class A.

Developing from this, a number of restrictions can be chained together using Description Logics to interrogate the ontology, with the relations 'some', 'only', 'and', 'or' and 'not'. In this way, a concept can be derived which is implicit in the available data but which may not be obvious.

In this ontology, the information is organised using a technique from TRIZ, a Russian system for solving technical problems. TRIZ is a collection of interrelated methods; the method around which the ontology is structured is probably the simplest of all of them. It owes much to a part of Hegelian Philosophy (which is taught early in the Russian education system) that is in turn based on the ideas of Plato and Heraclitus (thus establishing its credentials). A problem can often be characterised as due to the interaction of a pair of factors, one of which is the factor that is to be improved (Hegel's *thesis*) and the other, any perceived disadvantage of implementing the improvement (Hegel's *antithesis*). Resolution of the problem is achieved by making a change in the system, producing Hegel's *synthesis*. Genrich Altshuller, the inventor of TRIZ, used this construct to describe the action of patents. He drew up a list of factors that contribute the thesis and antithesis (currently, there are 39) and of the changes (Inventive Principles. Currently, there are 40) that produce the synthesis. Since each factor can be a thesis or an antithesis it is possible to arrange them along two sides of a matrix, with the appropriate synthetic Inventive Principles within the body of the matrix. We thus obtain a matrix (called the Contradiction or Conflict Matrix) containing 1,521 cells, most of them populated with up to four Inventive Principles and representing the resolution of a problem whose definition and solution were obtained from published patents. The matrix thus represents best practice of engineering (and related technologies) and so is a convenient baseline for comparisons.

The introduction of biology into this system is managed by considering the problems solved by living organisms in the same way that a patent is considered: an organism has a problem, what is its solution? There is a difficulty. Although, given a problem, it is logically simple to show that the solution has been derived from the definition of the problem using Hegel's approach, it is logically impossible to prove in the other direction. For instance, given the answer '42' it is impossible to state with certainty the question that led to that particular answer. Indeed, there is an infinity of questions. An organism is a compendium of solutions to biological problems that may have been solved millions of years ago by an organism living under conditions very different from the present. That the solution worked is evidenced by the survival of the organism; the nature of the problem has to be inferred, making this the most difficult part of generating a believable ontology. Ultimately, believability comes from the recognition of internal patterns that result from the logic of the ontology (which comes from the individual and independent decisions made during the assembly of the data) rather than some subliminal desire to impose a pattern on the data. I have tried very hard to make all judgements as objective and independent as possible (see next paragraph). Ultimately, it needs a small team of people who can introduce such independence with ease.

The starting point is the abstract of a published, peer-reviewed, scientific paper chosen from as wide a field of biology as possible, covering molecular biochemistry to ecosystems and behaviour. The only proviso is that, like a patent, the paper presents the analysis and solution of a biological problem. Consider the problem of making a skeleton from bone. Calcified bone is approximately twice as dense as other tissues, so it is important to minimise the size of the skeleton, but this implies increasing stress on the bones and the potential for fatigue fracture. If muscle mass increases in linear proportion to bone mass, extending a bone's fatigue life by increasing its cross-sectional dimensions may not be effective because the inertia of bigger bones would require larger muscles with increased skeletal loads. Thus, bone remodelling to remove fatigue damage may be essential for the existence of relatively large, long-lived vertebrates. The problem can be defined as the desirable strength and stiffness of bone (Feature number 14) but the undesirable force needed to move it because of its weight (Feature number 1). Relevant Inventive Principles (IPs) are then allocated by making judgements on their applicability. The choice has been made considerably easier by allocating typical functions to each of the IPs; Table 11.1 shows those developed for IPs 25 and 40, both relevant to the solution of the problem of the skeleton. Some 450 such functions have been identified, each of them unique to a single IP. Other IPs for this problem are IP31 (use porous material; introduce lightness) and IP34 (discarding and recovering material—for example, remodelling the bone). These IPs are then compared with the groups of IPs in the Conflict Matrix, each group defined by a pair of conflicting features (thesis and antithesis) looking for a minimum of 3 out of 4 of the Inventive Principles which biology would use to resolve the strength/weight duality. This is a check to see

Table 11.1 Relevant entry in the ontology

IP25: self-service
Remodel
Self-assemble
Self-clean
Self-regulate
Self-repair
Use passive feedback
Use proprioception
Use waste
IP31: porous material
Fill pores or cells
Filtration system
Introduce emptiness
Introduce pores or cells
Use cellular material
IP34: discarding and recovering
Discard and recycle
Discard objects
Recycle
Regenerate
Resharpen
Restore used or consumed objects
IP40: composite materials
Introduce chemical heterogeneity
Increase number of layers
Make composite
Make heterogeneous
Use fibrous composite
Use particulate composite
Use structural heterogeneity
Use a composite system

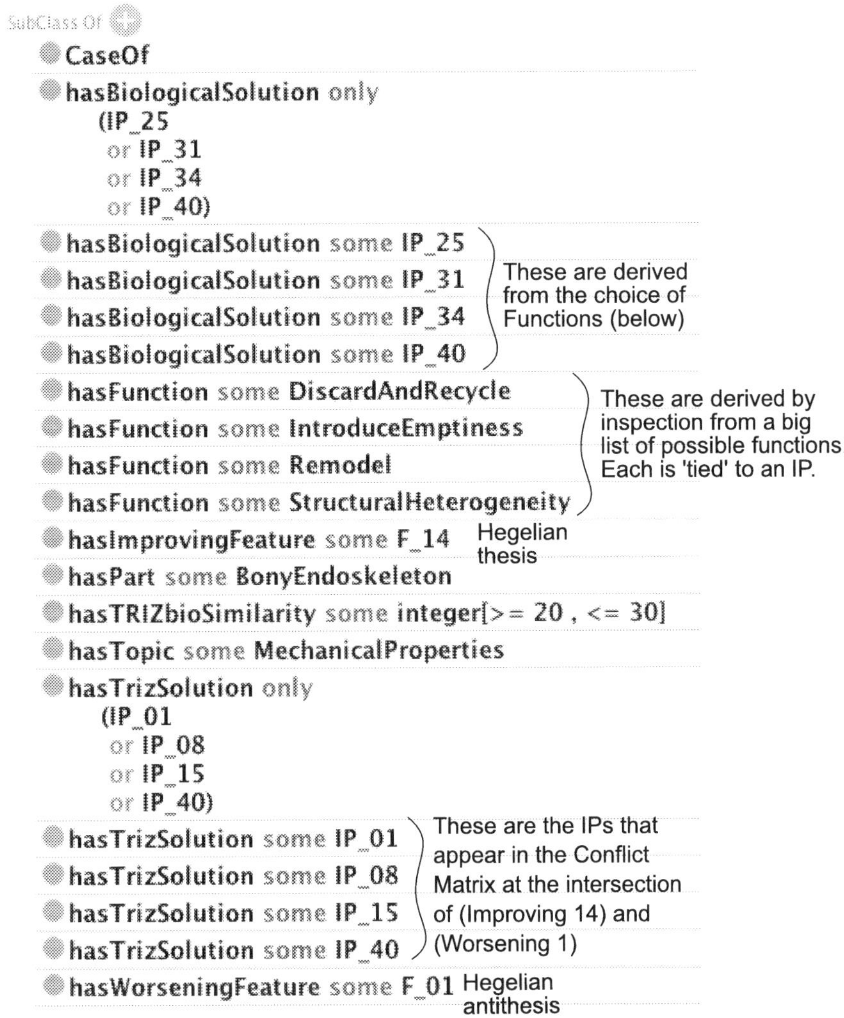

Fig. 11.4 The ontology entry for a bony skeleton and the effort required to move it

whether TRIZ has the same group of IPs as biology, but representing a different pair of Features. If the Matrix reveals such extra groups, the Features proposed as definition of the problem are reviewed and may be changed. Otherwise, the original pair of Features is retained, and the IPs which they define are used in the ontology. These two sets of IPs (from biology and from TRIZ) represent the degree to which biology and TRIZ are comparable. In this instance, there is a big difference—it is possible in biology to repair the skeleton while it is being used. This is not normally the case in an engineering context. The relevant entry in the ontology is shown in Fig. 11.4.

It is now possible to ask more interesting questions of the ontology. For instance, if the problem is to do with weight (which might be a positive or negative aspect) then the Description Logics feature of Protégé enables you to ask for the relevant examples to be listed. More interesting is to ask at the same time which examples in the ontology have the biological solution to weight as the introduction of porosity. Unsurprisingly, these turn out to be lightweight skeletons and bird bones. But if the question is raised—which of those examples that have weight as a problem would have used porosity had the problem been solved by a more engineering approach, the examples flagged up are shark tail and sycamore seed. Shark tail is more concerned with hydrodynamics, since the tail contributes to an upward force as the fish swims. However, the sycamore seed reduces its rate of descent (and hence a functional reduction in density as measured by its apparent response to the force of gravity) by autorotation rather than by being feather-light. While this is a mechanism present in the autogyro and, to a limited extent, in the helicopter, it is not necessarily the first solution that comes to mind.

Using the TRIZ system to define solutions to problems is probably useful in many realms, especially engineering. But biology is as much descriptive as anything, and it turns out that this aspect can be included in the ontology, using the Inventive Principles and their underlying Functions. The comparison with engineering is not then available, since there is no thesis/antithesis pair generated, and the ontology becomes more like a database of effects. This aspect has the advantage that features of biological materials which are common to them all (and which therefore will be less likely to be used as a medium for change), such as that they are (as far as I know) all composites and all hierarchical, can be catalogued. At the time of writing, this addition to the ontology is only just beginning, and the analysis presented here is based on the thesis/antithesis/synthesis data.

11.4 Analysis

Over a period, I have accumulated such results for rather more than 260 examples of biological problems. The resulting set of descriptions forms a pool of data that can be mined to discover general principles and trends that differentiate biological and technical solutions to problems, thus generating a set of design rules for biomimetics that can be used in the absence of biological knowledge. This analysis is preliminary and I present here only some basic simple statistics and cluster analysis. However, the differences and similarities are rational and continue the direction of results that the graphs shown in Fig. 11.1 have suggested. The current version of the ontology of biomimetics, together with limited instructions and comments, is available at http://wiki.bath.ac.uk/display/OOB/.

The similarity between the technical and biological solutions to problems can be calculated from the coincidence between the TRIZ and biological solutions. This ranges from zero to 100 (total identity) and is shown as a cumulative total (Fig. 11.5). The straight line is not necessarily significant. The proportion of

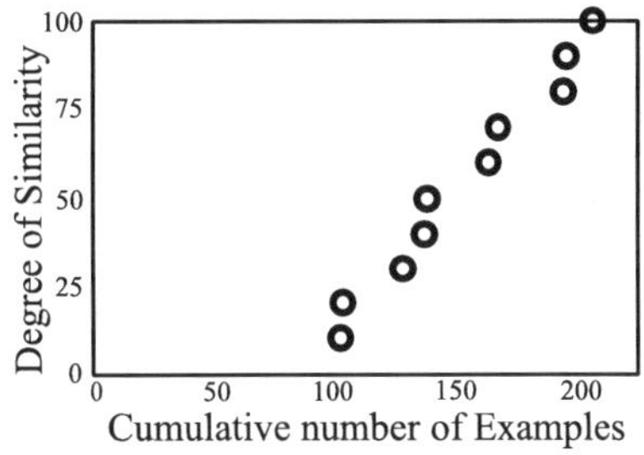

Fig. 11.5 The similarity between biology and technology in terms of the Principles used to solve biological problems

biological problems whose solution is not provided by TRIZ is about 60 %. About 5 % are totally identical to the TRIZ solution. This range covers the estimate of 12 % similarity (made from a 6 × 6 matrix generated in an independent study, using the six factors of Fig. 11.1) (Vincent et al. 2006) and 18 % (made in an independent study of the adaptations of insect cuticle) (Vincent 2005). Clearly, the degree of difference between biology and technology depends on the method of measurement, which should not be surprising. And seemingly the greater the level of detail at which the comparisons are made, the less the difference becomes.

The novelty which biology can introduce to technology seems to be evenly distributed across the size range (Fig. 11.6), which shows the novelty at the nine levels of hierarchy used in this study. This is comforting for those practising biomimetics, since it suggests that the concepts can be transferred to a wide range of technologies, from molecules to road systems. However, the data are still rather sparse and would benefit from more biological examples.

Things become more interesting when we look at the detail.

The Inventive Principles are fractionated into the six classes of function outlined above, primarily to allow an independent test of the results of Fig. 11.1. The decisions as to which class the Principles were placed, made by Olga and Nikolay Bogatyrev, were based on the original definitions expressed in Russian by Altshuller. The results are presented (Figs. 11.7, 11.8, and 11.9) as distributions showing the relative number of Inventive Principles used in the solution of problems in biology and technology. The number has been normalised against the total number of Principles used. Thus, a line plotted from (0,0) to (100,100) or its equivalent divides those Principles mainly used by technology from those mainly used by biology. This is shown on all the graphs.

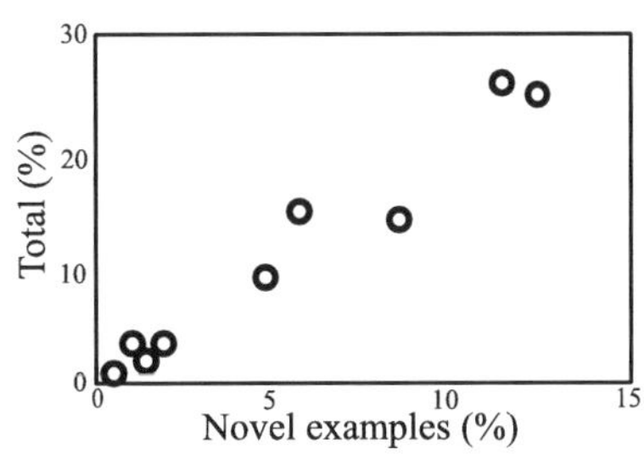

Fig. 11.6 The proportion of novel solutions is independent of the level of hierarchy

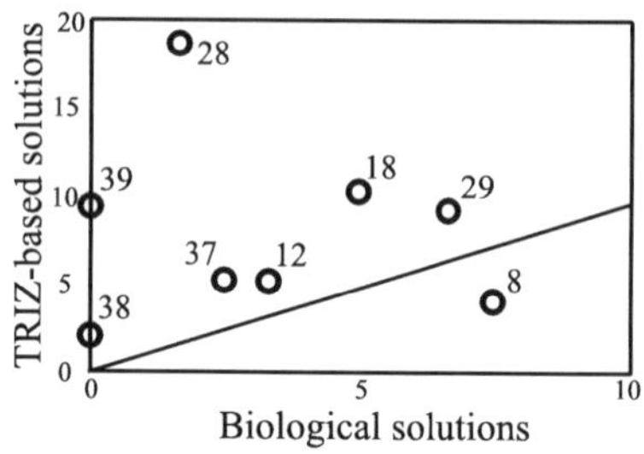

Fig. 11.7 The relative occurrence of energy-based Principles in the biological and technological solutions to biological problems. *8* Anti-weight; *12* Equipotentiality; *18* Mechanical Vibration; *28* Mechanics substitution; *29* Pneumatics and Hydraulics; *37* Expansion; *38* Strong oxidants; *39* Inert atmosphere

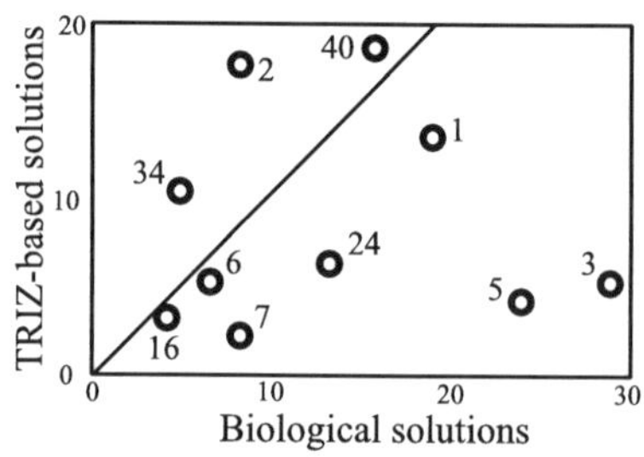

Fig. 11.8 The relative occurrence of structure-based Principles in the biological and technological solutions to biological problems. *1* Segmentation; *2* Extraction; *3* Local quality; *5* Consolidation; *6* Universality; *7* Nesting and Hierarchy; *16* Partial or Excessive Action; *24* Mediator; *34* Discarding and Recovering; *40* Composites

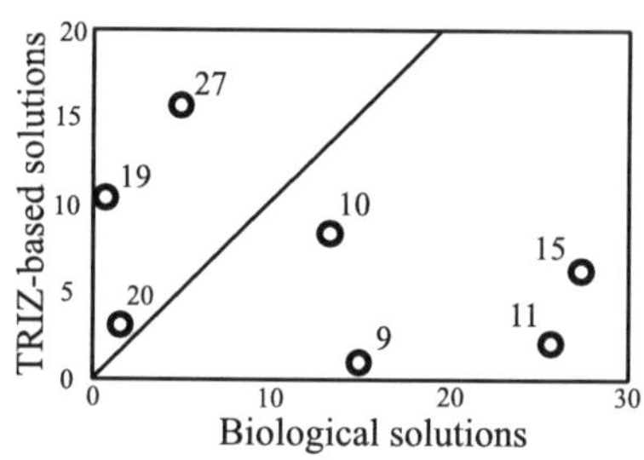

Fig. 11.9 The relative occurrence of time-based Principles in the biological and technological solutions to biological problems. *9* Prior Counteraction; *10* Prior action; *11* Prior Cushioning; *15* Dynamics; *19* Periodic Action; *20* Continuity; *27* Designed Obsolescence

I have chosen only three of the possible six subfactors of Fig. 11.1 to illustrate the differences between biology and technology that this study has generated. These differences, because the Inventive Principles are suggestions for the changes

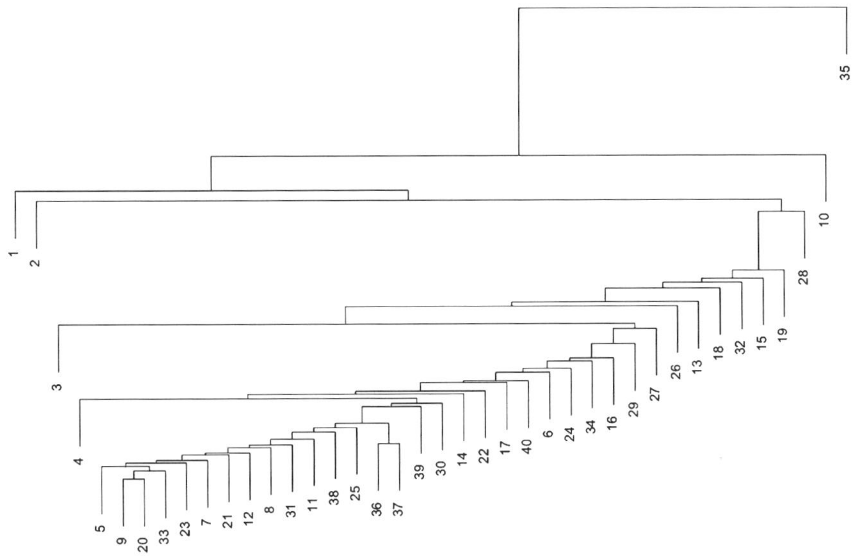

Fig. 11.10 Cluster analysis of the TRIZ Conflict Matrix

required to solve a problem, represent the changes that would move the design process towards biology and hence achieve apparently counter-intuitive goal of allowing designers to be biomimeticists without knowing anything about biology.

Considering *Energy* (Fig. 11.7), the only IP used more by biology is 'Anti-weight'. The suggestion that TRIZ makes in the definition of this IP is 'merge a heavy object with one that provides lift or use aerodynamic, hydrodynamic, buoyant and other forces'. More important is that this result places seven of the Energy Principles in technology and only one in the biology, mirroring the differential in importance of energy as a control parameter shown in Fig. 11.1. Thus, this result has been confirmed using an independent metric.

Considering *Structure* (Fig. 11.8), the balance between technology and biology shown in Fig. 11.1 is once again reflected, with three Principles in technology and 7 in biology, once again confirming, in more detail, the general picture presented in Fig. 11.1. Note that the Principle 'Composites' (IP40) is commoner in technology, possibly because composite materials are ubiquitous in biology and so do not constitute a novelty that could introduce change. The same could be said of 'Nesting and Hierarchy' (IP7) which, although in the biology half of the diagram, is not outstanding. Nearly, all biological materials and structures are hierarchical, being formed from the molecule up. Local Quality (IP3) makes up for some of this by suggesting heterogeneity, zonation, etc. Remember, however, that the inference taken from Fig. 11.1b is that *Information* is directly responsible for *Structure*. Unfortunately, as indicated by Fig. 11.1a, information is almost literally trashed by technology, being squozen out at the level of materials processing. This is reflected in the poor showing of information in the list of IPs—it contributes only four of

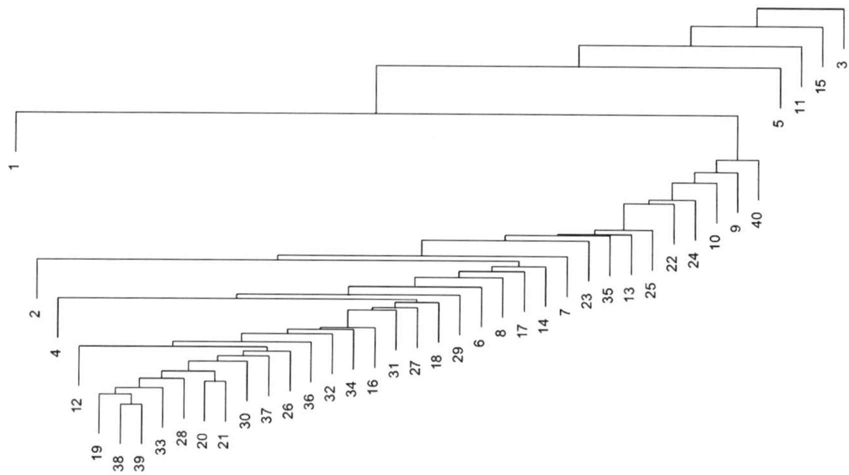

Fig. 11.11 Cluster analysis of biological examples with no TRIZ equivalent

them, whereas, statistically, it should contribute six or seven. This point will be returned to. Meanwhile, the other two important structural factors are Segmentation and Consolidation.

Time is marginally more important in biological than technological problem solving (compare Fig. 11.1a, b), and more detail (Fig. 11.9) shows that Dynamics and Prior Cushioning are important. In fact, 'Prior' is an important factor in the more biological Principles, suggesting that biology designs its structures and mechanisms in a way that we might regard as introducing redundancy. In biology, this preparedness for action, and the ability to adapt which 'Dynamics', shows that biological organisms are far more flexible and adaptable in their response to their surroundings and possible requirements.

11.4.1 Cluster Analysis

The questions to be asked are not just which are the commonest IPs in use, but do they show any sort of relation with each other. The four or so IPs associated with each solution (the crossing points in the Conflict Matrix of the pairs of thesis and antithesis) are compared as a group. Principles which are related, and which will tend to occur together in the synthesis of a problem, will be classified similarly and belong to the same cluster. There are several methods for calculating and presenting clusters—the one used here produces a tree. The analyses were performed using the public domain language R. Unfortunately, I have not yet been able to analyse and compare these results statistically, so what follows is a preliminary inspection.

Figure 11.10 is derived from the entire TRIZ Conflict Matrix, which is taken as providing the technology baseline. The order in which the IPs are numbered bears no relation to their recommendation for the design or to the principles by which they work. The first four Principles occur separately on the left edge of the cluster, remote from the rest of the tree. This shows that they have only very weak similarity to the rest of the Principles or to each other. It also seems reasonable that the most obvious Principles would be the first to be identified, so we can say with some certainty that these isolated Principles are mutually independent and strong. To the right of the tree, IP10 is also rather isolated, so we can say something similar about that one. IP35 (which in effect says 'If there is anything you can change, then change it'!) is by far the commonest in the TRIZ Matrix and is obviously the least related to any other Principle. It is also the Principle that has the most associated Functions (3 or 4 times more than any other Principle). It is well accepted amongst the TRIZ community that IP35 is a dump for a large number of solution strategies that could not be shoehorned into any other Principle. The last five Principles in the Matrix (making up the total of 40) were probably added at a much later date. This brief and rather casual analysis (which will be expanded and improved) exposes one of the basic problems with the TRIZ system—the Inventive Principles are not equally weighted; IP35 is grossly overweight.

However, so long as all the analyses are performed from the same starting position with the same assumptions, comparisons are still valid.

Figure 11.11 is the cluster tree for 117 biological examples from the ontology whose deduced Inventive Principles have no counterpart in the TRIZ matrix. There are a few similarities with the TRIZ Conflict Matrix cluster—notably, the position of IP1. IP3 is the dominant one (the same result came from a study on insect cuticle) (Vincent 2005), and IP35 has, in effect, been lost. Principles 5, 11 and 15 are also very important. These four Principles are the strongest indicators of biologically inspired design.

The thought occurs that, even though there is a significant number of examples whose Inventive Principles are more or less what TRIZ would predict (given the thesis/antithesis definition of the problem), the biological solution has nonetheless selected a set of results which does not resemble the cluster results for the entire TRIZ Conflict Matrix. This turns out to be true (Fig. 11.12) in that IP35 has been greatly reduced in its significance. Otherwise, this cluster seems to be not too different from the basic TRIZ cluster.

11.5 Systematic Omissions

There are some clear gaps when the clusters are compared with ideas at the beginning of this chapter. Most notably, there is no explicit mention of *Information*, which Fig. 11.1b shows clearly is a major component in the biological solution of problems. The answer appears in Fig. 11.1a—in our technology, we

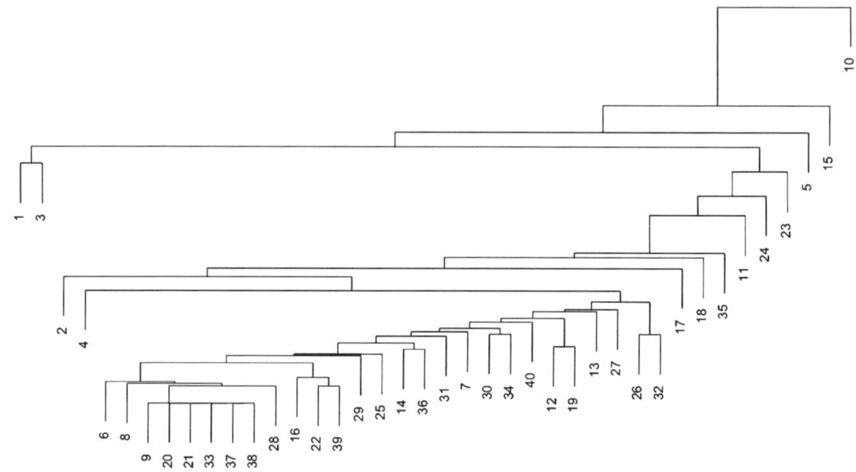

Fig. 11.12 Cluster analysis of biological examples with strong TRIZ identity

have very little use for information—at least, this was true during the period, about 30 years ago, when TRIZ was being formulated. The result is that information is not considered as an important part of the solution of problems. Unfortunately, many people treat TRIZ as a set of ultimate truths, so this lack tends not to be perceived by the TRIZniks who are most expert in this field. The very least that Fig. 11.1 shows is that technology is definitely not an assemblage of techniques of Best Practise. Biology (the most sustainable and arguably, therefore, effective way of doing things on this planet) differs in most respects. These two graphs represent the first real challenge to technology in that they not only show where it is wanting, but also show what is needed. And the recognition of the importance and use of information in modern technology is crucial and must be admitted into the model.

To be fair, information at the molecular level is being developed in the science of nanotechnology. In order to see how it should be used within the process of biologically inspired design, as developed here, we need to understand information from the molecular level. This brings in topics such as thermodynamics, liquid crystals and self-assembly, water as an assembly medium. This is a distinct lack and should go straight to the top of the biomimeticist's wish list. What other ways do we have of manipulating information that can be expressed in such a way that they can be incorporated into TRIZ system (and thus into the system presented here)?

A second clear gap is the materials used. This chapter has addressed problems of how to design structures and mechanisms but has said little about how they are to be realised. The assembly process is probably covered by information, but the materials are integral to this. Biological systems are assembled in a watery environment that provides the context for the types of interaction that drive the assembly process. Hydrophobicity is the driving force for molecular interaction—its outcomes are summarised in the design processes suggested by Principles 3 and 5—Local Quality and Consolidation. The advantage of the biological system is that the materials are easily recycled, but this is merely another way of saying that all the molecular processes of biology are provided by a relatively small range of chemical processes that are thus easily integrated. The recycling of metal is an equally well-closed loop, except (of course) that it requires large amounts of energy. Unfortunately, commercial pressures, commercial secrecy, the profit motive, even self-aggrandisement, all militate against the generation of a fully integrated system of materials. Metals are generally recyclable only because there is not an awful lot you can do with a material which has to keep in more or less its elemental state in order to be useful. Plastics come under commercial pressures and so are highly varied and generally incompatible and so difficult to recycle. Perhaps therefore this TRIZ-based system needs some design principles to aid the choice of materials.

A third apparent omission is that the pervasive appearance of composite materials and hierarchy in biology is not salient in this study. Presumably, this is because all biological materials and structures are made using these principles. Therefore, they cannot be introduced, although they can be fine-tuned. Even so, they appear in the Principles of Local Quality and Consolidation.

References

Craig S, Harrison D, Cripps A, Knott D (2008) BioTRIZ suggests radiative cooling of buildings can be done passively by changing the structure of roof insulation to let longwave infrared pass. J Bionic Eng 5:55–66

Hitzler P, Krötzsch M, Rudolph S (2010) Foundations of semantic web technologies. CRC Press, Boca Raton

Mak T, Shu L (2008) Using descriptions of biological phenomena for idea generation. Res Eng Des 19:21–28

Robinson PN, Bauer S (2011) Introduction to bio-ontologies. CRC Press, Taylor and Francis Group, London

Vincent JFV (2005) Deconstructing the design of a biological material. J Theor Biol 236:73–78

Vincent JFV, Bogatyreva OA, Bogatyrev NR, Bowyer A, Pahl A-K (2006) Biomimetics—its practice and theory. J R Soc Interface 3:471–482

Chapter 12
Evolution and Bio-Inspired Design: Natural Limitations

Frank E. Fish and John T. Beneski

Abstract Biomimetics is the incorporation of novel structures and mechanisms from nature into the design and function of engineered systems. Promotion of biomimicry has been justified on the basis that evolution has modified structures and functions in organisms to achieve optimal solutions and maximize performance. Such justifications reflect an incomplete understanding of evolution and constraints imposed on biology. Evolution is not a conscious or predictive process and does not drive toward perfection. Organisms are not optimal with regard to any one specific function. Where a biological feature will out-perform available technologies, these features can be targeted for assimilation into bio-inspired designs. For engineers and entrepreneurial investors interested in a biomimetic approach, an understanding of evolution and the limitations and constraints that have shaped biological organisms are necessary to avoid unsupportable and overzealous claims.

Keywords Evolution · Darwin · Natural selection · Ffitness · Adaptation · Phenotype · Genetic variation · Design · Reproduction and replication · MantaBot · Autonomous underwater vehicle (AUV) · Muscle

12.1 Introduction

The field of biomimetics and bio-inspired design has become an important source of innovative ideas. Biomimetics attempts to produce engineered systems that possess characteristics, resemble, or function like living systems (Vogel 1998). The biomimetic approach seeks technological advancement through a transfer of

F. E. Fish (✉) · J. T. Beneski
Department of Biology, West Chester University, West Chester, PA 19383, USA
e-mail: ffish@wcupa.edu

A. K. Goel et al. (eds.), *Biologically Inspired Design*,
DOI: 10.1007/978-1-4471-5248-4_12, © Springer-Verlag London 2014

innovation from natural to engineered systems. By emulating these biological characteristics in those instances where the performance of living organisms is superior to manufactured devices, the performance of engineered systems may be improved through biomimetics.

It has been a long-standing idea that new technologies can be developed from nature (Fish 1998; Vogel 1998; Lu 2004; Bar-Cohen 2006). Animals and plants have served as the inspiration for various technological developments. Copying biological organisms by the biomimetic approach attempts to seek common solutions from engineering and biology for increased efficiency and specialization (Vincent 1990; Ralston and Swain 2009). Bio-inspiration extends biomimicry by expanding and improving the original biological concept (Ralston and Swain 2009). Engineers that target the diverse specializations exhibited by organisms for technology transfer can effectively reduce the time for the development of innovative technological solutions.

Biologists have turned to physics and engineering for the answers when seeking explanations for the function of various adaptations involved with physiology and structural anatomy. The biology of adaptation is thus inferred through reverse-engineering. However, with biomimetics, the engineers must look to the biologists to provide working examples and mechanisms of physical phenomena. Novel approaches by living organisms, therefore, form the basis of the development of new technologies.

Biological organisms are in their own right machines. They are fully autonomous, self-powered, and self-repairing. The component structures that comprise these entities are adapted for particular functions that allow the organism to occupy a specialized environment space, or niche. The spectacular diversity of life-forms has produced a variety of mechanical and physiological solutions for interacting with both the biotic and abiotic environment. Inspired by such solutions, engineers are attempting to build machines that mimic the functions of organisms, but these machines and structures are limited and are nowhere near as sophisticated or versatile as real biological forms (Denny and McFadzean 2011).

Because biological designs result from the evolutionary process of "natural selection" (Darwin 1859), biological organisms are considered to have already performed the "cost-benefit-analysis," optimizing particular designs for specific functions. In this sense, biology has provided a design prototype (Allen 2010). Over the course of millions of years, different lineages of organisms have, in effect, experimented with various combinations of morphologies, physiologies, and behaviors to enhance performance. The planet is thus considered to be an enormous natural laboratory, where an infinite number of experiments have been attempted over the eons (Lu 2004; Bar-Cohen 2012). The results of these experiments are the diverse assemblages of animals, plants, fungi, protozoans, and bacteria that inhabit the Earth today. They are considered the evolutionary winners. The losers are the organisms that have gone extinct or never came into existence. They were assigned to an evolutionary trash heap of maladaptation. It is "survival of the fittest." However, such an appreciation of evolution is patently false. This erroneous rationale is too often employed as a natural justification for the perfection of nature and the power of the biomimetic approach. Many human-contrived inventions are thought to be the result of biomimicry (i.e., net spider web, submarine dolphin, wheel rolling round fruit; Lu 2004), although they may only be analogies with no cause and effect.

The purpose of this chapter is to provide a fundamental understanding of how evolution works and its importance with respect to biomimetics and bio-inspired design. The chapter has been written primarily for engineers as the mechanism of evolutionary change is generally known to biologists. However, biologists can also benefit from this chapter in understanding the inherent differences in the disciplines and seeing how a productive biomimetic design may be developed through a synergy of biology and engineering. By examining the relationship between evolution and biomimicry, limitations and constraints can be elucidated for nature-based technologies. It will then be possible to better target biological designs for technology transfer and more quickly bring biomimetic products to market.

12.2 Evolutionary Mechanics

Although the general principles of organic evolution are widely known, its implications are often misinterpreted or misdirected. This is unfortunate because evolution is arguably one of the most profound tenets of modern biology in that it provides a solid, unifying concept for all of biology's disciplines and subdisciplines as well as a common thread for continued investigation. At its core, evolution is a conceptual framework for the process by which living systems change over time. Because the time frame for evolutionary change is geological, it is difficult if not impossible to observe these changes directly. Instead, evolutionary biologists must rely on the results (or products) of evolution, which include all living systems past and present.

The genesis of modern evolutionary theory has a rich and colorful history involving numerous personalities. However, its formalization, development, and ultimate acceptance within the scientific community are generally attributed to Charles Darwin as documented in his seminal work "The Origin of Species" (Darwin 1859). In developing his theory, Darwin sought to explain the bewildering diversity of life by answering two basic questions: (1) How do organisms change over time? and (2) how do new types of organisms originate? Darwin's concept of evolution can be summarized as a series of observations and deductions (Fig. 12.1).

In contrast to his contemporaries, Darwin's extensive observations of nature led him to conclude that evolution operated at the level of the population rather than the individual. Darwin further concluded that populations evolved (changed over time) by differential reproductive success; that is to say, individuals with favorable traits leave more offspring than individuals with less favorable traits, thereby increasing the frequency of the favorable traits in the next generation. The measure of an individual's ability to pass its traits into the next generation is termed fitness.

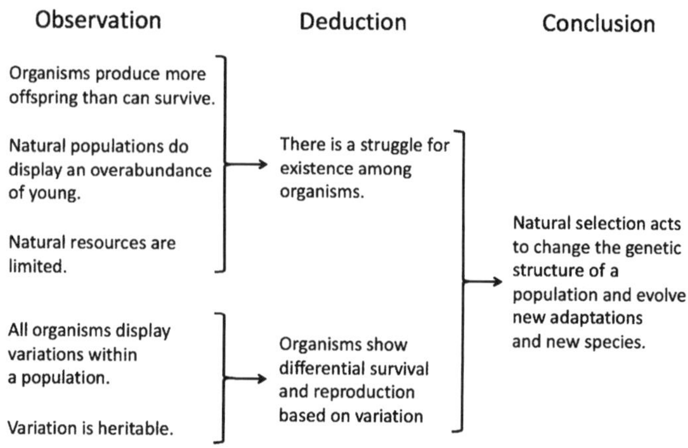

Fig. 12.1 The observations and deductions of Darwin (1859) that lead to the idea of evolution by natural selection

Individuals with a high fitness pass more of their traits into the next generation than individuals with a lower fitness. The process by which favorable traits are preferentially passed from generation to generation is termed natural selection.

Although Darwin's theory provided an elegant framework for the process by which populations change over time (natural selection), it lacked an adequate mechanism. Darwin's theory of natural selection relied on the passing of favorable traits from parent to offspring. However, the work of Gregor Mendel and therefore the genetic basis of inheritance were unknown to Darwin, and he therefore was unable to adequately explain how the traits of parents could be transferred to their offspring. R. A. Fisher, J. B. Haldane, Sewell Wright, and T. H. Huxley finally wed Darwinian evolution with the principles of genetic inheritance. The resulting elaboration of Darwinian evolution is variously referred to as neo-Darwinism, the Synthetic Theory of Evolution, or the New Synthesis. One of the first outcomes of this union was the field of population genetics.

According to the New Synthesis, evolution can be defined as the heritable changes that occur in a gene pool over time due to differential reproduction (i.e., ability to produce more offspring due to a heritable trait). This simple definition encompasses three key concepts: change, time, and populations. In studying evolution, it is important to distinguish between statements that apply to populations and statements that apply to individuals. Evolution occurs at the level of the population (populations evolve, individuals do not evolve). By contrast, natural selection occurs at the level of the individual by favoring the reproductive success of the fittest phenotypes. Phenotypic traits are characteristics of individuals that are coded for by specific genes (e.g., hair color, body size, wings). Therefore, if the frequency of a gene within a population changes, the frequency of the corresponding phenotypic trait in that population will also change. As such, evolutionary change does not occur within individuals, but occurs in a population over time as a consequence of which genes are passed from generation to the next generation. Evolutionary fitness is a measure of an individual's ability to pass genes on to the next generation. The more genes an individual passes on to the next generation, the higher the fitness of that individual.

It is important to note that by definition fitness is relative (there is no absolute fitness scale) and to recognize that fitness is not differential survival. To measure fitness, we compare the reproductive success of all phenotypes and then assign the phenotype with the highest reproductive success a fitness value of (1.0). For all other phenotypes, we assign a fitness value that reflects their reproductive success relative to the phenotype with the highest reproductive success. For example, if a second phenotype produced 80 % as many offspring as the fittest phenotype, it would be assigned a fitness value of (0.8); if a third phenotype produced 50 % as many offspring as the fittest phenotype, it would be assigned a fitness value of (0.5).

It is also important to note that selection operates on phenotypes, not the underlying genetics in that different combinations of genes or alleles can yield the same phenotype. Phenotypic traits include any morphological, physiological, behavioral, or other definable characteristic of an individual. Natural selection determines which phenotypic traits are favored and which are not. To be favored, a trait must increase the fitness (reproductive success) of an individual with that trait over an individual without it. Favored traits are often described as adaptive traits, so that adaptation is the process of acquiring favorable traits.

As a result of evolution, populations change through time (as traits change in response to changing gene frequencies) with the direction of change determined by natural selection. Evolution only works by selecting between alternative phenotypes already present in the population. This makes variation the raw material for evolutionary change. Yet, the process of evolution itself works by eliminating variation through natural selection. Continued evolution is therefore dependent on additional mechanisms, which add variation to a population.

Genetic variation is added to populations by several mechanisms. New genes can arise through mutations, genetic recombinations, or can be introduced (or reintroduced) by gene flow from neighboring populations (gene pools). In addition, the distribution and availability of genes within a population can be altered by various mechanisms or by chance (mating systems, genetic drift, genetic bottlenecks, and founder effects). For example, mating systems driven by female choice can lead to males and females being phenotypically different. Such species are referred to as being sexually dimorphic and the process leading to sexual dimorphism is termed sexual selection. Darwin recognized sexual selection is a special case of natural selection in which males and females are under different selective pressures. Because each of the different mechanisms for altering genetic composition can differ from population to population, each population of a species can be on a different evolutionary pathway. If these pathways diverge sufficiently, the original species can split into two new species.

As indicated by the fossil record, the rate and direction of evolutionary change are highly varied by both over time and taxonomic group. This variability is driven by multiple factors including changes in the range and scope of genetic variation, shifts in the adaptive landscape, and shifts in selective pressures. Various combinations of these factors over geological time have produced a wide range of evolutionary change that span from minor phenotypic adjustments to the apparent sudden appearance of new phenotypes (or disappearance of existing phenotypes). Because these factors are in constant flux, evolution is a continuous process and populations never stop changing.

One of the complications to studying evolution is that evolutionary change is not always evident. Although we define evolution as heritable changes in a gene pool, not all genetic change is translated into phenotypic change. In addition, the phenotypic consequences of genetic change are rarely predictable. Whereas some genetic mutations will have little if any effect on the phenotype, others can have profound effects depending, in part, on whether or not a gene codes for a functional or non-functional protein, whether the protein is a structural or regulatory protein, and the timing of protein expression.

Evolutionary change is limited or constrained by historical, developmental, and logistical factors. From a historical perspective, evolution implies a continuity of genetic information through time in the form of ancestor–descendant relationships. This ancestor–descendant relationship constrains evolution (descendants are limited by their ancestors). However, this relationship also provides a test for evolution in that character evolution should map to natural classifications, known as phylogenies, which in turn should reveal evolutionary trends. A phylogeny is the evolutionary history of an organism that shows its relationship to its ancestors and related species. Examination of evolutionary trends indicates that character evolution can be progressive or retrogressive and that generalized characters can provide a template for the radiation of more specialized characters across different groups of organisms.

Morphological evolution has strong ties to developmental programs. The specific area of biology that investigates this relationship is referred to as evo–devo (Arthur 2002; Gilbert 2003). According to evo–devo, developmental programs consist of a mosaic of interacting modules (Kuratani 2009; Breuker et al. 2006). Collectively, these modules add up to a specific body plan with a specific set of interdependent characteristics. The development of each module is influenced by the development of each other module both spatially and temporally. Modularity preserves the integrity and cohesiveness of each morphological unit while allowing for adjustments throughout the course of development in response to the interactions between the modules. When they are required for proper development, the interactions between modules can conserve existing morphological expression, thereby constraining the direction and magnitude of morphological change. However, temporal alterations in the expression of modules (heterochrony) can produce large, coordinated changes in morphology in a relatively short period of time. Therefore, developmental modularity can either canalize morphological change through spatial interactions or accelerate morphological change through temporal changes in expression or duplication of modules.

From a logistical perspective, not all evolutionary change is feasible. Each organism is a mosaic of interacting, interdependent characteristics. As such, individual characteristics do not evolve in isolation, but evolve in concert with all other characteristics. Whereas evolutionary modification to any particular character may have positive fitness effects on some characteristics, it may also have negative fitness effects on other characteristics. In these situations, evolution must settle for compromise rather than an optimized solution for each affected characteristic. Compromise selection can be driven by additional factors including seasonal variation, habitat variation, and ontogeny (development and life history). Each of these factors may provide competing selective pressures that vary across time, location, and age and thus further constrain the reach of evolution.

Despite the continuous march of evolutionary change and the numerous factors that influence, drive, and alter its direction and timing, the process of evolution is not restricted to greater and greater phenotypic divergence. It is not unusual for multiple evolutionary pathways to independently arrive at a common solution. These common solutions may be shared by unrelated groups (convergent evolution) or related groups (parallel evolution). Convergence not only demonstrates that there is more than one pathway to the same end point, but also demonstrates the power of natural selection to find a favorable solution from a variety of starting points.

12.3 Culture Clash

Evolution is the cornerstone of modern biological thought. It has propelled biology from a mere description of nature to a predictive science. The products of evolution are the adaptations. Adaptations are incorporated into the design of organisms to solve problems and to provide functions involved with survival and reproduction. In this case, nature has developed a technology through evolution. Biomimetics unites the natural technology of biology with physical technology of engineering. By incorporating both biology and engineering, the biomimetic approach requires that evolution be fully considered.

Effective transfer of technologies from biology to engineering requires the union of researchers working in these different fields. As simple as this may seem, there are differences between the two fields and their approaches to problem solving. The different approaches manifest themselves as cultural differences between the disciplines. Biologists and engineers often work apart. Whereas college biology curricula include courses in physics, engineering students are not necessarily required to take biology courses. Engineers may have little knowledge of biology beyond everyday experiences and evolution is often outside these experiences. Similarly, biologists have little appreciation for engineering issues as biological examples were not emphasized in physics classes and much of modern biology is chemistry-oriented (e.g., genetics, molecular biology) rather than physics-oriented (e.g., biomechanics).

A major difference between biologists and engineers is the concept of "design." For the biologist, "design" refers to a description of the physical structure of component or a whole organism in relation to the environment that it must interact.

In this sense, "design" implies only a functionally proficient arrangement of the parts composing an organism, which are the result of natural selection (Vogel 1988). For example, the gazelle has a design for running swiftly through open terrain, and the dolphin has a design for efficient movement in the ocean. The engineer's concept of "design" not only encompasses the structure of a system, but also includes the process by which the system is conceptualized for a particular function. In this sense, design is a human endeavor that implies anticipation and purpose (Vogel 1988). The design process involves discovery, planning, development, construction, evaluation, and invention (Vogel 1998). The difference in how biologists and engineers view "design" leads to misunderstandings in the interpretations of the origin of biological structure and function.

Without a combined understanding of biology and engineering, oversimplified assumptions can retard the development of viable biomimetic applications. For example, the development of the airplane, although inspired by birds, required abandonment of a bird model. Airplanes do not flap their wings like birds to simultaneously produce lift and thrust. Such a mechanism is impractical in modern aircraft due to limitations from scaling and the high speeds necessary to remain aloft by commercial and military jets. Lift generation at a size and speed scale that is sufficiently large to carry a human requires steady-state aerodynamics rather than the unsteady flapping of a bird wing (Jakalb 1990; Harris 1989). As a result, the design of aircraft has advanced beyond the size and capabilities of birds for level flight.

Harris (1989) argued that slavish adherence to the bird as a model system for early airplanes held back design improvements through the early 1900s. Birds did serve as the inspiration for flight and the early development of wing design (Lilienthal 1911; Jakalb 1990). However, large aircraft do not emulate the design of the wings and their control for birds and other powered flyers, such as bats and insects. Today, interest has focused on the agility of birds to perform complex aerial maneuvers. Flying with the agility of birds is not presently possible as there is an absence of detailed information on sensory input and control of the complex mechanical linkages, which are associated with the motion of the wings. Indeed, even the sensing and regulation of airflow over the wing and control surfaces is not understood. In regard to maneuverability and agility, birds continue to demonstrate superior performance to manufactured aircraft.

If biomimetic products are to be produced, the difference in cultures between biologists and engineers needs to be recognized. Each group works with systems that are parameterized in fundamentally different ways (Vogel 1998; Fish 2006).

The development of machines which have been the focus of engineered systems is relatively large in size compared to their biological counterparts. Skyscrapers, jumbo jets, and ships are of a scale that dwarfs termite mounds, birds, and whales, respectively. Conversely, microscopic cells and viruses are smaller than machines. These biological entities are not only capable of performing particular functions but also have the capacity to change their programming for new or modified functionality. The new approach of nanotechnology seeks to work at a level smaller than the cell to produce biomolecular machines, and nanomaterials for structural components and surface textures (Ummat et al. 2006; Zhang et al. 2006). Differences in size between current engineered and biological structures affect the forces experienced by these systems. While large vehicles traveling through a fluid medium experience an environment that is dominated by inertial and gravitational fluid forces, small organisms may be more affected by viscous forces. Water striders (Gerridae) can cruise across a water surface supported by molecular cohesion and surface tension (Hu et al. 2003), whereas ships operate at the water surface by buoyancy from the mass of water displaced by the hull. Grappling with such disparate criteria requires a designer to possess a working knowledge of both the relevant biomechanics and the engineering issues associated with adapting the biology to a machine for some targeted application.

The forces related to the scale of the systems dictate the construction materials used. Engineered systems are composed of rigid materials. These materials include metals, ceramics, and hard plastics. Even where compliance is required, hard materials are used (e.g., spring steel). The choice of these rigid materials is a matter of practicality for simple durability considerations as non-living systems do not possess the capacity for self-repair. Biological systems are generally constructed from materials based on organic molecules. Metals are of limited availability as construction materials in biological systems. Iron, copper, and magnesium are broadly used, but only in respiratory processes and only as one atomic component of a significantly larger molecule. Iron and copper are part of the molecules hemoglobin and hemocyanin. These molecules are essential to capture oxygen for transport through the bodies of animals. Magnesium is part of the basic structure of chlorophyll and functions to absorb radiant energy for plants. Magnetite (Fe_3O_4) is found capping the small teeth of the grinding radula in mollusks (Gordon and Joester 2011). A class of organic compounds, called siderophores, is used by marine bacteria to complex with iron and collect the metal from an environment, where this material is limited (Martinez et al. 2003).

Proteins, carbohydrates, and lipids form structures and components of simple cells to complex plants and animals (Wainwright et al. 1976; Vincent 1990). These materials are created and often function in an environment that is wet. Although biological skeletons can be formed from ceramics (e.g., vertebrate bones, mollusk shell, sponge spicules), these structures are composites with varying amounts of organic molecules (Vogel 1988). Generally, organisms are primarily composed of biological materials that are compliant (Wainwright et al. 1976; Vogel 1988). These compliant structures are part of a design that permits the body to bend (Wainwright 1988). Where motion is restricted due to a rigid skeletal framework, the compliant materials permit flexibility at joints. Furthermore, compliant materials allow for the storage of elastic energy when stressed. Release of the elastic tension can be used for energy recycling in repetitive motion. The springing ligament in the horse's leg aids in reaccelerating the hoof when running. The Achilles tendon in the kangaroo is stretched when hopping and recycles enough energy in each step to maintain a nearly constant metabolic effort as speed increases (Alexander 1988).

Engineered systems generally use rotational motors. The energy to power these devices is derived from radiant energy, chemical transformation, or thermal and nuclear sources. The energy for movement and manufacturing in biology is restricted to chemical catabolism, although ultimately radiant energy from the sun is used to produce the chemical compounds. High-energy chemical bonds are broken to release energy. This energy is then transferred to intermediate compounds (e.g., ATP, creatine phosphate), which can be transported around the cell. With the exception of bacteria, rotational motors have not evolved in biological systems (Fig. 12.2). Movement in plants is powered by growth and fluid pressure. Animals utilize translational movements that are activated by a chemical motor, that is, muscle. Muscles contract while exerting a force to do work. As muscles cannot lengthen on their own, muscles are typically arranged as antagonistic pairs. As one muscle contracts, the other is lengthened either passively or while exerting a force. This results in oscillatory motions of the body of an animal or its appendages, but not rotary motion. Some energy is lost by cyclical accelerations.

Complex neural networks with multiple sensory inputs control animal systems, whereas engineered machines are controlled by simple computational systems with limited sensory feedback. A human brain of approximately 1.5 kg is composed of 100 billion cells with one quadrillion synaptic connections (Denny and McFadzean 2011). The large number of nerve cells (neurons) and neural connections is associated with a large range of behavioral responses. Whereas autonomous machines must be pre-programmed to produce an appropriate response to a particular known stimulus, animals can be plastic in their response.

Biological organisms are functionally multifaceted (i.e., they move, feed, remove wastes, and reproduce) and must compromise optimal solutions for specialized functions to perform adequately rather than maximally (Katz and Jordan 1997; Webb 1997). A machine is constructed with a particular and defined purpose or a mission that it was designed to fulfill (Denny and McFadzean 2011). Having an engineered system with a single purpose increases the maximal efficiency of the targeted operation. As biological organisms are multitasking, they must balance any one function with a number of other functions that compete for energetic resources within the body, but are necessary to maintain life (Fish 2006). Ultimately, all biological organisms are driven by three primary motivations of obtaining sustenance (i.e., food and water), security (i.e., avoid being killed, self-preservation), and sex (i.e., reproduction) (Denny and McFadzean 2011). These are criteria that are seldom included in engineering schematics.

Perhaps the greatest difference between biological and engineered systems is that biological systems are capable of reproduction. Mechanical systems can be manufactured in large lots with strict control for exact duplication. However, for organisms, the ability to reproduce is not confined to merely making copies of individual units. Indeed, new mutations and recombinations of genetic material increase variation, which is the raw material of evolution. Reproduction allows for changes that can lead to new evolutionary solutions or improvements in functional efficiency. Machines cannot autonomously replicate themselves. Improvements in

Fig. 12.2 Improbable and real fish propulsion systems. A mechanical fish model with a rotational screw propeller (*top*), based on the art of Dr. Seuss on display at Universal Orlando. Lateral view of a fish exhibiting fins for propulsion and stability (*middle*). Dorsal view of a shark swimming by undulations of the body and caudal fin (*bottom*)

machine design and function only come about from tinkering with established designs or scrapping the old design for a completely new approach.

Internal reproduction of the cells of the body affords organisms the ability to grow and self-repair. Growth permits change in a body. This change in size may afford the reproducing adult the capability of maximizing the production of young by increasing the space available for storage of gametes or housing and protecting the young prior to parturition. Over the course of growth and maturation of the young, their small initial size will allow them to feed on different foods and occupy different microenvironments, which fully mature individuals would not be able to exploit. The assemblage of different species at varying stages of development and size would enhance the stability of a given ecosystem.

The ability of organisms to self-repair is not merely associated with wound healing. It is also associated with immunity in fighting disease, invasion by foreign bodies, and with coagulation in limiting damage to the body. Organisms can prevent damage to delicate tissue by generating new cells to replace those cells, which have been damaged or abraded away. Another mechanism is to coat an irritant, rendering it harmless. For example, mollusks such as bivalves (e.g., clams, oysters) can secrete a smooth nacre material around a grain of sand to produce a pearl. The pearl prevents abrasion to the soft tissue (mantle) responsible for secreting the animal's shell. Literally, sand in the gears of a machine would generate enough friction to stop operation. Filters and lubrication must be designed into machines to reduce contaminants from wear or external sources.

Engineered systems are built with an assessment of potential for failure and with a plan for maintenance that is based on failure of prototypes. In the development of an engineered design, such considerations are known as failure criteria and dictate the physical limits for the form, function and materials used (Petroski 1996). Autonomic healing is only recently being attempted with battery technology (Blaiszik et al. 2012).

Besides the inherent differences between biological and engineered systems, a cultural disparity in the way engineers and biologists view the data in their respective disciplines. Engineers strive to limit the number of variables of any mechanical system, especially in the design of structures or devices with targeted functionality. Biologists consider a large number of variables. Furthermore, engineers analyze all the errors associated with a system in an attempt to control and reduce variation. Biologists study variation for each of the parameters that control a system. Indeed, variation in biological systems is the foundation of the evolutionary process, which is at the core of modern biological thinking.

12.4 MantaBot: Example of Biomimetics

The biomimetic approach demands careful observation of the whole biological system to identify the principles and attributes of the system that are appropriate to the function that is to be emulated. Thus, major limitations and constraints of any biological design must be defined before translation to an engineered system. As biological systems are the product of evolutionary mechanisms with their limitations, it is possible to improve on the design where the biology is constrained.

In the area of autonomous underwater vehicles (AUV), there is a need for AUVs that can be deployed quickly and can be adapted for a variety of missions (e.g., surveillance, search and rescue, sentry duty, logistics support, and chemical or biological agent detection) (Bandyopadhyay 2005; Colgate and Lynch 2004; Fish et al. 2003, 2012; Low 2011; Moored et al. 2011a). Parameters that are important for the next generation of AUV include efficiency, maneuverability, stability in high-energy environments, operation in the littoral zone and open ocean, station holding, and ability to follow bottom terrain. Design considerations include a rigid hull and minimum of control and propulsive surfaces. The animal kingdom has a number of organisms that meet these qualifications and provide a viable solution. Because of these performance and design characteristics, particular attention has focused on the development of a biomimetic autonomous undersea vehicle (BAUV) fin propulsor that mimics the biological principles and kinematics of the myliobatoid rays with particular interest in the manta (*Manta birostris*, Order Myliobatiformes, Family Mobulidae). The biological role of the manta and other myliobatoid rays conforms to the design space of a BAUV.

Batoid fishes (skates and stingrays) represent a group of over 500 elasmobranch species that have evolved dorsoventrally flattened bodies with reduced or whip-like tails and expanded pectoral fins that are fused to the head to form a broad flat disk (Rosenberger 2001; Douady et al. 2003). This deviation from the typical torpedo-shaped bodies of bony fish and sharks is an adaptation to living on the ocean bottom. The batoids swim solely by movements of their two greatly expanded pectoral fins (Breder 1926; Klausewitz 1964; Heine 1992; Rosenberger 2001). The pectoral fins have triangular, wing-like planforms with an aspect ratio (the ratio of span to chord) of 3.5 (Fish et al. 2012). The cross-sectional geometry of myliobatoid has a streamlined appearance. The lateral pectoral fins display symmetrical cross-sectional profiles reminiscent of engineered foils (Abbott and von Doenhoff 1959).

The manta ray (*Manta birostris*) is phylogenetically one of the most highly derived species of batoid fishes. Manta rays inhabit tropical seas of the world. They are adapted to a pelagic life in the open ocean, but may live inshore, and are found along reef fringes near deep water (Deacon et al. 1997). Mantas are filter feeders, preying on crustaceans and small fish. Other rays feed on benthic prey. The manta and other highly derived pelagic rays (e.g., *Myliobatis*, *Rhinoptera*, *Aetobatus, Manta*) swim by oscillatory locomotion (mobuliform mode). This mode of swimming consists of a small undulatory component (the wavelength of the undulation is greater than the chord length of the pectoral fin), and the pectoral fins are flapped dorsoventrally, analogous to the flight of birds (Breder 1926, 1964; Klausewitz Heine 1992; Rosenberger 2001).

The geometry of the fins and their kinematics for thrust production indicate a high-efficiency propulsive system. The streamlined shape of the body and fins of myliobatoids indicates a low drag profile (Fig. 12.3; Webb 1975; Vogel 1994; Fish and Lauder 2006). Both spanwise and chordwise flexibility are apparent as the pectoral fins are oscillated (Fig. 12.3; Klausewitz 1964; Heine 1992; Rosenberger 2001; Schaefer and Summers 2005; Fish et al. 2012). Spanwise and chordwise flexibility are associated with enhancing propulsive efficiency and thrust production. Spanwise flexibility prevents the total loss of thrust at the reversal of an oscillatory stroke (Liu and Bose 1997). Chordwise flexibility at the trailing edge of

Fig. 12.3 Flexibility of the propulsive fins of batoids (*top*, manta; *bottom*, cownose ray)

the fin could increase the efficiency by up to 20 % with only a moderate decrease in the overall thrust (Katz and Weihs 1978, 1979; Bose and Lien 1989; Bose 1995; Prempraneerach et al. 2003). Actively swimming rays with flexible fins may have higher propulsive efficiencies compared to values predicted for models of rigid lifting surfaces. Hydrodynamic computations performed by Heine (1992) showed efficiencies of 0.7–0.9 for swimming rays. Efficiencies in this range are considered high, because few engineered propellers achieve efficiencies higher than 0.7 (Larrabee 1980; Liu and Bose 1993). In addition, oscillating biological hydrofoils with flexibility maintain high efficiency over an extended operational range (Fish and Lauder 2006). Standard fixed-pitch marine propellers have a maximum propulsive efficiency in only a very narrow range of operational speeds. Thus, oscillating hydrofoil propulsion as demonstrated by batoids is in keeping with the requirements for a BAUV.

The size of myliobatoid rays is appropriate as a model system for a BAUV. The manta is reported to be over 6 m wide and weigh over 1,580 kg, although other related species, such as the cownose ray (*Rhinoptera bonasus*), are smaller (Deacon et al. 1997; Compagno 1999). Scaling issues are inconsequential as the size and speed of these rays correspond to design and operation of AUVs. Thus, the hydrodynamics of the biological and engineered systems are equivalent as defined by important variables, the Reynolds number, *Re*, and the Strouhal number, *St*.

The Reynolds number represents the ratio of inertial forces to frictional forces and defines the flow conditions (laminar or turbulent) around a submerged body. A manta ray swimming at one body length/s would have a value of *Re* over three million, and data for the cownose ray from Heine (1992) indicate a *Re* of approximately 90,000. These high values indicate that inertial forces dominate.

Propulsion is provided by production of a caudally directed momentum jet of fluid. The momentum jet is generated by oscillatory motions of the enlarged pectoral fins (Fig. 12.4), leaving a wake of staggered alternately rotating vortices. The rate at which these vortices are shed and the efficiency of the propulsive movements are related to *St*. *St* is the ratio of inertial forces from local acceleration to inertial forces from convective acceleration and represents the degree of unsteadiness in the flow. *St* is the product of the propulsive frequency and vertical excursion of the fin divided by the velocity of the animal (Rohr and Fish 2004). The cownose ray has $St = 0.19$ (Heine 1992). This value is close to the Strouhal number range of 0.2–0.4, which is stated to be where the propulsive efficiency is maximal (Triantafyllou et al. 1993, 2000; Triantafyllou and Triantafyllou 1995; Rohr and Fish 2004).

There have been a number of attempts to develop a bio-inspired batoid AUV (Moored et al. 2011a, b). Robots have been constructed to mimic the oscillatory swimming of rays by motors (Gao et al. 2007; Yang et al. 2009; Zhou and Low 2010),

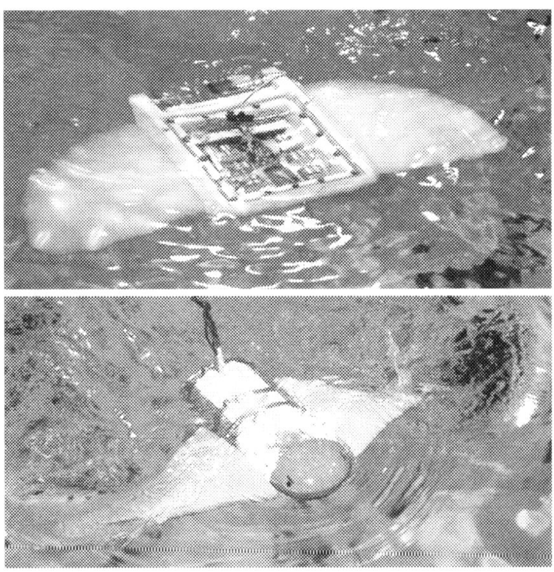

Fig. 12.4 Bioinspired robots (MantaBots) based on the design and swimming kinematics of rays. The mechanical rays were built by Princeton University (*top*) and the University of Virginia (*bottom*)

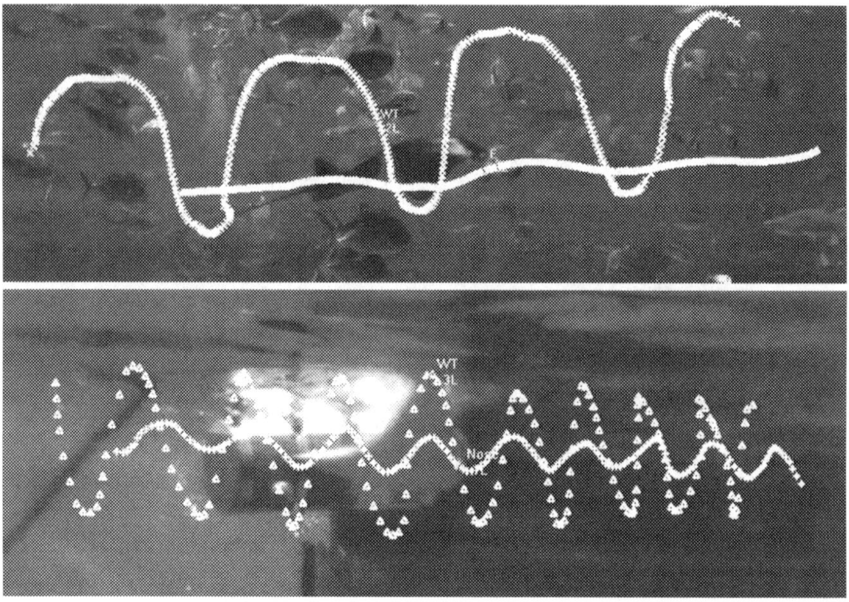

Fig. 12.5 Kinematic traces of the body and propulsive pectoral fins of a swimming manta (*top*) and MantaBot (*bottom*). The large oscillations are from the pectoral fins, and the small oscillations are indicated for the anterior end of the manta and the MantaBot. The manta shows asymmetrical movements of the fin strokes at lower frequency than for the MantaBot. Vertical motions of the anterior end were relatively greater for the MantaBot than the manta

pneumatic pectoral fins (Brower 2006; Suzumori et al. 2007; Cai et al. 2010), fluidic muscles (Festo 2008), and ionic polymer–metal composites (Takagi et al. 2006).

Recently, two prototypes (MantaBots) were developed and tested (Fig. 12.5; Pennisi 2011). The MantaBots were based on biologically derived data, which detailed the design geometry and kinematics of living batoids. Each MantaBot used different mechanisms to actuate the pectoral fins. In one version produced at Princeton University, four metal rods were actuated with a servomotor. The rods heaved a flexible plastic fin, producing varying degrees of spanwise and chordwise curvature (Moored et al. 2011b; Pennisi 2011). A second MantaBot was constructed at the University of Virginia and used a mobile tensegrity structure to move a propulsive elastomer fin (Pennisi 2011). The tensegrity structures consisted of truss-like structures, which acted like a skeleton tendon internal framework (Moored et al. 2011a, b; Fish et al. 2012). The rigid elements of the tensegrity structure were articulated. Two cable elements could generate tension to give integrity to the structure to support large loads and actuate movement.

While both MantaBots were capable of swimming and could perform elaborate maneuvers (Pennisi 2011), the propulsive movements of the pectoral fins were similar but not an exact duplication of a batoid (Fig. 12.4). The kinematics of the MantaBot pectoral fins was more symmetrical on the up- and down-strokes with respect to the longitudinal axis of the body than for live batoids. Living rays can swim faster and are more agile and maneuverable than the MantaBots (Parson et al. 2011). The MantaBots swam at just over 0.5 body lengths/s, and manta rays can swim at 0.8 body lengths/s. The amplitude of heave at the rostrum was 27 % of the fin tip amplitude of the MantaBot, whereas the heave amplitude was 12 % for manta (Fig. 12.4). *St* were 0.6 and 1.6 for the Princeton University and University of Virginia MantaBots, respectively. Such values were outside the range for maximum efficiency and above the value of *St* for the cownose ray (Heine 1992; Triantafyllou et al. 1993, 2000; Triantafyllou and Triantafyllou 1995).

For the MantaBots, engineering has derived separate solutions that provide similar motions, but are still limited compared to the actual rays. The skeletal, neural, sensory, and muscular systems of living organisms are still more complex than robotic systems. This complexity permits organisms to work within a large set of behavioral and performance responses. As further refinement to robotic systems is built into the design, emulation of biological systems will converge with the engineering. However, there are limitations to biological systems from evolutionary constraints that may reduce performance. Emulation of biological systems should only be taken to a point in which performance is maximized, but then can be enhanced through engineering of a bio-inspired design.

12.5 Evolutionary Constraints

Caution must be exercised when using evolution as a justification for the development of biomimetic products. Nature has served as inspiration for various devices for centuries, but the theory of evolution, as currently understood, has only been around since the mid-1800s. With the arrival of biomimetics and bio-inspired design as fields that could produce products and increase funding and investment, there has been an attempt to validate these fields with a natural justification. Evolution is seen as an iterative process that arrives at the best design. Therefore, the evolutionary process becomes imperative to highlight the advantages of natural designs. However, evolution does not hold all the answers or solutions to problems that affect our daily lives.

Evolution is a process that over the history of the Earth has produced a multitude of differing species that have developed solutions (adaptations) to local environments. In some cases, evolution has lead to different but analogous solutions to similar environments in different phylogenetic groupings. Consider swimming by a fish and a squid. The fish wags (undulates) its caudal fin from side to side to accelerate a mass of water into its wake, and thereby gain forward momentum (Fig. 12.2). The squid, however, uses jet propulsion to push the animal backward. Both animals can move rapidly through the water, but the mechanics of the propulsive systems differs. Flexibility of the vertebral column of the fish permits an undulatory wave to move posteriorly down the body to laterally flex the caudal fin. Because the body of the squid is held rigid by a non-flexing internal skeleton (i.e., pen), such movements are not possible in the squid, necessitating jetting. These differing solutions to the problem of rapid movement through water have further consequences. The fish can push a large volume of water at slow speed, thereby moving with a high efficiency, whereas the squid pulses a jet of fluid at a much higher speed to gain an equivalent momentum, resulting in a lower propulsive efficiency (O'Dor and Webber 1986).

For a non-biologist using evolution to justify biomimicry, the greatest mistake is inferring that evolution has an optimal design goal. Evolution is neither conscious nor predictive. Evolution is not visionary. Construction and organization of an organism's features have not evolved toward a specific goal, and evolution does not drive toward perfection. Evolutionary change by natural selection does not provide a "perfecting" principle, only a "better than" principle (Luria et al. 1981). Organisms evolve features not to be optimal or perfect, but merely to perform adequately (Katz and Jordan 1997; Webb 1997).

Genetic algorithms are used to mimic the process of evolution by natural selection. Genetic algorithms start with a population of randomly generated individuals [i.e., primordial ooze of hundreds or thousands of computer programs; Koza (1994)]. The individuals are evaluated iteratively by a fitness function, which represents a predetermined solution set. The solution set is encoded as a finite-length string that is composed of elements with a finite number of possibilities (Whitley 1994; Barrett 2002). By iteration and selection, a singular, predetermined optimal solution can be produced. However, biology is not driven by a predetermined optimal solution. The evolution of living organisms is shaped by the interactions of an organism with its environment, its phylogenetic history, and the genetic mechanisms that promote genetic diversity (e.g., mutation, recombination). Furthermore, evolution can work in the opposite direction from genetic algorithms to produce a diversity of solutions from a singular ancestral type. Each solution or species divides up the environment into various ecological niches.

Biological organisms are multitasking entities. An organism is a mosaic of integrated structures and functions to achieve evolutionary success (i.e., survive and reproduce). Some of these components may be at odds with other features of an organism. As a result, organisms must compromise optimal solutions for the necessity of having an integrated system that can perform a number of simultaneous functions. The integrated parts of an organism must share the limited metabolic energy available for maintenance and function. Increased allocation of limited resources to one component of a body may improve function, but be to the detriment of another component. Natural selection acts on an entire organism and not its individual parts (Luria et al. 1981). Despite the contrary argument (Bar-Cohen 2006; Allen 2010), evolution rarely leads to solutions with a maximal performance and with an economy of resources.

An example of how optimal design is lacking in biological organisms can be found in the structure and performance of the propulsive mechanics of fishes. Fish carry with them a large amount of muscle. Most of the muscle mass is not used during routine swimming, such as when a fish is cruising or migrating. Routine swimming is accomplished using muscle composed of slow oxidative (red) fibers

(Alexander and Goldspink 1977). These fibers use aerobic means of generating energy for muscle contraction and are highly efficient. As long as oxygen is available, red fibers can repeatedly contract over an extended period of time, but at a relatively slow rate. However, the bulk of the propulsive muscle mass is composed of fast glycolytic (white) fibers. These fibers contract faster and more powerfully than red fibers, but cannot sustain repeated contractions over a long time and have a very low efficiency. For fish that migrate long distances at low-to-moderate speeds, it makes more sense for these fish to have a muscle mass composed of a higher proportion of red fibers than white fibers. Indeed, carrying a large mass of white muscle seems detrimental. The inactive white fibers continue to metabolize nutrients, and the extra mass encumbers additional energy costs to move the body. Why then is so much of the musculature composed of white fibers? While not utilized all the time, the white fibers are advantageous during those brief instances when life and death are on the line. When a fish has to chase prey for food or to escape being preyed itself, it is advantageous to have a large mass of white fibers to generate the forces required to accelerate quickly. At these times, efficiency is not as important as rapid acceleration to close the distance on prey or increase the distance away from a predator. In addition, the fish must carry other organs that add to its mass, thereby impeding performance. Carrying large gonads and reproductive products can reduce swimming speed and survival, although reproductive organs are necessary for evolutionary fitness.

Not all possible structures and processes are available to organisms. As mentioned previously, biological organisms do not use metal as a framework for a physical support. The structures of organisms can only work within the constraints of materials based on organic molecules. The formation of these molecules is directed through recipes encoded on the DNA molecule and manufactured by living cells. Metabolic processes are only possible within a narrow range of temperatures for the formation of complex molecules. Above critical temperatures, proteins denature and cellular systems fail. Alternatively, synthetic manufacturing systems can use high temperatures and pressures to produce new molecular configurations or meld materials together in construction.

The wheel may be considered one of the greatest inventions by humans. It allows for a reduction in the energy cost for movement by reducing friction and eliminating oscillatory motions. The wheel is free to rotate continuously around an unattached central axis. Although the wheel is ubiquitous in engineered mechanical systems, it is rare in natural systems. With the exception of bacteria, multicellular organisms have not evolved wheels. Rotation movements translated over 360° are found in whole body rolling maneuvers by organisms, such as tumbleweeds, caterpillars, stomatopods, and desert spiders (Full et al. 1993; Armour and Vincent 2006). The energy for rolling is derived from air and water currents or from gravity. Self-actuating movements by animals are powered by muscles, which only allow a limited degree of rotation. A physical connection is made by muscles and stabilizing ligaments across a rotational joint. It is this construction of rotational joints that makes the evolution of wheels in animals impossible. As continuous rotation is not possible in the musculoskeletal system of animals, locomotion requires oscillatory movements. These oscillations are characterized by reversals in the direction of body and appendage movements. The energetics is impacted by operating in such a fashion. Periodic accelerations occur over a propulsive cycle. Thus, kinetic energy varies greatly and the efficiency is decreased (Alexander 1983).

Another constraint imposed by evolution is that the environment is in a non-equilibrium state, which places design criteria in a state of constant flux. Organisms must be plastic enough to deal with the constant changes in the environment. For any geographic location, the physical parameters of temperature, pressure, humidity will vary with time, season of the year, and across geological eons. Additional factors include the availability of food and shelter, distribution of predators, and prevalence of diseases. Changes in these factors can be unpredictable. This state of constant flux means that no design ever lasts indefinitely (Van Valen 1973). What is good today may not be good tomorrow. Furthermore, what is observed today may be on the way out. All species eventually go extinct.

The dinosaurs represent a successful group of medium to large reptiles that were the dominant land vertebrates on the planet for approximately 150 million years. However, in a geologically short period of time, the entire lineage went extinct. The only remnant of the dinosaurs is the offshoot that gave rise to the birds. Although dinosaurs might have evolved adaptations that solved a number of problems, their total extinction was due to a sudden catastrophic environmental change that overwhelmed the genetic capabilities of these animals to evolve to meet the ecological insult. The environmental changes that occurred afterward may have made it impossible for dinosaurs to reestablish themselves. Furthermore, the mammals usurped the ecological niches formerly occupied by the dinosaurs or formed new niches.

Organisms themselves can change the very environment in which they live, and not always for the better. Expectations of harmonious living between individuals and species (Benyus 1997) may be merely wishful thinking. Whereas harmonious symbiotic relationships have occurred such as in the origin of eukaryotic cells (Margulis and Sagan 1997) and organisms like lichens and corals, competition is fierce between individuals and species. Natural selection works at the level of the individual. Because there is an inherently selfish interest, life does not work for the best interests of the community. Adages such as "kill or be killed" and "Nature, red in tooth and claw" have an element of truism to them. Individual organisms can degrade the physical environment to their own benefit and the detriment of others. Certain plants will release secondary compounds that negatively affect the growth of neighboring plants. This strategy is referred to as allelopathy. The black walnut (*Juglans nigra*) releases a phenolic compound from its leaves, stems, branches, and roots. The toxic effect extends up to 27 m from the trunk of the tree (Lambers et al. 1998). Even the beaver (*Castor canadensis*) will modify its environment, so that various species are negatively impacted. By damming small streams, the creation of a pond will flood areas to the detriment of various terrestrial plants and animals and organisms that inhabit faster flowing water.

Lastly, in regard to the constraints imposed on evolutionary systems, design is constrained by evolutionary history. Species are merely terminal points on branches of an evolutionary bush. Organisms evolved along lines of common descent with shared genetically regulated developmental patterns, confining the design space. For example, the insulation of birds is fixed to a morphology that uses feathers, whereas mammals utilize fur. Both hair and feathers function similarly entrap an insulative air layer against the skin and maintain an elevated body temperature. These structures are derived from the keratinized scales of reptilian ancestors, although from different parts of the scales and with different molecular conformations (Vincent 1990). Once taking separate evolutionary pathways, species are limited with respect to developmental options. Although originally used for insulation, the feathers became co-opted for flight in the structure of the wings. Mammals cannot grow feathers. Therefore, the evolution of flight in mammals required a wing constructed of a flexible skin membrane, rather than a feathered wing. Whereas feathers have sufficient strength to maintain the wing surface and deal with aerodynamic forces generated by the flapping wing, the skin membrane in bats must be reinforced by elongation of the bony digits.

Phylogenetic history can impede performance by a species. Most fish breath solely using gills, which allows them to remain submerged and away from the surface of the water. This means that the drag on a swimming fish can be minimized. Moving in the proximity of the water surface incurs additional drag due to the formation of waves (Fish et al. 1991). Dolphins and whales (i.e., cetaceans) secondarily returned to the sea. These marine mammals evolved a body design analogous to fish for reduced drag (Fish and Hui 1991). However, dolphins and whales are restricted to the mammalian body plan, which uses lungs for gaseous respiratory exchange. Although it would be more efficient to possess gills and remain submerged, cetaceans must come to the surface to breathe. As the additional drag at the surface can be as high as five times the submerged drag (Fish and Hui 1991), the phylogenetic history of cetaceans can impose severe energetic penalties. Cetaceans have had to develop behavioral and physiological adaptations, such as porpoising and prolonged apnea, to avoid surface effects.

12.6 Final Comments

It is tantalizing to consider the development of new and superior technological designs for enhanced performance based on biological systems. However, such innovations have been elusive (Fish 1998, 2006). The commercial production of biomimetic products has been rare (Vogel 1998). There has been greater success from bio-inspiration. Strict adherence to biological designs in biomimicry rarely produces any practical results and in some cases can impede the development of engineered systems (Vogel 1994, 1998; Fish 1998). Yet, the fields of biomimetics and bio-inspired design are generating excitement. Biology can provide fresh solutions to more conventional approaches and even suggest new avenues of research. For engineering, the designs, structures, and materials of biology represent potentially untapped resources. Entrepreneurs, venture capitalists, and corporations speculate on the future of new designs (Petroski 1992; Vogel 1998). Each is hedging on getting a lead in the marketplace. The number of biomimetic-based patents has increased faster than the total of all US patents between 1985 and 2005 (Bonser 2006). "Biomimicry," "biomimetics," and "bio-inspired design" have become terms that draw attention to an emerging field of study and a source of investment capital. However, if these terms are not to be merely buzzwords that focus attention without comprehension, then clearer definitions and a conceptual biological framework are required.

As this chapter has emphasized, it is necessary to understand evolution with its inherent limitations to all possible designs. The technology that nature has evolved is not always ahead of the technology of human ingenuity (Vogel 1998). Only by understanding evolution and how organisms have adapted to their present and past environments can one avoid the pitfalls of overstatement regarding biomimicry. Besides an understanding of evolutionary mechanisms, practitioners of biomimicry must be aware of the limitations of biology for transition to engineered systems, and differences in the culture of biologists and engineers.

How can the process of bio-inspired design be accomplished to live up to the expectations for the development of new products? To address this question, one must first ask the question, "What needs to be improved?" Engineering is a goal-directed, applied science. The technological problem must be outlined to determine the direction of design for improvement in functionality. Next, biological structures or processes need to be identified that perform better than the current technology or solve the problem from a direction different from the current design path. The input of a biologist, who has knowledge of the workings of the system, is requisite. Simplified explanations of biological phenomenon from the popular literature and general media are inadequate to base development of new designs. This step is followed by research on the mechanism of action by the biological structure or process. In this analysis, considerations of the limitations of the biology are necessary. Any limitations due to energy efficiency, construction of materials and morphology must be evaluated with respect to a cost-benefit analysis. For example, engineered systems can be economically constructed as the performance of the materials used and the forces to be encountered are highly predictable. Alternatively, biological systems require increased safety factors due to the unpredictability of the environment and variation in the structural design of the organism (Alexander 1998), where the "safety factor" is the ratio of a component's strength or performance to the maximum expected load during operation (Diamond 1998). The cost to introduce the biological advancement or replicate it may be uneconomical. Finally, the biomimetic approach requires a coordinated effort of biologists, engineers, industrial designers, and business people to produce future bio-inspired products.

Acknowledgments We would like to thank Dr. Anthony Nicastro and Janet Fontanella for their comments on the manuscript. We also great appreciate the cooperation of Drs. Hilary Bart-Smith, Keith Moored, Hossein Haj-Hariri, Tetsuya Iwasaki, and Alexander Smits on the robotic manta project. This chapter is based in part on research performed with support from the Office of Naval Research grant no. N000140810642 to FEF.

References

Abbott IH, von Doenhoff AE (1959) Theory of wing sections. Dover, New York

Alexander R McN (1983) Animal mechanics. Blackwell Scientific Publications, Oxford

Alexander R McN (1988) Elastic mechanisms in animal movement. Cambridge University Press, Cambridge

Alexander R McN (1998) Symmorphosis and safety factors. In: Weibel ER, Taylor CR, Bolis L (eds) Principles of animal design: the optimization and symmorphosis, Cambridge University Press, Cambridge, pp 28–35

Alexander RMN, Goldspink G (1977) Mechanics and energetics of animal locomotion. Chapman & Hall, London

Allen R (2010) Bulletproof feathers. University of Chicago Press, Chicago

Armour RH, Vincent JFV (2006) Rolling in nature and robotics: a review. J Bionic Eng 3:195–208

Arthur W (2002) The emerging conceptual framework of evolutionary developmental biology. Nature 415:757–764

Bandyopadhyay PR (2005) Trends in biorobotic autonomous undersea vehicles. IEEE J Ocean Eng 29:1–32

Bar-Cohen Y (2006) Biomimetics: biologically inspired technology. CRC, Boca Raton

Bar-Cohen Y (2012) Biomimetics: nature-based innovation. CRC, Boca Raton

Barrett DS (2002) Optimization of swimming locomotion by genetic algorithm. In: Ayers J, Davis JL, Rudolph A (eds) Neurotechnology for biomimetic robots. MIT Press, Cambridge, pp 207–221

Benyus JM (1997) Biomimicry: innovation inspired by nature. Perennial, New York

Blaiszik BJ, Kramer SLB, Grady ME, McIlroy DA, Moore JS, Sottos NR, White SR (2012) Autonomic restoration of electrical conductivity. Adv Mater 24:398–401

Bonser RHC (2006) Patented biologically-inspired technological innovations: a twenty year view. J Bionic Eng 3:039–041

Bose N (1995) Performance of chordwise flexible oscillating propulsors using a time-domain panel method. Int Shipbuild Progr 42:281–294

Bose N, Lien J (1989) Propulsion of a fin whale (*Balaenoptera physalus*): why the fin whale is a fast swimmer. Proc Roy Soc Lond B 237:175–200

Breder CM Jr (1926) The locomotion of fishes. Zoologica 4:159–297

Breuker CJ, Debat V, Klingenber CP (2006) Functional evo-devo. Trends Ecol Evol 21:488–492

Brower TPL (2006) Design of a manta ray inspired underwater propulsive mechanism for long range, low power operation. Master's thesis, Tufts University

Cai Y, Bi S, Zheng L (2010) Design and experiments of a robotic fish imitating cow-nosed ray. J Bionic Eng 7:120–126

Colgate JE, Lynch KM (2004) Mechanics and control of swimming: a review. IEEE J Ocean Eng 29:660–673

Compagno LJV (1999) Systematics and body form. In: Hamlette WC (ed) Sharks, skates, and rays: the biology of elasmobranch fishes. The John Hopkins University Press, Baltimore, pp 1–42

Darwin C (1859) On the origin of species by means of natural selection. John Murray, London
Deacon K, Last P, McCosker JE, Taylor L, Tricas TC, Walker TI (1997) Sharks and rays. Fog City Press, San Francisco
Denny M, McFadzean A (2011) Engineering animals: how life works. Belknap Press, Cambridge
Diamond JM (1998) Evolution of biological safety factors: a cost/benefit analysis. In: Weibel ER, Taylor CR, Bolis L (eds) Principles of animal design: the optimization and symmorphosis. Cambridge University Press, Cambridge, pp 21–27
Douady CJ, Dosay M, Shivji MS, Stanhope MJ (2003) Molecular phylogenetic evidence refuting the hypothesis of Batoidea (rays and skates) as derived sharks. Mol Phylogenet Evol 26:215–221
Festo M (2008) Fluidic muscle brochure. http://www/festo.com
Fish FE (1998) Imaginative solutions by marine organisms for drag reduction. In: Meng JCS (ed) Proceedings of the international symposium on seawater drag reduction, Newport, Rhode Island, pp 443–450
Fish FE (2006) Limits of nature and advances of technology in marine systems: what does biomimetics have to offer to aquatic robots? Appl Bionics Biomech 3:49–60
Fish FE, Blood BR, Clark BD (1991) Hydrodynamics of the feet of fish-catching bats: influence of the water surface on drag and morphological design. J Exp Zool 258:164–173
Fish FE, Hui CA (1991) Dolphin swimming—a review. Mamm Rev 21:181–195
Fish FE, Lauder GV, Mittal R, Techet AH, Triantafyllou MS, Walker JA, Webb PW (2003) Conceptual design for the construction of a biorobotic AUV based on biological hydrodynamics. In: Proceedings of the 13th international symposium on unmanned untethered submersible technology. Autonomous Undersea Systems Institute, Durham New Hampshire
Fish FE, Lauder GV (2006) Passive and active flow control by swimming fishes and mammals. Ann Rev Fluid Mech 38:193–224
Fish FE, Smits AJ, Haj-Hariri H, Bart-Smith H, Iwasaki T (2012) Biomimetic swimmer inspired by the manta ray. In: Bar-Cohen Y (ed) Biomimetics: nature-based innovation. CRC, Boca Raton, pp 495–523
Full R, Earls K, Wong M, Caldwell R (1993) Locomotion like a wheel? Nature 365:495
Gao J, Bi S, Xu Y, Liu C (2007) Development and design of a robotic manta ray featuring flexible pectoral fins. Robotics and Biomimetics, ROBIO 2007. IEEE International Conference, pp 19–523
Gilbert SF (2003) The morphogenesis of evolutionary developmental biology. Int J Dev Biol 47:467–477
Gordon LM, Joestner D (2011) Nanoscale chemical tomography of buried organic-inorganic interfaces in the chiton tooth. Nature 469:194–197
Harris JS (1989) An airplane is not a bird. Invent Tech 5:18–22
Heine C (1992) Mechanics of flapping fin locomotion in the cownose ray, Rhinoptera bonasus (Elasmobranchii: Myliobatidae). Ph.D. Dissertation, Duke University
Hu DL, Chan B, Bush JWM (2003) The hydrodynamics of water strider locomotion. Nature 424:663–666
Jakab PL (1990) Visions of a flying machine. Smithsonian Institution Press, Washington
Katz J, Weihs D (1978) Hydrodynamic propulsion by large amplitude oscillation of an airfoil with chordwise flexibility. J Fluid Mech 88:485–497
Katz J, Weihs D (1979) Large amplitude unsteady motion of a flexible slender propulsor. J Fluid Mech 90:713–723
Katz SL, Jordan CE (1997) A case for building integrated models of aquatic locomotion that couple internal and external forces. In: Tenth international symposium unmanned untethered submersible technology: proceedings of the special session on bio-engineering research related to autonomous underwater, Durham, NH, pp 135–152
Klausewitz W (1964) Der lokomotionsmodus der Flugelrochen (Myliobatoidei) Zool Anz 173:111–120

Koza JR (1994) Genetic programming as a means for programming computers by natural selection. Stat Comp 4:87–112
Kuratani S (2009) Modularity, comparative embryology and evo-devo: Developmental dissection of evolving body plans. Dev Biol 332:61–69
Lambers H, Chapin FS III, Pons TL (1998) Plant physiological ecology. Springer, New York
Larrabee EE (1980) The screw propeller. Sci Am 243:134–148
Lilienthal O (1911) Birdflight as the basis for aviation. Longmans, Green, London
Liu P, Bose N (1993) Propulsive performance of three naturally occurring oscillating propeller planforms. Ocean Eng 20:57–75
Liu P, Bose N (1997) Propulsive performance from oscillating propulsors with spanwise flexibility. Proc R Soc A 453:1763–1770
Low KH (2011) Current and future trends of biologically inspired underwater vehicles. Def Sci Res Conf Expo. doi:10.1109/dsr.2011.6026887
Lu Y (2004) Significance and progress of bionics. J Bionics Eng 1:1–3
Luria SE, Gould SJ, Singer S (1981) A view of life. Benjamin/Cummings, Menlo Park
Margulis L, Sagan D (1997) What is sex?. Simon and Schuster, New York
Martinez JS, Carter-Franklin JN, Mann EL, Martin JD, Haygood MG, Butler A (2003) Structure and membrane affinity of a suite of amphiphilic siderophores produced by a marine bacterium. Proc Nat Acad Sci 100:3754–3759
Moored KW, Fish FE, Kemp TH, Bart-Smith H (2011a) Batoid fishes: inspiration for the next generation of underwater robots. Mar Tech Soc J 45:99–109
Moored KW, Dewy PA, Leftwich MC, Bart-Smith H, Smits AJ (2011b) Bioinspired propulsion mechanisms based on manta ray locomotion. Mar Tech Soc J 45:110–118
O'Dor RK, Webber DM (1986) The constraints on cephalopods: why squid aren't fish. Can J Zool 64:1591–1605
Parson JM, Fish FE, Nicastro AJ (2011) Turning performance of batoids: limitations of a rigid body. J Exp Mar Biol Ecol 402:12–18
Pennisi E (2011) Bio-inspired engineering: manta machines. Science 332:1028–1029
Petroski H (1992) The evolution of useful things. Alfred A Knapf, New York
Petroski H (1996) Invention by design. Harvard University Press, Cambridge
Prempraneerach P, Hover FS, Triantafyllou MS (2003) The effect of chordwise flexibility on the thrust and efficiency of a flapping foil. In: Proceedings of the Thirteenth international symposium on unmanned untethered submersible technology: proceedings of the special session on bio-engineering research related to autonomous underwater vehicles, Lee, New Hampshire, Autonomous Undersea Systems Institute
Ralston E, Swain G (2009) Bioinspiration-the solution for biofouling control? Bioinsp Biomim 4:015007. doi:10.1088/1748-3182/4/1/015007
Rohr JJ, Fish FE (2004) Strouhal numbers and optimization of swimming by odontocete cetaceans. J Exp Biol 207:1633–1642
Rosenberger LJ (2001) Pectoral fin locomotion in batoid fishes—undulation versus oscillation. J Exp Biol 204:379–394
Schaefer JT, Summers AP (2005) Batoid wing skeletal structure: novel morphologies, mechanical implications, and phylogenetic patterns. J Morph 264:298–313
Suzumori K, Endo S, Kanda T, Kato N, Suzuki HA (2007) A bending pneumatic rubber actuator realizing soft-bodied manta swimming robot. In: Robotics and Automation, IEEE International Conference on 2007, pp 4975–4980
Takagi K, Yamamura M, Luo Z, Onishi M, Hirano S, Asaka K, Hayakawa Y (2006) Development of a rajiform swimming robot using ionic polymer artificial muscles. In: Intelligent Robots and Systems, IEEE/RSJ International Conference, pp 1861–1866
Triantafyllou GS, Triantafyllou MS, Grosenbaugh MA (1993) Optimal thrust development in oscillating foils with application to fish propulsion. J Fluids Struct 7:205–224
Triantafyllou GS, Triantafyllou MS (1995) An efficient swimming machine. Sci Am 272:64–70

Triantafyllou MS, Triantafyllou GS, Yue DKP (2000) Hydrodynamics of fishlike swimming. Ann Rev Fluid Mech 32:33–53

Ummat A, Dubey A, Mavroidis C (2006) Bio-nanorobotics: a field inspired by nature. In: Bar-Cohen Y (ed) Biomimetics: biologically inspired technologies. CRC Press, Boca Raton, pp 201–227

Van Valen L (1973) A new evolutionary law. Evol Theory 1:1–30

Vincent J (1990) Structural biomaterials. Princeton University Press, Princeton

Vogel S (1988) Life's devices. Princeton University Press, Princeton

Vogel S (1994) Life in moving fluids. Princeton University Press, Princeton

Vogel S (1998) Cat's paws and catapults. W. W. Norton, New York

Wainwright SA (1988) Axis and circumference: the cylindrical shape of plants and animals. Harvard University Press, Cambridge

Wainwright SA, Biggs WD, Currey JD, Gosline JM (1976) Mechanical design in organisms. Princeton University Press, Princeton

Webb PW (1975) Hydrodynamics and energetics of fish propulsion. Bull Fish Res Bd Can 190:1–158

Webb PW (1997) Designs for stability and maneuverability in aquatic vertebrates: What can we learn. In: Tenth international symposium unmanned untethered submersible technology: proceedings of the species on bio-engineering research related to autonomous underwater vehicles, Durham, NH, pp. 85-108

Whitley D (1994) A genetic algorithm tutorial. Stat Comp 4:65–85

Yang S, Qiu J, Han X (2009) Kinematics modeling and experiments of pectoral oscillation propulsion robotic fish. J Bionic Eng 6:174–179

Zhang S, Yokoi H, Zhao X (2006) Molecular design of biological and nano-materials. In: Bar-Cohen Y (ed) Biomimetics: biologically inspired technologies. CRC Press, Boca Raton, pp 229–242

Zhou C, Low K (2010) Better endurance and load capacity: an improved design of manta ray robot (RoMan-II). J Bionic Eng 7:S137–S144

Author Biographies

Wojciech Bejgerowski received the MS degree in mechanical engineering from Poznan University of Technology, Poznan, Poland in 2003. He received his Ph.D. degree in mechanical engineering from the University of Maryland, College Park in 2010. His research interests include In-mold Assembly Processes, Integrated Product and Process Design and Manufacturing Process Optimization.

Dr. John T. Beneski is a Professor of Biology at West Chester University of Pennsylvania, where he has been on the faculty since 1986. He earned a BA in Zoology (1978) and a MA in Biology (1982) from Humboldt State University. He received his Ph.D. in Evolutionary Morphology from the Department of Zoology, Washington State University in 1986.

Dr. Beneski's research emphasizes the evolution and adaptations of primitive terrestrial vertebrates. Recent projects include the ontogeny, kinematics, evolution and functional morphology of prey capture and intra-oral transport, the biomechanics of tail autotomy, the evolution and functional significance of tooth shape, the quantitative and qualitative analysis of tissue samples, and the 3-dimensional modeling of biomechanical systems. Primary analytical techniques include scanning electron microscopy, image analysis, and computer modeling. He has been funded by the National Science Foundation, the PA Department of Education, Endo Pharmaceuticals and the PA State System of Higher Education.

Janine Benyus is a biologist, innovation consultant, and author of six books, including Biomimicry: Innovation Inspired by Nature. In Biomimicry, she names an emerging discipline that seeks sustainable solutions by emulating nature's designs and processes (for instance, solar cells that mimic leaves). Since the book's 1997 release, Janine has evolved the practice of biomimicry, consulting with sustainable businesses and conducting seminars about what we can learn from the genius that surrounds us. Over the past 14 years, the Biomimicry Guild and the Biomimicry Institute have developed the practice of nature-inspired innovation and introduced it to the world. We have reached hundreds of thousands through TED talks, feature films and documentaries, and media such as Time, Newsweek, Forbes, Fortune and Wired. We started the world's first bio-inspired consultancy, bringing nature's ideas to the design table and boardroom of 250+ companies

A. K. Goel et al. (eds.), *Biologically Inspired Design*,
DOI: 10.1007/978-1-4471-5248-4, © Springer-Verlag London 2014

including Nike, HOK, GE Procter and Gamble, Levi's and General Mills. In 2011, we combined our staffs and administration services into a single entity, to create Biomimicry 3.8.

Dr. Amaresh Chakrabarti is a Professor at Centre for Product Design and Manufacturing, Indian Institute of Science (IISc), Bangalore. He has a BE (Mechanical Engineering) from University of Calcutta, India, ME (Mechanical Design) from IISc, and a Ph.D. (Engineering Design) from University of Cambridge, UK. After completing his Ph.D., he led for ten years the Design Synthesis team at the EPSRC Centre for Excellence Engineering Design Centre at University of Cambridge, before joining IISc. He works in design synthesis, creativity, eco-design, sustainability, and knowledge engineering/management. He has authored or edited 9 books, authored over 250 peer-reviewed articles, and has 8 patents granted or pending. He co-authored DRM, a methodology used widely around the world as a framework for carrying out engineering design research. He is an Associate Editor of AI EDAM Journal (CUP), Area Editor for Research in Engineering Design (Springer), Regional Editor for Journal for Remanufacturing (Springer), and Advisory Editor for seven International Journals. He is on the Advisory Board of Design Society, UK, and a member of the CII National Committee on Design, India. He founded IDeASLab – the first laboratory in India for research into design, and ICoRD – the first conference series in design research. He is an Honorary Fellow of the Institution of Engineering Designers, UK.

Hyunmin Cheong is currently completing his Ph.D. at the University of Toronto. He obtained his MASc and BASc in Mechanical and Industrial engineering at the University of Toronto. His research interests include applying natural-language processing techniques to support biomimetic design and understanding designers, analogical reasoning, and creativity.

Jon-Michael Deldin is a software developer for Biomimicry 3.8. He maintains the current version of AskNature and is developing its next evolutionary leap. Additionally, he is the lead programmer of the Biomimicry Student Design Challenge and Biomimicry Education Network websites. His professional interests include designing developer-friendly APIs, building highly-usable products, and applying machine learning techniques for insight into data. He holds a master's degree in computer science from the University of Montana-Missoula, and undergraduate degrees in biology and chemistry from the University of Great Falls. In his spare time, he can be found riding mountain bike trials, hiking, and pursuing assorted side-projects in programming and data visualization.

Dr. Frank Fish is a Professor of Biology at West Chester University of Pennsylvania, where he has been on the faculty since 1980. He received a BA in Biology from SUNY Oswego in 1975. He completed a MS in 1977 and a Ph.D. in 1980 from the Zoology Department of Michigan State University.

Dr. Fish has served as chairperson of his department at West Chester University. He was the chairperson of the Division of Vertebrate Morphology of the Society for Integrative and Comparative Biology. He is currently an associate editor for the journal Marine Mammal Science, and he served on the editorial board of the journal Bioinspiration and Biomimetics.

Dr. Fish's research is focused on the energetics and hydrodynamics of aquatic locomotion by vertebrates and on biomimetic applications. He has been funded by the Defense Advanced Research Projects Agency, the National Science Foundation, and the Office of Naval Research. Recent projects have included examinations of the evolution of swimming modes in aquatic mammals, energetics and maneuverability of aquatic animals, hydrodynamic design of biological control surfaces, and analysis of manta swimming for biomimetic AUV design.

John Gerdes is currently working as a Mechanical Engineer in the Vehicle Technology Directorate of the US Army Research Laboratory. He received his M.S. in mechanical engineering in 2009 and is currently pursuing his Ph.D. in mechanical engineering, both from the University of Maryland, College Park. His research interests include biological robotics and systems engineering.

Dr. Ashok K. Goel is a Professor of Computer and Cognitive Science in the School of Interactive Computing at Georgia Institute of Technology in Atlanta, USA. He is Director of the School's Design and Intelligence Laboratory, and a Co-Director of Georgia Tech's Center for Biologically Inspired Design. He serves on the Boards of Biomimicry 3.8 Institute and BioInspired! digital magazine. Born and brought up in India, Ashok immigrated to the U.S. for graduate studies. He lives in mid-town Atlanta with his two sons. Ashok conducts research into artificial intelligence and cognitive science, with a focus on computational design, modeling and creativity. His recent TEDx talk http://www.youtube.com/watch?v=wiRDQ4hr9i8 summarizes some of his research on AI, biologically inspired design, and environmental sustainability. An interactive tool for supporting some aspects of biologically inspired design developed by his laboratory is available at http://dilab.cc.gatech.edu/dane/.

Dr. Satyandra K. Gupta is a Professor in the Department of Mechanical Engineering and the Institute for Systems Research at the University of Maryland. He was the founding director of the Maryland Robotics Center. His interest is in automation problems arising in Computer Aided Design, Manufacturing Automation, and Robotics areas. The current focus of his research is on automated planning for human robot collaborations in assembly tasks, autonomous unmanned sea surface vehicles, autonomous unmanned ground vehicles, and automated optical micro-manipulation. He has authored or co-authored more than two hundred fifty articles in journals, conference proceedings, and book chapters. He has received many awards for his research contributions including a

Presidential Early Career Award for Scientists and Engineers (PECASE) and five best paper awards in conferences. For description of his latest projects please visit http://terpconnect.umd.edu/~skgupta/.

Michael Helms holds a BA in Finance from the University of Illinois, Urbana-Champaign, an MS in Computer Science from DePaul University in Chicago, and is currently a Ph.D. candidate in Computer Science at the Georgia Institute of Technology working in the Design and Intelligence Laboratory in the School of Interactive Computing. He is an instructor in Biologically Inspired Design (BID) at Georgia Tech, and has taught BID courses at NASA and Tohoku University, Japan, through his affiliation with the Center for Biologically Inspired Design at Georgia Tech. His forthcoming Ph.D. thesis investigates problem formulation and problem-solution coevolution in the context of Biologically Inspired Design. Prior to his work in artificial intelligence and design theory, Michael provided business information strategy consulting to Fortune 500 financial services companies.

James K. Hopkins currently works in the Propulsion and Power Engineering Department at the U.S. Naval Air Systems Command. He received his BS degree in mechanical engineering from Ohio University in 2001. He received his MS degree in engineering management from Florida Institute of Technology in 2004. He is pursuing his Ph.D. in mechanical engineering from University of Maryland at College Park. His research interests include design and mechanics of snake-inspired robots.

Dr. Venkat Krovi is an Associate Professor of Mechanical and Aerospace Engineering and the Director of Automation, Robotics and Mechatronics (ARM) Laboratory at the State University of New York at Buffalo. His research interests are in the lifecycle treatment (design, modeling, analysis, control, implementation and verification) of novel biorobotic and biomechatronic systems. His work has been funded by NSF, DARPA, ARO, state and industrial grants and has received multiple awards, including the National Science Foundation (NSF) CAREER Award, Petro-Canada Young Innovator Award, several best poster and journal paper awards. See http://mechatronics.eng.buffalo.edu for details.

Dr. Lengfeng Lee is currently working as a post-doctoral researcher in Department of Kinesiology at Univ. of Massachusetts, Amherst since 2011. He completed his Ph.D. (2011), MS (2005) and BS (2003) degrees in Mechanical Engineering from State University of New York at Buffalo. His research interests include design and control of articulated multi-body systems, motion planning and musculoskeletal analysis.

Dr. Julie S. Linsey is an assistant professor at the Georgia Institute of Technology in the Woodruff School of Mechanical Engineering. She earned a Ph.D. and MS in mechanical engineering from The University of Texas at Austin and a BS in mechanical engineering from the University of Michigan. Her research focus is on systematic methods and tools for innovative design with a particular focus on concept generation and design-by-analogy. Her research seeks to understand designers' cognitive processes with the goal of creating better tools and approaches to enhance innovation. She has co-authored over forty technical publications including four book chapters and she holds two patents.

Dr. Daniel A. McAdams is currently an associate professor Department of Mechanical Engineering at Texas A&M University. His research interests are in the area of design theory and methodology with specific focus on functional modeling; innovation in concept synthesis; biologically inspired design methods; inclusive design; and technology evolution as applied to product design. Dr. McAdams leads the Product Synthesis Engineering Lab (http://www.prosedesign.org/). He has served in various scholarly leadership roles including Technical Committee Chair for ASME Design Theory and Methodology Committee and serving on the editorial board of the Journal of Engineering Design. Dr. McAdams served as an associate and assistant professor at the Missouri University of Science and Technology. He received his Ph.D. from the University of Texas at Austin.

Dr. Frank Mendel is the Vice Chairman and Associate Professor in the Department of Pathology and Anatomical Sciences at the State University of New York at Buffalo, NY, USA. He received his MA and Ph.D. degrees in anthropology from the University of California, Davis in 1971 and 1975, respectively. His research interests include forensic pathology of extant and extinct big cats and cat-like carnivores (saber-toothed forms), developing virtual clinical training platforms and determining best practices for managing sprains and other soft tissue injuries.

Dr. Jacquelyn K. Nagel is an Assistant Professor in the Department of Engineering at James Madison University. Dr. Nagel has seven years of diversified engineering design experience, both in academia and industry, and has experienced engineering design in a range of contexts, including product design, bioinspired design, electrical and control system design, manufacturing system design and design for the factory floor. In 2012, Dr. Nagel was recognized by the National eWeek Foundation and IEEE-USA as one of the New Faces of Engineering for her pioneering work in using biological systems as models for sensors, instrumentation and processes. She earned her Ph.D. in Mechanical Engineering from Oregon State University, and her MS and BS in Manufacturing Engineering and Electrical Engineering, respectively, from Missouri University of Science and Technology. Her research interests include biomimicry, sensors, engineering design theory, sustainable energy, and manufacturing.

Madusudanan Sathia Narayanan received the BEng degree in mechanical engineering from the College of Engineering Guindy (Anna University Main Campus), India in 2006. He received the MS degree in mechanical engineering in

2008 from the State University of New York at Buffalo, Buffalo, NY, USA. He is currently pursuing his Ph.D. degree in mechanical engineering since 2009 from the same institution. His research interests include kinematics, haptics and medical robotics.

Megan Schuknecht is the Director of University Education at the Biomimicry 3.8 Institute. She works with faculty and administrators to incorporate biomimicry tools and concepts into university and college classrooms, curricula, and degree programs. She promotes and fosters formal relationships between the Biomimicry 3.8 Institute and universities to establish biomimicry degree programs through the Institute's Biomimicry Affiliate Program. She also develops and evaluates biomimicry teaching and learning materials, and shares those materials with educators and students through workshops, in-person and online courses, and curricula development. Megan orchestrates both the Institute's annual student design challenge and its annual Biomimicry Education Summit. She played a lead role in developing the AskNature website and continues to contribute to the site's growth, with an eye on how best the site can evolve to meet student and faculty needs.

Dr. L.H. Shu, Associate Professor in the Department of Mechanical and Industrial Engineering at the University of Toronto, received her SM and Ph.D. in Mechanical Engineering from the Massachusetts Institute of Technology. She has held research fellowships at the Technical University of Denmark, AlliedSignal Aerospace Canada Inc., Eastman Kodak, Naval Ocean Systems Center, and Naval Training Systems Center. She received the CIRP (International Academy for Production Engineering Research) F. W. Taylor Medal and numerous best paper awards for her work on biologically inspired design, and held leadership positions in both the Design Society and the ASME Design Theory and Methodology community. Dr. Shu's research interests are in creativity in conceptual design, systematic identification and application of biological analogies in biomimetic design, affordances, and identifying and overcoming obstacles to personal environmentally significant behavior.

Dr. Robert B. Stone is a Professor and the Interim Head of the School of Mechanical, Industrial and Manufacturing Engineering at Oregon State University. Dr. Stone s research interests include design theories and methodologies, specifically automated conceptual design techniques and concept generation techniques such as biologically inspired design. He has authored chapters on product architecture and reverse engineering techniques in product design texts. He leads the Design Engineering Lab (http://www.designengineeringlab.org). Previously, Stone utilized his design background to assist in creating the design-focused Interdisciplinary Engineering degree program at the University of Missouri-Rolla. At Rolla he served as Director of the interdisciplinary Student Design and Experiential Learning Center from January 2004 through June 2008 where he oversaw the design competition activities of ten teams and guided the Center's new engineering design and experiential learning initiative. He was a Distinguished Visiting Professor at the US Air Force Academy during the 2006–2007 academic year. Prior to initiating his graduate work, Dr. Stone worked in the Missions Operation Directorate of NASA-Johnson Space Center as a Space Shuttle Flight Controller for the Guidance, Navigation and Control Section. Stone completed his Ph.D. in Mechanical Engineering from The University of Texas at Austin.

Dr. Craig Tovey is the David McKenney Family Professor in the School of Industrial and Systems Engineering at Georgia Tech. He received his AB in applied mathematics from Harvard College in 1977 and both an MS in computer science and a Ph.D. in operations research from Stanford University in 1981. Tovey's principal research and teaching activities are in operations research and its interdisciplinary applications to social and natural systems, with emphasis on sustainability, the environment, and energy. He co-founded and co-directs Georgia Tech's Center for Biologically Inspired Design. His current research concerns inverse optimization for electric grid management, classical and biomimetic algorithms for robots and webhosting, the behavior of animal groups, political economy, and sustainability measurement.

Dr. Tovey received a Presidential Young Investigator Award in 1985 and the 1989 Jacob Wolfowitz Prize for research in heuristics. He was granted a Senior Research Associateship from the National Research Council in 1990, was named an Institute Fellow at Georgia Tech in 1994, and received the Class of 1934 Outstanding Interdisciplinary Activity Award in 2011.

Dr. Swaroop S. Vattam is a Research Scientist at the School of Interactive Computing at Georgia Institute of Technology. He has degrees in computer science and artificial intelligence and recently completed his Ph.D. degree in Computer Science from the Georgia Institute of Technology. His research interests lie in artificial intelligence, cognitive science, and human-centered computing. He is primarily interested in the development of novel artificial intelligence systems that espouse the potential of emerging technologies, but that are also grounded in theories of human cognition and studies of communities of practice. He is a firm believer that computational thinking enables us to conceptualize, modularize, and approach discovery and innovation in fundamentally different ways than we might consider otherwise. He is also a reviewer of several international journals and conferences in the areas of design computing and cognition.

Dr. Julian Vincent began his career as an entomologist with a Ph.D. studying insect hormones acting on cuticle, then a DSc on the mechanical design and properties of the extensible cuticle of the female locust. This got him interested in materials and mechanical properties, a topic which he developed to cover aspects of plants and animals and food, concentrating specifically on fracture mechanics. All this time he was at the University of Reading in the department of zoology, collaborating with George Jeronimidis and Jim Gordon in the department of

Engineering. Fascinated by the idea of turning biological phenomena into engineering, he became interested in biomimetics, eventually being appointed Professor in the department of Mechanical Engineering at the University of Bath. He has retired, but continues writing and lecturing around the world, and works part time for Swedish Biomimetics 3000, the only company in the world specialising in the development of biomimetic products.

Dr. Vimal Viswanathan is a post-doc at at the Georgia Institute of Technology in the Woodruff School of Mechanical Engineering. He earned a Ph.D. and MS in mechanical engineering from Texas A&M University. He had co-authored over twenty technical publications.

Dr. Marc Weissburg is Professor of Biology and Co-Director of the Center for Biologically Inspired Design. He is an ecologist/sensory biologist who examines the mechanisms and consequences of information transfer via aquatic chemical signals. He uses multidisciplinary approaches and field ecological investigations to study the structure of aquatic plumes and the dynamics of fluids in the marine environment and to behaviorally analyse the sensory strategies of aquatic organisms and their capability to rely on turbulent chemical plumes for guidance and navigation. He has used biological principles to devise artificial sensory processing strategies for autonomous navigation in chemical plumes. He has applied principles of ecological organization to human infrastructure in his search for more sustainable practices. He has co-taught Biologically Inspired Design for seven years to a variety of audiences ranging from undergraduates to professional engineers and scientists.

Bryan Wiltgen is a Ph.D. student in Computer Science at the Georgia Institute of Technology, where he is advised by Dr. Ashok K. Goel. Bryan conducts research in the Design & Intelligence Laboratory at Georgia Tech. Bryan holds a BS with Honors in Computer Science from Kennesaw State University and a MS in Computer Science from the Georgia Institute of Technology. Bryan has research interests in the areas of analogical reasoning, knowledge representation, conceptual change, and distributed cognition. Please go to http://www.cc.gatech.edu/~bwiltgen/ to find more information about Bryan and his research.

Dr. Jeannette Yen is the Director of the Georgia Institute of Technology's Center for Biologically Inspired Design (CBID). The goals of CBID are to bring together a group of interdisciplinary faculty who seek to facilitate interdisciplinary research and education for innovative products and techniques based on biologically-inspired design solutions. The participants of Georgia Tech's CBID believe that science and technology are increasingly hitting the limits of approaches based on traditional disciplines, and Biology may serve as an untapped resource for design methodology, with concept-testing having occurred over millions of years of evolution. Experiencing the benefits of Nature as a source of innovative and inspiring principles encourages us to preserve and protect the natural world rather than simply to harvest its products. Jeannette team-teaches the interdisciplinary course in biologically inspired design [http://www.youtube.com/watch?v=PMlvUJ9_GSk].

Jeannette Yen's Ph.D. is in interdisciplinary environmental science of biological oceanography where she studies how fluid mechanical and chemical cues transported at low Re flow serve as communication channels for micro-aquatic organisms, primarily zooplankton: key link in aquatic food webs. She has been to all 7 continents, including Antarctica for her research and education. She is a Professor of Biology and has been at Georgia Tech since 2000.

Index

A. K. Goel et al. (eds.), *Biologically Inspired Design*,
DOI: 10.1007/978-1-4471-5248-4,